DICK POLWARKOW

JAN 6 1987

MEASUREX
MX P.C. DESIGN GROUP

Printed Circuit Board
Precision Artwork Generation
and Manufacturing Methods

Printed Circuit Board Precision Artwork Generation and Manufacturing Methods

PREBEN LUND

*Quality Manager, Radiometer A/S,
Copenhagen, Denmark*

The Innovators
Bishop Graphics, Inc.

This is a revision of Preben Lund's, *Generation of Precision Artwork for Printed Circuit Boards,* originally published in Great Britain in 1978 by Wiley-Interscience, a division of John Wiley & Sons.

Copyright © 1986 by Bishop Graphics, Inc.

All rights reserved. No part of this book may be reproduced or transmitted in any form or by any means electronic or mechanical, including photocopying, recording, or by any information storage and retrieval system, without permission in writing from the publisher.

Printed in the United States of America
First Printing, May, 1986
ISBN 0-9601748-7-7
Library of Congress Catalog Number 85-73135
Reorder No. 10011

The Innovators
Bishop Graphics, Inc.
® 5388 Sterling Center Drive
P.O. Box 5007
Westlake Village, California 91359-5007, U.S.A.

Contents

Preface to the
Revised Edition

There currently exists very little technical literature on the preparation of precision artwork for printed circuit boards, and that which is available provides only scanty information. It is therefore hoped that this book, dealing as it does, both with manual methods, and with automated drafting and computer-aided design techniques, will fill a need for a comprehensive description of rational working methods for the preparation of precision artwork.

The first edition was published in 1978, but since then a rapid development within the area of computer-aided design has taken place. Very effective systems for computer-aided design and automated drafting have appeared, for which reasons the old Chapter 9 on automated generation of artwork/filmwork has now been updated and rewritten. In order to better understand the mode of operation, the new Chapter 9 also gives a detailed and easily understood explanation of the algorithms that form the basis of all computer-aided design systems.

Although the computer-aided design systems have taken over a very large part of the layout work not least the more tedious tasks, it is still the PCB designer or draftsman who decides whether the layout is to become a success. The author has therefore attached much importance to giving a deeper insight into the board manufacturing processes so that the PCB designer or draftsman will be able to set up reasonable design criteria for the preparation of artwork and avoid the common pitfalls which beset the manufacturer of the boards.

A new Chapter 11 on the design of boards for automatic testing has been incorporated. Particularly when testing assembled and soldered boards, contact problems can arise but proper design of the boards, primarily the introduction of test pads, can eliminate these problems to a very high degree.

Acknowledgments

As in his other book *Quality Assessment of Printed Circuit Boards,* the author wishes to thank Mr. Louis N. Jones, Santa Maria, California, for his painstaking efforts to correct the English, and the publisher, Mr. Robert Meisterling, of Bishop Graphics, Inc., for publishing this book and for his personal interest in the project.

In the "Acknowledgments" stated in the first edition, the author has expressed his thanks to quite a number of friends and colleagues who have assisted in many ways, so in the second edition the author wishes to extend the list by singling out the names of Mr. Arne Christensen, Print-tegn, Mr. Tom Hertz and Finn Vesthede, Dansk Data Elektronik A/S, and Mr. Finn Andersen, Radiometer A/S, to whom the author is grateful for inspiration and contributions in different ways to this book.

The author wishes to repeat his sincere and heartfelt thanks to Mr. Johan Schroeder, Director and General Manager of Radiometer A/S, for his kind permission to publish this book, which to a high degree is based upon the author's experience of preparing precision artwork for printed circuits boards over many years at Radiometer A/S.

It is also a matter very near to my heart to express my heartfelt thanks to Ellen, my dear wife, for her encouragement during the work of revision.

Preben Lund
Copenhagen
February, 1986

1

Introduction

Many discussions with printed circuit board manufacturers in various countries have confirmed the author's view that much of the artwork submitted to them by their customers is of less than acceptable quality. This is regrettable since it is no more difficult or time consuming to generate acceptable artwork than to produce artwork embodying faults which, by the nature of the process, will repeat themselves in all the boards manufactured and if not leading to complete rejection will at least cause an unnecessary degree of re-work and scrap following inspection of the manufactured product.

The author sets out to show how artwork faults occur and gives a detailed description of a novel copying technique which enables the designer to produce manual artwork of notably higher quality and to do this more quickly than is possible by following the methods which have been in general use up till now. A chapter on automated drafting methods and computer-aided design (CAD) has been included to encourage the reader who usually prepares artwork manually to try other methods. Special attention has been given to describe the preparation of a documentation package for a service bureau. The final two chapters deal with the design of printed circuit boards for automatic insertion of components and for automatic testing, respectively.

In short, improved artwork means less scrap, a consequent reduction in manufacturing costs, and lower prices to the user of the manufactured product; otherwise, production problems due to poor artwork can only be reflected in higher prices to the consumer than are really necessary. A conservative estimate of the benefits obtained by adoption of the design and layout methods described indicates that cost savings in the region of 10% can easily be realized. There is, therefore, little need to stress further the importance of these methods to every designer in this field.

In the early years of printed circuit board manufacture, the patterns were usually coarse, with plenty of room for wide tracks, larger pads, and wide spacing between the individual constituents of the pattern; there was no need for great accuracy and stability in artwork and its preparation was usually entrusted to a general drawing office. The introduction and rapid expansion in the use of semi-conductors and integrated circuits have lead to corresponding increases in the complexity of printed circuit boards resulting in much finer line structures and more intricate detail in the patterns which have to be produced. In consequence, highly accurate artwork permitting very close registration between the pads on the component side and those on the solder side of a double-sided board has become essential.

In spite of these more exacting requirements, more or less obsolete and inadequate techniques remain in general use for manual artwork preparation and the conditions for reasonably trouble-free manufacture are not met by the artwork produced. The following are some of the reasons for this state of affairs.

1. It has not been generally recognized by equipment manufacturers that the design of printed circuit boards is a highly technical activity calling for a group of specialists who are fully aware, not only of the technical problems themselves in the preparation of the artwork, but also of their impact on the manufacturing process. It is important that this knowledge extends not only to the various manufacturing processes but also to quality assessment, purchasing specifications and the problems which may occur in assembly and soldering. Only such a well-informed group working in its specialist technical environment is likely to be able to raise and maintain an adequate standard of artwork design in a situation where requirements and constraints are continually becoming more exacting.

   The above statement also applies to computer-aided design. Although many of the CAD systems offer automatic placement and routing routines, the knowledge of the designer (operator) is of great importance. Particularly in the case of interactive systems where he is engaged in a dialog with the computer, the designer can use his experience and ingenuity to solve layout problems that exceed the capacity of the computer.
2. The designer may not understand the requirements of the artwork and may be reluctant to adopt or develop new and improved methods for its preparation.
3. The designer or draftsman, though trained in general design work, may have no specific training for the design of printed circuit boards. In this field he is often self-taught and has had to learn mainly by experience through his own mistakes.
4. The limitation of the technical literature available on this subject is an additional obstacle for the designer, however willing he may be to learn.

The purpose of this book is to give a detailed and comprehensive description of the important disciplines to be mastered by the printed circuit board designer and draftsman when laying-out the pattern in accordance with the schematic diagram and taping the artwork. The discussion leads to the formulation of rational working methods which inherently lead to artwork of higher quality. Hence, another purpose of this book is to describe automated artwork-generation techniques: automated drafting and computer-aided design.

In the author's experience it is essential for the printed circuit board designer and draftsman to be familiar with the manufacturer's techniques of fabrication, especially in those areas which depend greatly on artwork quality. Printed circuit board design should, preferably, be the result of close cooperation between the customer and the manufacturer enabling many of the errors which become evident on the finished boards to be avoided. To achieve this, both parties must have a good understanding of the techniques involved.

Occasionally, insufficient importance is attached to the activities of the drawing office. Yet another aim of this book is, therefore, to show that artwork generation must be highly specialized if full consideration is to be given to the production processes which follow. Consequently, it has been necessary to introduce theoretical considerations concerning certain drawing-office procedures, in order to clarify the technical aspects. Complete command of a special field of activity can only be obtained when the background is fully understood. For this reason, the author has, in every instance, attempted to supply reasons for the views expressed whether the subject is, for example, the design of a special pad or the application of a special copying technique.

To impart a thorough understanding of the CAD systems' working principles to the designer (operator) is an important goal of the book, and the algorithms employed by the CAD systems have therefore been described by means of simple and easy to grasp diagrams showing how the CAD layout progresses.

This book, in contrast to others on printed circuit boards has, as its main subject, the drawing-office methods of artwork generation. The practical production processes used by the printed circuit board manufacturers are described only to a limited extent, the principal aim being to impart to the design personnel a basic knowledge of the production problems—not least the maintenance of tolerances. The author has purposely elected to cover the subtractive production processes, and to mention only briefly the additive and semi-additive processes, as these processes do not introduce any different requirements in the generation of acceptable artwork.

It is the author's intention to give the reader a good working knowledge of manual and automatic generation of artwork, for both single-sided and double-sided boards with plated through and non-plated through holes. He hopes, thus, to fill a need for the most up-to-date and comprehensive description of rational working methods for the routine generation of precision artwork. The procedures and rules of design are not to be construed as standards in an MIL (Military Standard) or an ESA (European Space Agency) sense, but they are very useful when designing ordinary commercial boards for use in professional equipment. An understanding of the principles for generation of precision artwork will, however, enable the designer or draftsman to extend and modify the rules of design to meet different customers' specifications and to cover the layout of flexible printed circuits and multilayer boards.

Since automatic insertion of components as well as automatic testing have become very common during the last years, two chapters state design recommendations to be observed by the designer in order to achieve boards that are easy to assemble and test automatically.

For the reader who is not familiar with the design and manufacture of printed circuit boards, a short survey is given below. In this way he will get a general outline of the various aspects of printed circuit boards before he embarks on the more detailed technical discussion.

An example of a printed circuit board (PCB) is shown in Fig. 1.1. It is clearly seen that the PCB itself is a board carrying a pattern consisting of conductors and pads. The electronic components are soldered to the pads and electrically interconnected by means of the conductors. In other words, the printed circuit board serves to carry and interconnect all the electronic components. It can be easily appreciated that the use of printed circuit boards results in a more uniform and error-free product if the board pattern is accurately and consistently reproducible. The methods described for the generation of a correct documentation package for use by the PCB manufacturer are designed to ensure that this reproducibility is realized in manufacture. The basic information required when starting to design a PCB comprises the schematic diagram and the component list together with mechanical dimensions and tolerances of the board. In the layout

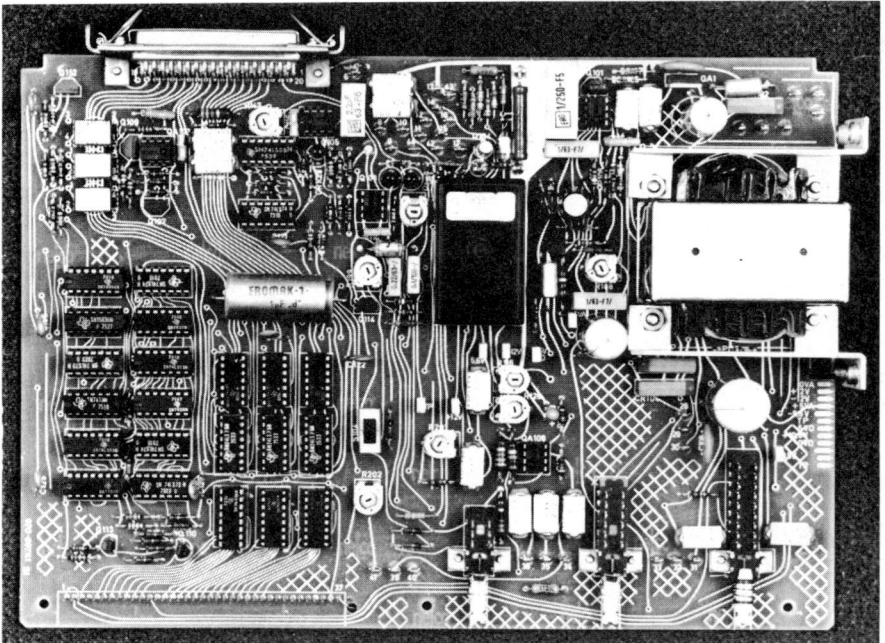

Fig. 1.1 Example of assembled printed circuit board.

stage, which is illustrated in Fig. 1.2., a sketch of the component location and the interconnection of the components is worked out. The next stage is the taping of the artwork, which means the preparation of a very precise picture of the pattern, very often made on a large scale, for example, four times full size. This is illustrated in Fig. 1.3. The taping is usually carried out using self-adhesive pre-cut symbols for the various constituents of the pattern, e.g., solder pads, contact fingers of edge connectors, and conductors. A close-up of the taping stage is shown in Fig. 1.4, the various symbols being easily recognized.

Fig. 1.2 The layout stage.

Fig. 1.3 The taping stage.

6

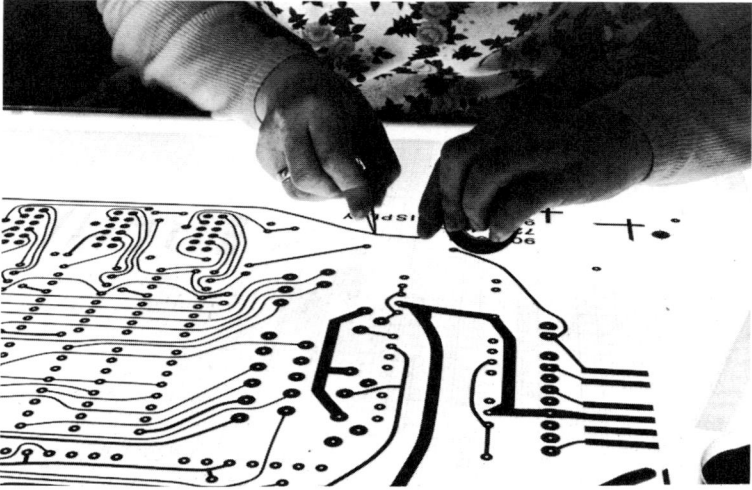

Fig. 1.4 Close-up of the taping.

The artwork is subsequently reduced to actual size by photography and the reduced copy used by the PCB manufacturer as the master pattern. For this reason it is not possible to indicate mechanical dimensions and other relevant specifications on the artwork itself. Instead, it is necessary to prepare an engineering drawing which contains a complete specification of the printed circuit board including the mechanical dimensions, hole diameters, tolerances, and surface treatment of the boards. In order to achieve correct and easy mounting of the components in the assembly department, the board is very often provided with a notation indicating the location of the various components. The artwork for this notation is also prepared by the PCB draftsman.

Nowadays, many electronics companies use automatic machines for soldering the assembled printed circuit boards. Unfortunately, this process sometimes creates unwanted solder bridges, i.e., small local short-circuits between adjacent parts of the pattern. In order to prevent solder bridging, it is a common practice to apply a resist on the solder side of the board. This solder resist leaves all the solder pads open and accessible for soldering but covers all other parts of the pattern, thus preventing solder bridging. The necessary artwork is also prepared by the PCB draftsman.

Now a few words about the manufacture of the boards: The raw material is a laminate consisting of a thin, rigid sheet of insulating material which is clad with a very thin copper foil on one or both sides. All the solder holes are drilled in accordance with the master pattern and the mechanical drawing. By selectively etching all unwanted copper areas away, only those parts which form the pattern, i.e., the solder pads and the conductors, remain on the finished board. This is a subtractive process, and thus it can be easily understood that an additive process

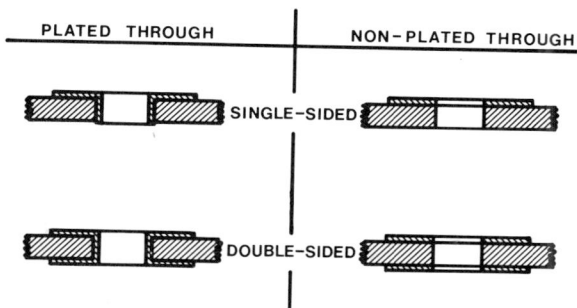

Fig. 1.5 Cross-sections of different types of printed circuit boards.

starts with an unclad base material on which the copper necessary tor forming the pattern is deposited. Whether the pattern is derived by the former or the latter process, its surface has to be protected in order to avoid oxidization of the copper which would lead to a serious reduction of the solderabililty. Simple single-sided boards are very often given an organic coating, for example, a colophony resin, but an electrolytic deposition of tin or a tin/lead alloy is used for the more elaborate plated-through boards. In these boards, it is possible to deposit copper and tin/lead on the walls of the holes so that the holes become conductive and solderable. These plated holes serve to make connection between conductors on opposite sides of the boards.

The above description shows that printed circuit boards come in several different types as illustrated by the cross-sections of Fig. 1.5. The two main types are the more expensive plated-through boards and the less expensive nonplated-through boards. The former are with a few exceptions always double-sided and the latter usually single-sided. Plated-through boards are chiefly used in professional electronic equipment such as measuring instruments, computers and military equipment, whereas the nonplated-through boards are mostly used for entertainment applications such as radios, television, amplifiers and the like. There is, however, a continuous growth in the use of plated through boards and, taking into account all the problems both in design and manufacture connected therewith, the author has found it justified to devote the greater part of this work to these boards.

2

Base Materials

2.1 Classification

A wide selection of copper-clad laminates is available to the manufacturer of printed circuit boards. These fall into two main groups: paper-base laminates and glass-base laminates. These materials are built from several layers of paper or glass cloth which are bonded together under heat and pressure to form rigid sheets. The binder is usually a phenolic resin, in the case of paper base, or an epoxy resin in the case of glass base, but sometimes polyester, silicone or Teflon* (PTFE) may be used. The laminate is clad on one or both sides with copper foil of standardized thicknesses. The characteristics of the finished printed circuit board depend upon the selected combination of the above elements as do the manufacturing processes, including the machining of the board.

2.2 Manufacture of Copper-clad Laminates

The base of the laminate can, as mentioned above, be either paper or glass-fibre cloth. First the base is passed through a vat filled with fluid resin (phenolic or epoxy) for impregnation, then between rolls to reduce the material to a uniform thickness and, finally, through a drying oven. Here the volatile elements of the resin are driven off and the semi-cured base, which is now called a 'prepreg', becomes tack free and ready for the laminating process.

The copper foil is produced by electroplating a thin layer of copper on a large rotating drum of stainless steel. As the drum turns, the deposited copper layer is peeled off and forms a continuous length which is coiled into rolls for use. To ensure good adhesion between the copper foil and the base material, it is common practice to oxidize the surface of the copper foil thereby roughening it.

The build up of the laminate must be done in a clean room in order to avoid electrostatic attraction of dust particles to the copper foil and prepreg material. Such particles could cause voids or dents in the copper foil when the laminate is pressed later. A piece of copper foil is placed on a polished steel plate followed by

* Trademark, E.I. du Pont de Nemours & Co., Inc.

the number of prepregs required to build the laminate to the desired thickness. For a double-sided laminate, the build up is topped with another piece of copper foil, and, finally, with another polished steel plate. For single-sided boards, the polished steel plate is placed directly on top of the prepregs and in this case is sprayed with a mould-release agent. The 'sandwich' is now placed in a large steam-heated hydraulic press where the laminate is pressed and cured to a rigid sheet which is then trimmed to size.

The heat applied in the curing process causes oxidation of the copper foil and it is, therefore, necessary to clean the surface of the copper after pressing by scrubbing, usually by means of a machine employing rotating brushes. A mild abrasive such as pumice powder is often used to help in the removal of the oxide coating. This may, however, have an unfavourable effect on the solderability of the finished printed circuit board as tiny particles of the pumice may adhere to the soft copper. Unless the PCB manufacturer takes measures against this contamination, for example by etching the surface, the customer may have some solderability problems.

Following manufacture, the product is inspected visually for scratches, pinholes, dents or other blemishes and tested for its electrical and mechanical properties.

2.3 Types of Base Material

As previously mentioned, many different types of copper-clad laminates exist. There are several type-designation systems, but, in order to limit the survey, the author has chosen the NEMA (National Electrical Manufacturers' Association, USA) specifications and will ignore the others, e.g. Military Standards or the IEC Recommendations. The board manufacturer will, however, find, from time to time, that he is asked to work to these and other specifications.

2.3.1 Phenolic Paper

This is a paper base impregnated with a phenolic resin. The colour is normally brown and the material is opaque. Phenolic paper boards are generally used for 'entertainment' applications for reasons of cost. Type XXXPC is a cold-punching material often used in very low-cost applications whereas XXXP is not punchable. When required, chemicals are added to the resin to produce a self-extinguishing material FR2 (FR, flame retardant).

2.3.2 Epoxy Paper

This is also a paper base but it is impregnated with epoxy resin. The colour is often yellow or white and more or less translucent depending on the thickness. The FR3 grade is self-extinguishing and is mostly used for entertainment applications.

2.3.3 Epoxy Glass

This material is widely used because it combines relatively high mechanical strength and good electrical properties. The strength is derived from the intrinsic strength of the glass fibres when supported by the cured epoxy resin. The material is semi-transparent, and the colour is usually green.

Type G10 is the grade in most common use. Type G11 uses a heat-resisting grade of resin and can withstand higher temperatures than type G10. Corresponding self-extinguishing types, FR4 and FR5, are available when the application calls for this property.

2.3.4 Copper Foil

The copper foil, which is used for cladding the base material, is produced by electrolysis and is at least 99.5% pure; it is available in several thicknesses. In the UK and USA the thickness is expressed in oz/sq. ft (ounces per square foot) and in other countries preferably in micrometres, usually written as 'μm' or just 'μ'. (1 μm = 0.001 mm.). However, 'μ' is not strictly correct. The most generally-used standard thicknesses are as follows:

$\frac{1}{2}$ oz/sq. ft ~ 17.5 μm
1 oz/sq. ft ~ 35 μm
2 oz/sq. ft ~ 70 μm

Recently, ultra-thin copper foils ranging down to 5 μm have been introduced. Their principal advantage is a considerable reduction of the undercut of the conductor edges. This aspect is described in more depth in Section 3.3.

2.4 Applications

As a rule, neither phenolic-paper nor epoxy-paper laminates should be used in professional electronic equipment, and, in particular, not in the case of plated through holes. Plated through boards will be discussed in later chapters but it can be mentioned at this stage that a thin layer of copper (approximately 25 μm = 0.001 in) is deposited on the wall of the holes. Since the coefficient of thermal expansion of copper is about 6 to 12 times less than that of phenolic-paper laminates, there is a certain risk of cracks in the hole walls if the board is subjected to a thermal shock such as that which occurs when the board is passed through a soldering machine. Obviously, this implies reduced reliability of the finished product when the plated through holes are used as a connection between the solder side and the component side of the board. Such holes are called interfacial holes or via holes, and often they will not be filled with solder after the soldering process so a crack results in a very unreliable connection.

In the case of epoxy-glass laminates the ratio of the coefficients of thermal expansion is about 3, and, therefore, the risk of cracked holes is considerably less. From the specification surveys which follow (Section 2.5) it will be evident that in all respects (apart from the price) epoxy-glass laminates are superior to paper-base laminates.

2.5 Mechanical Properties

In this section the mechanical properties of glass-epoxy and paper-phenolic laminates are compared. The figures should be taken as a general guide since there are significant variations in most of the characteristics from product to product. The differences between phenolic-paper and epoxy-glass laminates are, however, so marked that the variations within each type can be largely ignored in comparing the two types.

2.5.1 Flexural and Impact Strengths

The flexural and impact strengths of a printed circuit board must be sufficiently great to ensure that the board can carry heavy components without damage. Some constructions use, for example, power transformers soldered directly on to the printed circuit board which, therefore, has to support this relatively massive component. Under static conditions there will usually be no risk of damaging the board but, during shipment, the equipment can easily be subjected to very rough handling. Under unfavourable conditions the equipment can be exposed to very high acceleration forces, for example, $20g$. In such a case, a transformer weighing 0.25 kg (0.5 lb) would impose on the printed circuit board a force of 5 kg (10 lb) which could result in damage.

The flexural strengths of 1.6 mm (1/16 in) thick laminate are as follows:

Phenolic paper:	800 kg/cm^2 ~ 12,000 lb/in^2
Epoxy glass:	3500 kg/cm^2 ~ 50,000 lb/in^2

It is evident that the flexural strength of epoxy-glass laminates is $4 - 5$ times that of phenolic paper laminates.

The impact strengths are:

Phenolic paper:	7 kg cm/cm	(0.50 ft lb/in)
Epoxy glass:	10 kg cm/cm	(7 ft lb/in)

Therefore, the impact strength is approximately 15 times higher for epoxy-glass laminates than for phenolic-paper laminates.

The above values are stated only to give typical mechanical characteristics of the laminates. A comprehensive description of the test procedures to be followed in determining these characteristics for any material is given in various standards, e.g., ASTM, IEC, and reference should be made to the standards issued by these and other authorities.

2.5.2 Coefficient of Thermal Expansion

Generally, the coefficient of thermal expansion is of minor importance. For the laminates mentioned, it lies between 1×10^{-5} to 2×10^{-5} per 1°C, measured in the plane of the board. This means that if a printed circuit board having a length of 250 mm (10 in) is subjected to a temperature change of 40°C the length will increase by 0.1–0.2 mm (0.004–0.008 in) which is quite negligible. The normal manufacturing tolerance for the length of the said board would be around ± 0.2 mm (± 0.008 in).

On the other hand, it is true to say that the coefficient of thermal expansion is of importance if the board is to be punched. Although many paper-base laminates can be cold punched, heating the laminate will, in most cases, result in a cleaner fracture at the punched edges. By heating the above board to 120°C, the length will be changed by 0.3–0.6 mm (0.012–0.024 in) which has to be taken into account by the tool designer.

2.5.3 Mechanical Deformation under Load

It is common practice to use screws for fastening certain components to the printed circuit board or for mounting the board itself to the chasis. If, however, the board settles a little under the screw heads, the screws might become loose and, in the case of mobile equipment, even fall out.

Epoxy glass boards show a much lower deformation than epoxy and phenolic paper boards, the order of magnitude being 10 times less for epoxy glass boards.

Note:

It is recommended that lock washers be used with screws, preferably spring washers which produce a pre-determined pressure. Tooth lock washers and helical spring washers are not recommended since these, in time, will cut into the laminate thereby reducing the spring action.

2.5.4 Maximum Operating Temperatures

The maximum operating temperatures for the base materials previously discussed are:

Phenolic paper:	110 – 120°C	(XXXP)
Epoxy paper:	110 – 120°C	(FR3)
Epoxy glass:	130°C	(G10 and FR4)
Epoxy glass:	150°C	(G11 and FR5)

These temperatures are the highest which the laminates can withstand under continuous operation. It should be noted that operating at or around the maximum temperatures may cause discolouration and softening of the laminate. The mechanical strength of laminates such as G10 and FR4 is reduced by 90% of the value at normal room temperature whereas the reduction for G11 and FR5 is only 50%. Under all circumstances, it is recommended that the operating temperature should be limited to 100–105°C maximum and, in the case of still higher operating temperatures, the use of heat-resistant laminates is advisable, e.g. silicone glass or Teflon glass.

2.5.5 Soldering Time

It is important that printed circuit boards are able to withstand, without damage, the high temperatures which occur when components are being soldered to the board. This is especially so where automatic soldering machines are to be used. The soldering temperature can be 240–260°C which is about 130 to 150°C above the maximum operating temperature of the board. Fortunately, the soldering is done quite quickly and the damage to the board is negligible. Typically, these materials withstand a soldering temperature of 260°C for the following times:

Phenolic paper:	5 sec	(XXXP)
Epoxy paper:	10 sec	(FR3)
Epoxy glass:	20 sec	(G10, G11, FR4 and FR5)

2.5.6 Peel Strength

It is important that there is a good bond between the copper foil and the base material. The finer and more detailed and complicated the pattern of the printed circuit board is, the greater the risk that solder pads and conductors lose adhesion. The risk is particularly high when a printed circuit board has to be serviced or repaired, for example, if defective components have to be unsoldered and replaced.

To ensure satisfactory performance, a peel-strength test has been devised enabling the adhesion to be measured. This test measures the force required to peel off a strip of copper foil at a rate of approximately 50 mm (2 in) per minute in a direction perpendicular to the board. The peel strength of the strip is then converted to kilograms per 25 mm width of peel (lb/in width of peel). The previously mentioned laminates have peel strengths of the order of 3 kg (6 lb) but vary considerably with the type of laminate and the thickness of the copper foil.

2.5.7 Dimensions

2.5.7a Thickness of Laminate

Laminates may be purchased in various standard thicknesses. Table 2.1 is an extract of the NEMA specification Ll–l (1971) and shows the thicknesses which are generally available.

In general, the thickness tolerances are of no special significance in design, but, if an edge connector is used, the mating contacts of the fixed connector must accept the thickness tolerance of the board. Similarly, card guides used in equipment assembly must be able to accept boards on the upper thickness limit.

14

Table 2.1 Extract of the NEMA specification LI–1 (1971) on epoxy-glass laminate thickness and tolerance.

Nominal thickness	Tolerance on thickness*	
	Class I	Class II
0.8 mm (0.031)	±0.165 mm (±0.0065)	±0.10 mm (±0.004)
1.6 mm (0.062)	±0.19 mm (±0.0075)	±0.125 mm (±0.005)
2.4 mm (0.093)	±0.23 mm (±0.0090)	±0.175 mm (±0.007)
3.2 mm (0.125)	±0.30 mm (±0.0120)	±0.23 mm (±0.009)

* Values in parentheses are equivalents in inches. (Reproduced by permission from NEMA publ. No. LI 1–1971, Industrial Laminates.)

2.5.7b Sheet Size

There are no standards for the sheet sizes of laminates since they depend on the laminator's production equipment. The sheet size is also of no particular importance to the customer unless he plans to use exceptionally large boards. If so, there may be other limiting factors such as the production equipment of the PCB manufacturer. In such a case, it is necessary to consult both manufacturers before proceeding with the design. (Factors determining the board size are discussed in detail in Section 5.2.)

2.6 Electrical Properties

In discussing the electrical characteristics, we are dealing with quantities which vary widely in magnitude among types and within any one type. Any figures quoted must therefore be taken as typical only. Except in the case of boards used for high-frequency work and, occasionally, where exceptionally high insulation resistance is called for, these variations can be ignored. The laminate will normally be of such a high quality that the electrical performance of the assembled board is in no way affected.

2.6.1 Insulation Resistance

The insulation resistance between two points on a board is a combination of the surface resistance and the volume resistance. When a voltage is applied between two solder pads of a printed circuit board, a leakage current will pass across the surface. The magnitude of this current will be determined by the voltage applied and the surface resistance. Also, a leakage current will pass within the laminate itself, the value of which is determined by the voltage and the volume resistance.

The surface resistance is determined under prescribed test conditions, very often involving a three-terminal measuring set-up as described in the IEC Recommendation 249.

To give an idea of the order of magnitude, the data given below is typical of surface resistance after the boards have been conditioned at a temperature of 35°C and a relative humidity of 90% for 96 hours:

Phenolic paper:	1×10^3 MΩ
Epoxy paper:	1×10^3 MΩ
Epoxy glass:	1×10^4 MΩ

The surface resistance is, of course, very dependent on the relative humidity, so, before testing, the surface should be allowed to dry for one hour.

Other sources indicate that a reduction of the relative humidity from 90% to 70% could cause an increase of the surface resistance by a factor of 10.

It is worthwhile to note that the surface resistance is affected by dust particles and fingerprints on the surface. Also, chemicals used in the manufacturing process, if not completely removed by thorough rinsing and neutralization, will cause a reduction of the surface resistance.

The volume resistance is also determined under prescribed conditions, and the boards are pre-conditioned as described above. The following values of the volume resistance are typical:

Phenolic paper:	1×10^4 MΩ
Epoxy paper:	1×10^5 MΩ
Epoxy glass:	1×10^6 MΩ

The information given above does not give a comprehensive picture of the insulation resistance between parallel conductors on a printed circuit board. The insulation resistance is composed partly of the surface resistance between the conductors, and partly of the volume resistance between the conductors as well as between the plated through holes associated with the conductors. The values stated in the examples below give an impression of the order of magnitude of the insulation resistance as measured on a 1.6-mm (1/16-in) epoxy-glass board.

Between two plated through holes, diameter 0.8 mm (0.031 in) and a centre distance of 2.54 mm (0.1 in) the resistance ranged from 1×10^5 MΩ to 5×10^8 MΩ, depending on the make.

Between two conductors having a length of 50 mm (2 in), a spacing of 0.6 mm (0.024 in), and two plated through holes attached to each conductor, diameter 0.8 mm (0.031 in), the insulation resistance ranged from 10^4 MΩ to 10^7 MΩ, depending on the make. Clearly, such insulation resistance will have little effect on the rather low impedance circuits generally used today.

2.6.2 Dielectric Constant and Dissipation Factor

When dealing with high-frequency circuits, the capacity—and also to a certain degree the losses—between the conductors becomes more important and it is necessary to know the dielectric constant and the dissipation factor of the laminate. The values quoted below give an idea of these properties for engineering purposes but it is recommended that the manufacturer be consulted for specific information.

The values are normally specified for a frequency of 1 MHz and they vary slightly with the frequency. A drop of 5% in the dielectric constant can be expected when the frequency changes from 1 MHz to 1000 MHz.

Laminate	Dielectric Constant, ε (1 MHz)	Dissipation Factor, tan δ (1 MHz)
Phenolic paper:	5.3	0.050
Epoxy paper:	4.8	0.040
Epoxy glass:	5.4	0.035

An ordinary double-sided printed circuit board may very well be designed in such a way that it becomes suitable for high-frequency applications. For this purpose a stripline technique (derived from Microstrip Transmission Line) is used for designing the pattern as a transmission line. In this way, it is possible to achieve the desired values of the characteristic impedance, the propagation velocity and the attenuation. If a large frequency range is to be covered it is necessary to choose a laminate with a dielectric constant which shows only a small variation over this range.

3

Board Manufacture

3.1 Introduction

Before discussing the methods used by the PCB drawing office in preparing an artwork, we should spend some time investigating the sequence of manufacture of a printed circuit board in order to impart to the designer or draughtsman knowledge of the problems and difficulties experienced. Thus he will be able to perform the layout and taping of the artwork in such a way that these problems and difficulties are taken into account and so minimized in the subsequent manufacture. In addition to the personal satisfaction of a job well done there are two very important advantages to be gained: lower prices and more reliable deliveries.

If a PCB is so designed that the PCB manufacturer cannot achieve a yield greater than 80%, i.e. he has to begin with at least 125 PCBs in order to deliver 100, it is quite certain that the customer will, in the long run, have to pay for the excess boards in the form of higher prices. Alternatively, if the manufacturer begins with what might be regarded as a reasonable excess number, for example 5%, i.e. in the above case with 105 PCBs, it will not be possible to achieve full delivery since the yield may be only 84 PCBs. This may result in a break in the production rhythm of the customer's production department with consequent impairment in his delivery performance and profit.

It follows that we have to ensure that the artwork, which is sent to the PCB manufacturer, is debugged for the various production traps, and this can only be done if the designer understands the PCB manufacturer's production methods. This chapter, therefore, gives a qualitative description of commonly used production methods and draws attention to the aspects which are important when the board is being designed. A more detailed quantitative description of the production methods can, if desired, be obtained from other sources, e.g. *Printed Circuits Handbook* by Clyde F. Coombs Jr. (McGraw-Hill Book Company).

In this chapter we will discuss various production methods for both non-plated through boards and plated through boards. In order to achieve a better understanding of the flow diagrams given in Section 3.9 we will consider first the various part-processes such as image transfer, etching, plating and machining.

3.2 Image Transfer

The pattern is derived from the artwork which, by a photographic process, is transferred to a master film on a 1:1 scale. Now it is up to the PCB manufacturer to transfer the pattern in a very accurate manner to the copper foil of the laminate. In order to do this he can use several methods, each having advantages and disadvantages.

3.2.1 Screen Printing

The simplest method is a screen-printing technique taken from the printing industry. The pattern is transferred by screen printing a positive image on the copper foil of the board. A positive image means simply that the copper foil is covered with printing ink in the areas where the conductors and solder pads are going to be, the printing ink being resistant to the etchants used in the subsequent manufacturing stages. The principle of a screen-printing machine is shown in Fig. 3.1. It consists mainly of a hinged frame over which a fine fabric of nylon,

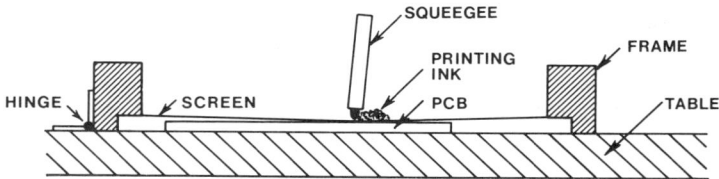

Fig. 3.1 Simplified cross-section of a screen-printing machine.

polyester or, even better, stainless steel is tightly stretched. There are several ways of preparing the screen, all based on a photographic process, but common to all is that the screen ends up as a sort of stencil having open meshes where the master film is opaque and closed meshes where it is transparent. Thus the open meshes correspond to the pattern. The board is now placed under the screen which is held 1–2 mm (0.040–0.080 in) above the board. The printing ink is placed at one end of the screen and, by means of a rubber squeegee, it is pushed across the screen. During the operation the screen is pressed down and the printing ink squeezed through the open meshes onto the surface of the board. The frame is now lifted a little and the squeegee and the printing ink brought back to the starting position. The board is then removed from the machine for drying.

As mentioned above, the screen has to be very closely meshed. It is common to use a 200–250 mesh fabric, i.e. 200–250 meshes per 25 mm (1 in) which corresponds to a mesh size of 0.07–0.08 mm (0.0027–0.0031 in). The mesh size should, theoretically, determine the highest resolution obtainable, but as the printing ink flows a little, the resolution is rather less, perhaps around 0.10–0.15 mm (0.004–0.006 in). This resolution is inadequate for boards with narrow conductors and close spacing, often called fine-line boards, and an alternative method is used. Usually the PCB manufacturers prefer not to screen print the pattern if the conductor widths and spacings are below 0.3–0.4 mm (0.012–0.016 in). Furthermore, they will assess the pattern in general and if the conductors run closely parallel for long distances so that there is a risk that the tracks may tend to merge, they will choose an alternative printing process and so avoid subsequent retouching—a relatively costly process.

Another limitation of screen printing is the loss of printing accuracy due to stretching of the screen by the mechanical action of the squeegee. In general, the accuracy obtainable when printing a 200–250 mm (8–10 in) long board is limited to 0.1 mm (0.004 in), depending on the manufacturer's equipment and the skill and care of the operator.

In the case of double-sided boards, the screen-printing process has to be repeated, i.e. one side is first printed and dried, followed by the second side.

As the PCB technician should be able to assess the screen-printing quality the following common faults are highlighted for his benefit:

(a) Voids or pinholes in the printed resist. If these are overlooked at inspection and retouching, they may result in undesirable etching of the copper or un-desirable plating depending on the process used. The voids or the pinholes are very often caused when the screen mesh is blocked by dust particles or by dried ink.

(b) Blurred or jagged printing. This causes uneven conductor edges on the finished boards. The reason is often incorrect ink viscosity or a worn screen.

(c) Very thin extensions in the printing. The result is often a blurred pattern with uneven conductors. Some of the causes are incorrect viscosity of the printing ink, an excessive distance between the screen and the board, insufficient squeegee pressure and relaxed tension of the screen.

(d) Hairlines in the printing. Such hairlines result in corresponding faults in the finished board, in extreme cases causing open-circuited conductors. The basic fault is scratches across the screen, arising from particles mixed with the prin-ting ink. It is an absolute requirement that screen printing is carried out in a clean and dust-free room.

Screen printing is advantageous when the number of panels is more than 25 and the pattern is not too fine. In the case of fewer boards, the initial costs, especially for preparing the screen, will be relatively too high and the PCB manufacturer will choose a photo-resist printing method which is described in the next section.

3.2.2 Photo-resist Printing

In order to achieve higher resolution than that obtainable by screen printing, a so-called photo-resist printing process employing photopolymer dry film is used. By the use of this process 'fine-line' conductors, i.e. conductors having widths and spacings down to a lower limit of 0.1 mm (0.004 in), can be produced.

A photopolymer resist is a light-sensitive organic material which is applied to the board as a thin film. The most commonly used photo resist is negative acting which means that it polymerizes on exposure to ultraviolet light, or in other words, it becomes insoluble to certain chemical solvents called developers. A developer dissolves the non-polymerized, i.e. unexposed areas, of the film, leaving all the polymerized areas as a resist on the copper surface. Photo-resist printing is done by placing the master film on top of the board to which a photo-sensitive film has already been applied. The pattern is now contact printed by means of ultraviolet light. Sharp definition requires a close contact between the master film and the photo resist, so a vacuum frame is always used for the contact-printing process.

In order to obtain a positive pattern on the board, the master film has to be a negative. In this way, all the conductors and solder pads will be exposed by the ultraviolet light which causes the photopolymer film to harden. This corresponds closely to screen printing where the conductors and solder pads are protected from the etching chemicals by means of the printing ink, and thus appear in copper when the resist is removed after etching.

Alternatively, a negative pattern is obtained if a positive master film is used for the contact printing. When the resist is developed after the exposure, all conductors and solder pads appear as bare copper, whereas all other areas of the copper surface will be covered by polymerized photo resist. The exposed copper is then electro-plated with a thin layer of metal, e.g. a tin/lead alloy. When the polymerized photo resist is stripped, all the unwanted areas will appear as uncovered copper. These areas can then be removed by etching using the tin/lead plating as an etch resist. The tin/lead-plated areas which remain after etching, constitute the required conductor pattern.

To make matters even more complicated, it is possible to obtain positive-acting photo resists. In this case, the process is reversed.

The prerequisites for high definition in contact printing are, firstly, that the photo resist is fine grained and, secondly, that the light source is collimated, i.e. a beam of parallel light. Even so, some of the light diffuses under the edges of the pattern as a result of scattering at those edges and reflection at the interfaces of the copper and the film. This is illustrated in Fig. 3.2.

The result is that the definition is imperfect since the division between the polymerized and non-polymerized areas becomes somewhat ill-defined. After development, the polymerized photo resist shows a slight rounding towards the copper foil. If the printed circuit board is to be etched directly, the widths of the conductors will increase, and if the board is to be plated (i.e. to deposit tin/lead—or first copper and then tin/lead—by an electrolytic process), a very slight nicking takes place at the transition between the photo resist and the copper foil. The thicker the photo resist, the more critical is the need for parallel copying light. Furthermore, it is important not to expose the photo resist to random ambient light, so copying rooms are usually shielded against daylight by means of yellow filtering foil and lit by yellow light which does not affect the photo resist.

In the past, liquid photo resists were used in a variety of ways, including dipping, spraying, roller coating or spinning, followed by drying. These methods were difficult to handle and results were not always uniform as many factors could affect the final result. Very often the thickness and evenness of the photo resist could vary which, of course, affected the quality of the contact printing.

Several years ago, the first photopolymer dry films appeared and they have now generally superseded fluid photo resists since they are far easier and safer to use. A dry film is constructed as a sandwich with the polymer photo resist placed

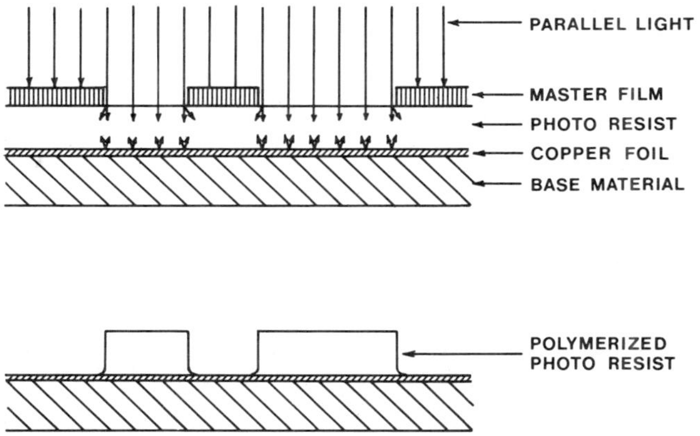

Fig. 3.2 Creeping of light in contact printing.

between two protective films. In order to make automatic application possible, the dry film is delivered in rolls. The application takes place under heat and pressure using a laminator. In this process, one of the protective films is removed while the other remains to protect the outer side of the dry film. A laminator can also be arranged to apply dry film simultaneously to both sides of a board. After exposure the film, which protects the photo-sensitive material, is removed and the dry film is developed as described above. A polymerized resist has now been applied to the copper foil and this resist will protect the copper which it covers during the sub-sequent manufacturing process whether this is etching or plating.

The process of printing a double-sided board using dry film is shown in Fig. 3.3. The master films of the two sides have to be aligned accurately with respect to each other to avoid misregistration of the conductor patterns on the finished board. In order to achieve a good contact between the emulsion of the master film

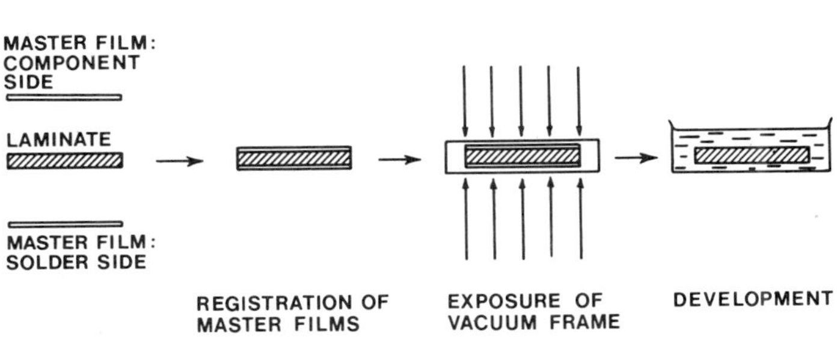

Fig. 3.3 Photo-resist printing of double-sided board.

22

and the dry film, the printing must be done in a vacuum frame. Following exposure, the master films and then the protective films are removed and the board developed.

Typical faults of the photo-resist printing method are as follows:

(a) Voids in the dry film itself which, as in screen printing, cause unwanted etching of copper or unwanted plating. If, however, a good quality film is used, the incidence should be relatively low.
(b) Dust particles or scratches in the photographic master film have the same effect as described in relation to screen printing.
(c) Insufficient development, i.e. failure to remove all the non-polymerized areas of the resist, causes etching faults or plating faults.

Dry films are being used more and more commonly and all leading PCB manufacturers have the necessary equipment and expertise. In the case of small batches, for example up to 25 panels, the dry-film printing method is generally found to be more economical than screen printing, regardless of the degree of detail in the pattern. For boards having relatively narrow conductors and small spacings (so-called 'fine-line boards'), the dry-film method is now universally used.

3.3 Etching

Etching is a chemical process whereby all the unnecessary copper is removed (dissolved) from the laminate while the copper which is protected by an etch resist is retained and forms the conductor pattern itself. The etching process is illustrated in Fig. 3.4 which shows a cross-section of a board with a transferred

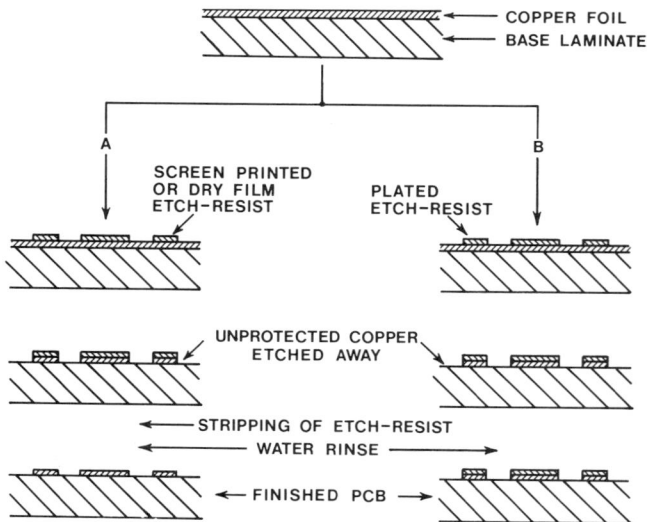

Fig. 3.4 Basic etching process.

image. The etch resist can be either an etch-resistant screen-printing ink or dry film (A) or an etch-resistant electro-plated metal layer, for example, tin/lead or pure tin (B).

Most commonly used is a spray-etching machine in which the etchant is forced under high pressure onto the surface of the board from a bank of nozzles. A double-sided board is usually etched simultaneously on both sides which presumes a double bank of nozzles. This avoids the danger of over-etching the side which would otherwise be etched first.

The type of etchant used depends on the type of board, i.e. whether it is a bare copper board, or plated with tin/lead, pure tin, tin/nickel, or gold. Commonly used etchants are ferric chloride, ammonium persulphate, cupric chloride and chromic-sulphuric acid.

After etching, a very thorough water rinsing is necessary to remove the etchant, followed by a neutralization of possible etch residues.

It is worthwhile noting that a board should not be profiled before etching takes place. The reason is that the etchant which, to a certain degree, can penetrate into small cavities and porous areas along the edges, is very difficult to remove by water rinsing and the problem is accentuated if the board has been blanked or sheared since the edges can be very rough as a result of the cutting action. It has therefore become accepted practice to profile the boards after the etching process.

In addition to the removal of unwanted copper depthwise, the etchant also attacks the copper sideways under the resist and causes a so-called undercut as illustrated in Fig. 3.5. The undercut is more or less proportional to the thickness of

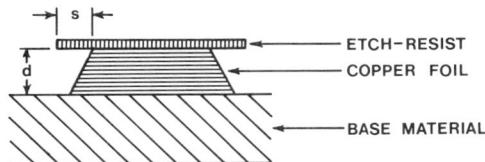

Fig. 3.5 Cross-section of etched conductor.

the copper to be etched, so an etch factor has been defined as the ratio d/s between the etching in depth (d) and sideways (s). The etch factor, which preferably should be well above 1, depends on the etching conditions, i.e. the etchant, the etching time, and the etching machine. In order to minimize the undercut, it is usual to etch as quickly as possible, preferably in an etching machine with high-pressure spraying.

Particularly in the case of 'fine-line' boards, it is necessary to achieve a high etch factor since the width of the conductors, which is, by definition, small, can be reduced to zero in extreme cases. It should be noted that a thick copper foil automatically leads to a larger undercutting effect. It is recommended, therefore, that the thinner copper foils should be used where undercutting is of significance. Preferably a 17.5 μm ($\frac{1}{2}$ oz/ft) copper foil, or perhaps an ultra-thin copper foil (5μm) should be selected. A different approach, whereby these problems can be overcome, is to deposit the total pattern on an unclad base material. This method is called an 'additive process' which is the opposite of the subtractive process described so far.

3.4 Plating

3.4.1 Introduction

Plating is the electrolytic or chemical deposition of metal on the board. The main purposes are, firstly, metallization of the solder holes in order to make them conductive, and, secondly, protection of the copper surface of the pattern against oxidation which would otherwise reduce the solderability and render the board unfit for use.

As indicated previously, the surface protection of conductors and solder pads may be a layer of tin/lead or pure tin usually with a thickness of 10–20 μm (0.0004–0.0008 in). In order to achieve sufficient protection against etching of the underlying copper, the tin/lead layer must be at least 6 μm (0.00024 in) thick. It is normal practice to specify 10 μm (0.0004 in) as minimum thickness. The tolerance is generally − 0%, + 100%.

Selected areas, for example the contact fingers of an edge connector, can be gold plated. In this case a thickness of 2–5 μm (0.00008–0.0002 in) is normal, sometimes with an underlying nickel layer of 5–10 μm (0.0002–0.0004 in).

It should also be mentioned that the protection of the copper against oxidation can be achieved by applying a coating of rosin, compatible with the flux system used in the soldering machine. The shelf life—which in this connection means the time within which satisfactory solderability is maintained—is normally about three months for rosin-coated boards and about six months for tin/lead-plated boards.

The metallization of the solder holes consists of an adherent electrolytic deposition of metal, initially copper, to a thickness of about 25 μm (0.001 in) followed by tin/lead or pure tin to a thickness of 10–20 μm (0.0004–0.0008 in) as required, for surface protection of the copper during the subsequent etching process.

3.4.2 Plated Through Holes

The purpose of the plated through holes (very often abbreviated to PTH) is partly to establish the so-called interfacial holes or via holes, i.e. holes intended as electrical connections between conductors on the component and the solder sides, and partly to create good anchoring holes for the component leads, resulting in increased reliability.

With growing complexity of electronic circuitry and progressive pressure for reduction of equipment size, plated through holes are used more and more widely. We shall now consider the plating processes in relation to the production of satisfactory plated through holes.

The first step is, of course, to drill the holes in the board. In order to achieve the correct diameter of the finished holes, the holes are drilled slightly oversize to allow for the internal metallization. This often amounts to 0.07–0.10 mm (0.0028–0.004 in) of the finished diameter, making room for 2×25–$35\,\mu$m (2×0.001–0.0014 in) copper plus 2×10–$20\,\mu$m (2×0.00004–0.00008 in) tin/lead. The tolerance of a plated through hole of 0.8 mm (0.031 in) is generally ± 0.1 mm (± 0.004 in).

As previously mentioned, phenolic-paper laminates are rarely used for plated through boards. Good through plating depends upon good hole-wall quality. A hole wall should be smooth and even. The surface copper foil should not have lifted nor should burrs appear at the edge of the hole. It is, therefore, not permissible to punch the holes as this results in localized crushing of the hole wall. In order to achieve smooth holes it is necessary to drill with a fast-running drill. To avoid epoxy smearing—which means epoxy softened by the drilling heat and smeared over the entire hole wall, including the edge of the copper foil—the drill feed has to be correctly related to the drill speed. Since the heating effect is minimized by the use of sharp drills, a record is kept of the number of holes drilled by individual drill bits so that they can be withdrawn for re-sharpening while the hole quality remains acceptable.

It is also important that the hole walls are drilled without cavities since small quantities of electrolyte may be trapped in these during the plating processes. At a later stage, such corrosive materials could cause serious trouble. Another danger is the breakdown of such areas during the etching process leaving voids in the plating of the hole walls. More and more PCB manufacturers are going over to the use of NC (numerically controlled) drilling machines. These machines are very expensive but they offer a very fast working speed and good accuracy. The justification for the use of these very costly machines is readily realized when it is considered that a board of 200×300 mm (8×12 in) can contain 2000 or more holes which all require a high positional accuracy.

As a rule, a number of boards to be drilled are assembled in an accurately registered stack so that all are drilled in one pass of the drill. To avoid drift of hole position on the lowest board due to flexure of the drill, the number of boards in the stack is normally limited to three or four depending on the thickness of the material. To avoid lifting of the copper foil on the underside of the lowest board when the drill breaks through, a sheet of cheap phenolic-paper laminate is always placed on the bottom of the stack.

After the holes have been drilled, the hole walls are made conductive by a copper-reduction process by which a coating about $0.5\,\mu$m (0.00002 in) thick is deposited. Without going into details of the chemistry, it can be mentioned that the process includes sensitizing by which stannous ions are adsorbed on the wall surface of the holes and the copper foil on both sides of the board. Catalysing then follows, through which the hole walls and the copper foil are covered with an extremely thin film of palladium. Finally, a chemical reduction of copper ions takes place and a thin film of metallic copper is deposited on the hole walls and the copper foil. The metallic copper film is very thin and delicate and can easily be damaged. Before the final plating is carried out it is, therefore, common practice to strengthen the copper film by a special and carefully performed copper plating in order to achieve a final copper thickness of 5 μm (0.0002 in). The board is now ready for the final build-up of the metallization of the hole walls as well as the conductors and solder pads. Two different methods are available to the PCB manufacturers:

(a) Pattern plating, which is a selective plating of the pattern, i.e. holes, conductors and solder pads.
(b) Panel plating, which is a plating of the entire board including the hole walls and the total copper foil area.

3.4.3 Pattern Plating

The method of pattern plating is shown in Fig. 3.6. The surface of the board and the hole walls have been made electrically conductive as described in the last section. A resist is now applied to the board which leaves the conductors and solder pads exposed while the rest is covered with screen-printing ink (Section 3.2.1) or with a dry-film photo-resist (Section 3.2.2). Metal is now deposited by electrolysis on the exposed areas including the hole walls. First a copper layer of minimum thickness 25 μm (0.001 in) is deposited followed by a tin/lead or pure tin layer of minimum 10 μm (0.0004 in). The plating resist is then stripped. As tin/lead or pure tin acts as an excellent etch resist, it is now possible to remove, by etching, the copper-plated areas previously covered by the plating resist, leaving only the required pattern of tin/lead plated copper.

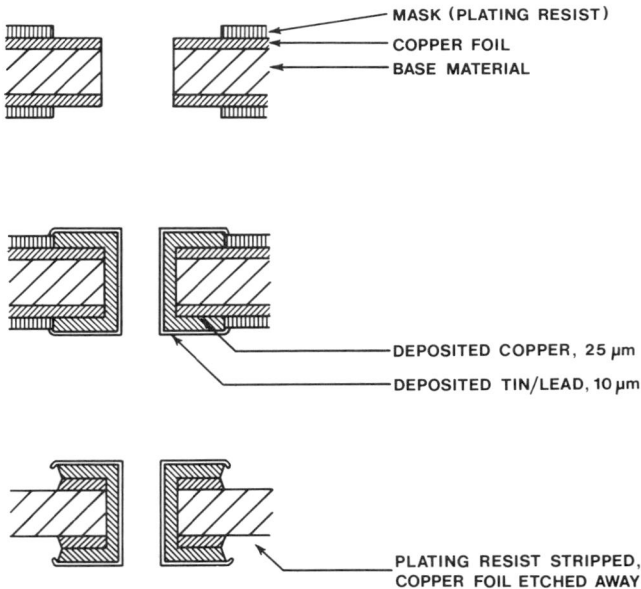

Fig. 3.6 Pattern plating. (Cross-section of hole.)

3.4.4 Panel Plating

The method of panel plating is shown in Fig. 3.7. The surface of the board, including the hole walls, has been made electrically conductive, exactly as before. The process starts with an electrolytic process by means of which copper is

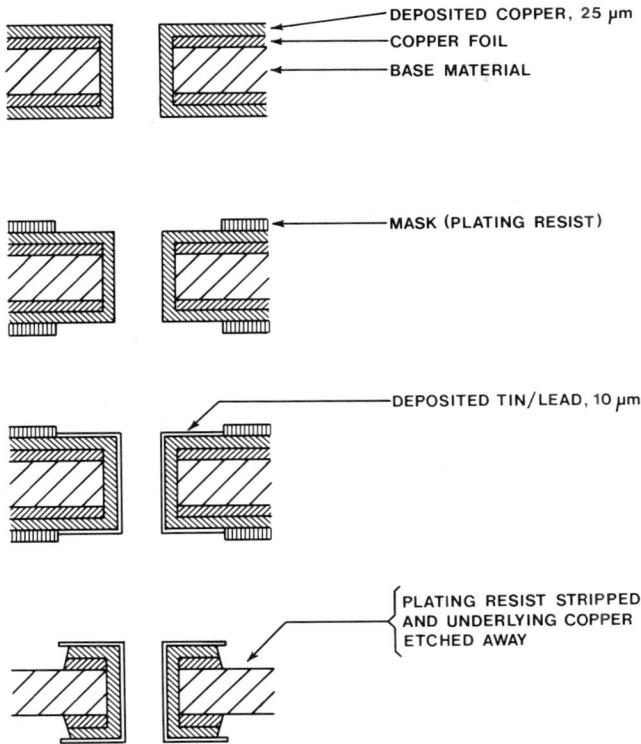

DEPOSITED COPPER, 25 µm
COPPER FOIL
BASE MATERIAL

MASK (PLATING RESIST)

DEPOSITED TIN/LEAD, 10 µm

PLATING RESIST STRIPPED
AND UNDERLYING COPPER
ETCHED AWAY

Fig. 3.7 Panel plating. (Cross-section of hole.)

deposited to a minimum thickness of 25 µm (0.001 in). In contrast to the former case, the copper is deposited all over the board, i.e. the surface of the two sides and all the hole walls. Afterwards, a plating resist is applied, exactly as described above, allowing the deposition of min. 10 µm (0.0004 in) tin/lead or pure tin on all areas which remain uncovered. The resist is now stripped and the board etched. As before, only the tin/lead- or the tin-plated areas stand the etching and remain on the board as the desired pattern.

It should be noted that the electrolytic processes are far more complicated than it would appear from this description. A great number of process steps is involved, including degreasing, water rinsing, and acid dips. Added to this is the necessary process control, including the adjustment and maintenance of the electrolytic solutions, the determination of the plating current and not least the control stages. The qualitative description given above should not lead the reader into believing that it is a fairly simple matter to produce plated through boards.

3.4.5 Comparison of Pattern-plating and Panel-plating Methods

An advantage of pattern plating is that the etching time is considerably shorter than for panel plating. As has previously been shown, deep etching results in heavy undercutting and for this reason the panel-plating method is less suited to the manufacture of 'fine-line' boards. Furthermore, observations can be made about the waste of resources and the effects which arise from the unfortunate fact that a thick layer of copper is plated then is removed and must be disposed of without pollution. On the other hand, it must be realized that the copper plating is less critical in the case of panel plating, since the surface to be plated is homogeneous and untouched by human hands, no resist being applied. The homogeneity implies that a higher plating-current density can be used which in turn allows faster production. In the case of pattern plating, a high current density could introduce 'burning' of solder pads and conductors in relatively open areas of the pattern.

In the following sections we will discuss in more detail the various forms of plating (copper, tin/lead, tin, gold, etc.) as a certain appreciation of these processes is necessary if one is to establish meaningful quality specifications for printed circuit boards.

3.5 Principle of the Electrolytic Process

3.5.1 Introduction

The plating itself takes place in a vat filled with an electrolyte. Figure 3.8 illustrates a copper-plating bath. The boards which are to be submerged in the bath are clamped in metal frames through which the electrical contact to the copper foil of the boards is established. The frames are hung on live rails giving contact to one pole of the current source. Electrodes, made of electrolytic copper, are placed on each side of the boards and all electrodes are connected to the other pole of the current source. The process takes place by oxidizing the electrode copper to copper ions. These migrate through the electrolyte, which consists of copper sulphate and sulphuric acid towards the printed circuit board. The board acts as a

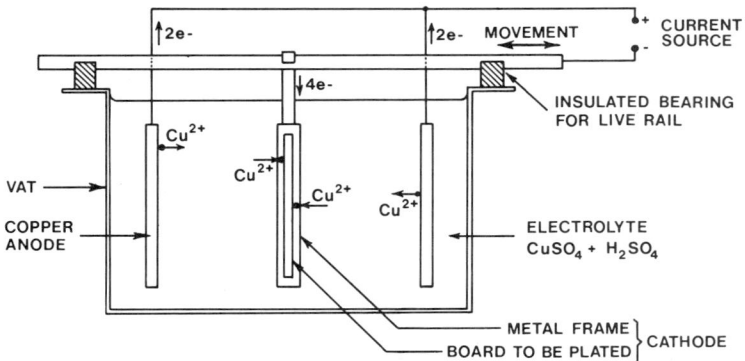

Fig. 3.8 Principle of the electrolytic process.

cathode and the copper ions are reduced to metallic copper and deposited on the surface and in the holes. In order to achieve good deposition within the holes themselves it is ordinary practice to move the frames backwards and forwards in the bath so that there will always be fresh electrolyte in the holes. An important process factor is the throwing power of the process which means the hole-to-surface plating ratio. This factor should never be below unity, otherwise it will be necessary to plate a fairly thick layer on the surface of the board to achieve the specified thickness in the holes.

3.5.2 Copper Plating

Since copper has a high conductivity, is reasonably easy to plate and is a good base for subsequent plating, it is invariably used as the basic plating material for printed circuit boards. Several types of copper baths exist, each type having its own advantages and disadvantages. By adding various chemicals it is possible to control the performance of the bath and to ensure that the copper deposit on finished boards has acceptable characteristics. With this in view, manufacturers use formulations based on their own experience. We shall now briefly discuss two of the most commonly used bath types.

Copper Pyrophosphate

This bath is among the most commonly used because it produces a fine-grained copper deposit which is smooth and dense, and reasonably ductile (i.e. plastic). The throwing-power factor usually approaches unity; in other words, the amount of copper deposited is not much less in the holes than on the surface of the board.

In certain cases, the PCB manufacturers add a so-called 'brightener' to the bath in order to obtain a bright deposit which is relatively resistant to oxidation and forms a good base for subsequent plating. A side-effect is that the copper deposit becomes harder and less ductile. The harder the copper, the greater is the risk that some of the hole walls will crack during the soldering process. The quantity used represents a compromise between the desirable and undesirable effects and has to be carefully controlled. This aspect was discussed in Section 2.4.

Copper Sulphate

This type of bath is also in common use. It is very simple in its make-up in that it contains only copper sulphate and sulphuric acid. The deposited copper is soft but by adding suitable additives it is possible, as in the case of the pyrophosphate bath, to produce a harder, brighter, and more fine-grained copper. Unfortunately, the throwing power is less than for the pyrophosphate bath, so that for a specified copper thickness in the holes, the deposit thickness on the two sides of the board is greater.

3.5.3 Hull Cell

The performance of a formulation can be checked by means of a Hull cell. This is, in a sense, a miniature plating bath in which the anode is placed at an angle to the test board as illustrated in Fig. 3.9. The current density is greatest where the anode is nearest to the board and decreases over the surface of the board as the separation increases. The deposit varies from a dark and somewhat spongy texture where the current density is highest, to a bright and even appearance at a lower current density. The cell is calibrated and enables the optimum operating current density for any particular formulation to be determined.

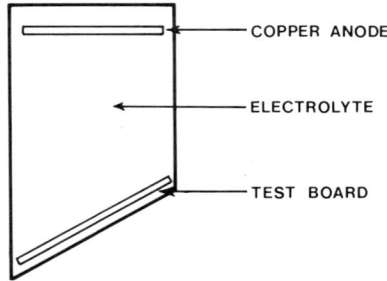

Fig. 3.9 Hull cell. (Seen from above.)

3.5.4 Tin/Lead Plating and Tin Plating

It is a fortunate coincidence that tin or tin/lead plating not only provides a surface which is readily soldered in subsequent use but also acts as a satisfactory resist to etchants which are used to remove unwanted areas of copper from the board. Although these materials do not retain their solderability indefinitely (due to slow oxidation of the surface) both are greatly superior to bare copper in this respect.

In the USA and UK and, to a certain degree, in France, tin/lead plating (solder plate) is preferred whereas pure tin plating is usually specified in West Germany and the surrounding countries. There is, however, an increasing preference for tin/lead plating since this finish can be reflowed as described in Sections 3.5.5 and 4.5.6. This process cannot be used on tin-plated boards because the melting point of pure tin is relatively high and the heat required to melt the finish would damage the base material.

The plating processes for tin and tin/lead are closely similar but since the deposition of a tin/lead alloy in controlled proportions of the two constituents presents certain problems which do not arise when plating pure tin, the following discussion deals only with the deposition of tin/lead.

The alloy composition of tin/lead solders in general use is 60% and 40% by weight of tin (Sn) and lead (Pb), respectively. The tin/lead-plating process is intended to produce a deposit of the same composition. The phase diagram (Fig. 3.10) shows how the melting and freezing temperatures of tin/lead alloys vary with varying proportions of the two constituents. Only at the eutectic composition, i.e. 63% Sn/37% Pb, do the two coincide—this at a temperature of 183°C (361°F). The cross-hatched areas in the diagram represent a condition in which the solid and liquid phases coexist and the metal is in a more-or-less plastic condition. The 60/40 alloy gives a melting point only slightly above that of the eutectic and a limited plastic freezing range. The choice of composition is necessarily a compromise aimed at keeping the soldering temperatures low to avoid damage to the base material of the board and ensuring that, in solidifying, a sound standard of soldering is achieved, i.e. the prevention of dry joints. Such joints are often due to small movements of the components during the freezing stage and fast freezing should, therefore, be avoided.

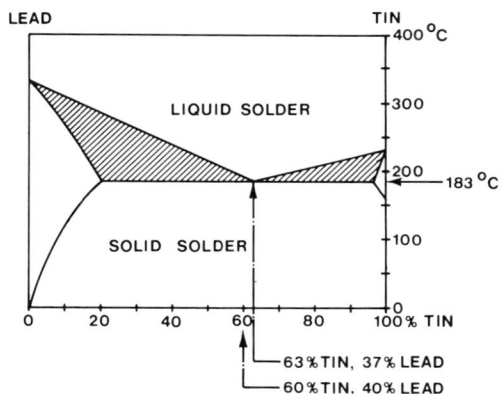

Fig. 3.10 Tin/lead phase diagram.

The composition of the alloy is, therefore, chosen away from the eutectic point, namely, at 60% tin and 40% lead. When the composition of the alloy is changed the melting temperature rises faster if the lead content is increased than for an equal increase of the tin content. This indicates that the tolerance of the tin content should not be $\pm 10\%$ as usually stated, but rather -5%, $+10\%$, i.e., from 55 to 70% tin, in order to obtain more or less the same melting temperatures at the ends of the tolerance range. It should be noted that the strength of a solder joint also depends on the composition of the alloy. The highest strength is found in the region 65–70% tin, but even at a 55% tin content the reduction in strength is insignificant.

The plating bath operates in the same way as the copper bath described previously. In this case, the anodes are made from tin/lead (60/40), or pure tin, depending on the type of plating, and the electrolyte is the fluoborate of these materials. As before, additives are used to ensure a smooth and fine-grained deposit.

The major plating problem is to maintain the composition of the alloy within the given tolerances. In particular, the current density is important since the tin content of the deposit increases with increasing density. If the melting point becomes too high, soldering problems can result because soldering machines are normally set to optimal operation for 60/40 solder alloys. The composition of the deposited alloy varies across the board depending on the uniformity of the pattern, and very often the highest tin content is found at the edge of the board.

For this reason, the so-called 'test coupons' which are nearly always located outside the boards and which are used for various tests or even as confirmation documentation of the quality of the board, are not proof of correct tin/lead composition. Also, the thickness of the deposit is subject to some variation due to the non-uniform density of the pattern. These problems are discussed in more detail in Sections 4.5.3 and 6.6.3, where suitable counter-measures to be taken by the designer are outlined. Tin/lead and pure tin finishes should be plated to a minimum thickness of 6 μm (0.00024 in) in order to achieve full etch protection of the underlying copper. A thickness of 10 μm (0.0004 in) with a tolerance of -0%, $+100\%$ is usually specified.

On occasions, the tin/lead plating becomes dull during the etching process. This can very easily be corrected by treating the boards in a bath containing a tin/lead brightener.

3.5.5 Edge Growth and Undercut

In the plating process the object, whether copper or tin/lead is being deposited, is to reach the thickness of the deposited layer. Lateral growth is generally prevented by the dry-film plating resist but should the thickness of the deposit exceed that of the resist, lateral growth will ensue at about the same rate as the increase in thickness of the deposit. This effect is illustrated in Fig. 3.11(A) showing the mushroom-like cross-section which results.

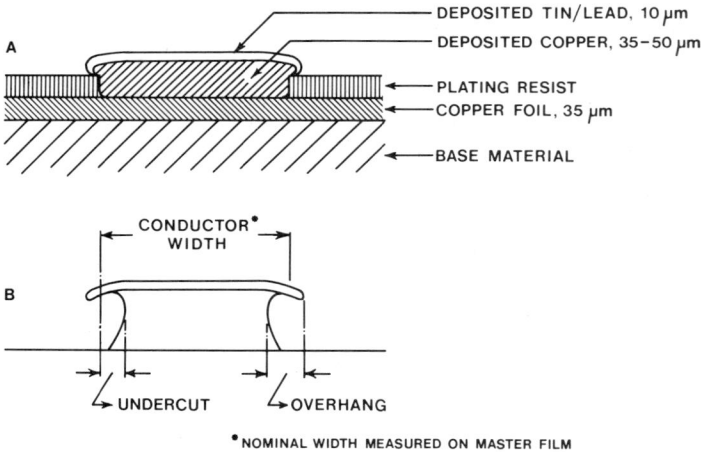

Fig. 3.11 Definition of undercut and overhang.

During etching, the conductor edges are subjected to undercut in which copper is removed from the exposed edges of the conductors leaving an overhang of tin/lead as shown in Fig. 3.11B. This overhang along a conductor is brittle and tends to break off in very thin slivers which can short-circuit adjacent conductors. The slivers are very often just 25–50 μm (0.001–0.002 in) thick and are difficult to detect with the naked eye.

By reflowing, or fusing, as described in Section 4.5.6, the surface tension of the molten tin/lead draws the material of the slivers back over the underlying copper as shown in Fig. 4.1. While this additional operation adds to the cost of the finished board, it is compensated for by the assurance of freedom from slivers and a reduction in the time required at incoming goods inspection.

3.6 Resist Removal

It is important that before the etching process is started, all the plating resist is removed from the board. Any traces of resist material that remain at this stage result in unwanted areas of copper remaining on the board, resulting in short-circuits or, at best, reduced clearances between conductors.

It has been shown in the description of the plating process how the plating resist, whether screen printed or photo-resist printed film, tends to be trapped by the overhanging plating deposit (Fig. 3.11A) thereby making the complete removal of the resist more difficult. The difficulty is less acute with screen-printed resists since these are fairly easy to remove by suitable industrial solvents. In contrast, dry-film photo-resists require a combination of chemical cleaning and mechanical scrubbing processes to ensure that no residues are present to cause the detrimental effects described above.

At the design stage this simply means that, in so far as other constraints permit, conductor spacings should be kept as wide as possible. In particular, where conductors run parallel over relatively long distances with small conductor-to-conductor gaps, the risk of short-circuiting is high and, consequently, the board will be more expensive.

3.7 Gold-plated Contacts

3.7.1 Introduction

In many cases, external connections to PCBs are made through contact fingers spaced along an edge of the board and integral with the circuit pattern. These contact fingers engage with mating contacts in a suitable socket in the equipment in which the board is to be used. Where cost is a prime consideration, and the highest standards of reliability are not called for, the contact fingers may simply be left tin/lead plated and in many cases will give a quite satisfactory performance. Where however, a high standard of reliability is required—such as in most professional equipment—it has been found necessary to use gold-plated fingers mating with gold-plated contacts. The economical advantages of tin/lead-plated contacts are obvious since their use involves no additional processing in manufacture. In contrast, use of gold-plated contact fingers introduces a number of considerations—directed at the integrity of the contact in manufacture and use—which place constraints on the manufacturing processes. These relate to the avoidance of corrosion affecting the contact surfaces and to securing a gold layer which will withstand the wear and tear of a reasonable number of insertions into, and withdrawals from, the mating contacts.

3.7.2 Corrosion in Gold-plated Contacts

Gold is a so-called noble metal, i.e. it does not form oxides and is not subject to attack by any corrosive agents likely to occur in everyday use. In addition, it has a high electrical conductivity and is, for these reasons, the ideal contact material. It is, however, very expensive and it is therefore important to minimize the quantity used.

Corrosion, when it occurs in gold-plated contacts, arises from the underlying metal and not from the gold itself. The corrosion arises from pollutants found in industrial environments of which the most serious are hydrogen sulphide and the oxides of sulphur and nitrogen. Hydrogen sulphide attacks copper in the absence of moisture and the others combine with atmospheric moisture to form acids which readily attack copper. It is, therefore, clear that avoidance of corrosion implies complete integrity of the gold layer covering the base metal, i.e. the complete absence of pinholes, microcracks and scratches which would provide sites for its initiation. Corrosion started at such sites (usually sulphates, sulphides or oxides) eventually spreads over the surface of the gold, resulting in poor contact due to increased electrical resistance or even complete open circuits. In the case of the acid corrosives, the process is encouraged by electro-chemical reactions at the exposed junction of the gold and the underlying metal.

The thickness of the gold deposit and the surface characteristics of the underlying metal are the main factors influencing the quality of the gold-plated contacts. The cost of the material limits the thickness that can be used, and the preparation of the contact surface prior to plating is, therefore, very important. Abrasive cleaning and etching both leave a somewhat pitted surface which tends to produce a granular and porous deposit of gold and it is therefore advantageous to plate a thin layer of nickel over the copper before the gold. The nickel tends to fill the roughnesses in the copper and leaves a highly polished surface on which a relatively thin gold deposit can be produced free of pinholes and porosity.

Nickel also has the advantage of being relatively stable against hydrogen sulphide. Within certain limits, it is possible to achieve the same degree of corrosion resistance by using either a thick gold layer on top of a thin nickel layer or vice versa. The latter combination is, of course, cheaper. Thicknesses of the gold (Au) and the nickel (Ni) shown by experience to give good results are tabulated below:

Hard Gold

Either 1–3 μm (0.00004–0.00012 in) Au on top of 10–12 μm (0.0004–0.00048 in) Ni

or

2–4 μm (0.00008–0.00016 in) Au on top of 1.5–2.5 μm (0.00006–0.0001 in) Ni

Soft Gold

Either 0.5–2 μm (0.00002–0.00008 in) Au on top of 3–4 μm (0.00012–0.00016 in) Ni or

1.5–4 μm (0.00006–0.00016 in) Au with no underlying Ni

Figures 3.12–3.16 which stem from an investigation made by the Danish Research Centre for Applied Electronics, show how corrosion occurs on gold-plated contacts after exposure to corrosive atmospheres. These samples were tested under more severe conditions than would be found in practice. The purpose of these tests was, chiefly, to emphasize the effect of pinholes and microcracks.

3.7.3. Hardness of Gold Plating

The durability of gold plating depends directly on the thickness of the gold and the hardness of the materials forming the mating contacts. Also, the difference in hardness of the gold plating on the mating surfaces is of particular importance. Hardness is expressed in kilograms per square millimetre (kg/mm^2) and is usually measured using a Vickers Hardness Tester. The measurement is carried out by indenting the metal under test with a diamond pyramid under a specified load and measuring the size of the impression produced. The impression is very small, for example, 1–2 μm (0.00004–0.00008 in) deep within a corresponding diagonal in the range 10–15 μm (0.0004–0.0006 in).

Hard gold is produced by alloying soft gold with various other metals such as cobalt (e.g. 1%) or nickel plus indium (e.g. 3%). The Vickers Hardness for hard gold is typically within the range 140–240 kg/mm^2, while for soft gold it ranges from 60 to 80 kg/mm^2. The longest life of a mating contact system will usually be achieved by choosing two materials of different hardness. As the contacts of the female connector are very often plated with hard gold, the contact fingers of the board's edge connector should, preferably, be plated with soft gold. Another argument in favour of soft gold is that it provides better protection than hard gold, probably because it is not alloyed with any base metals. On the other hand, the wear on the edge connector will not be excessive, as the board will normally be re-inserted only a few times. Very often it is difficult to obtain female connectors with high-quality, smooth, individual contacts. Therefore, it is rational to choose hard gold for the plating of the edge connector as this gives higher security against mechanical tearing or scraping of the gold layer. It should, however, be noted that certain types of hard gold show somewhat higher contact resistance than the soft gold but, in general, no problems arise from this.

The points made above, and in the preceding section, lead the author to recommend a minimum gold-plating thickness of 2.5 μm (0.0001 in) hard gold with a Vickers Hardness of 140–160 kg/mm^2, over a nickel layer of 5 μm (0.0002 in).

3.7.4 Gold Plating of Edge Connectors

The gold plating is carried out in essentially the same way as that described for copper and tin/lead plating; that is, an electrolytic deposition. There is however a difference because the printed circuit board is to all intents and purposes, finished,

Fig. 3.12 The corrosion products are seen as green crystalline flakes. The corrosion is spreading over the majority of the surface, but it is a far-advanced point corrosion which started in pinholes and microcracks. The test conditions are a sulphur dioxide atmosphere with a concentration of 100 p.p.m. SO_2, a relative humidity of 87%, and a temperature of 40°C, for 96 hours. (Courtesy of the Danish Research Centre for Applied Electronics.)

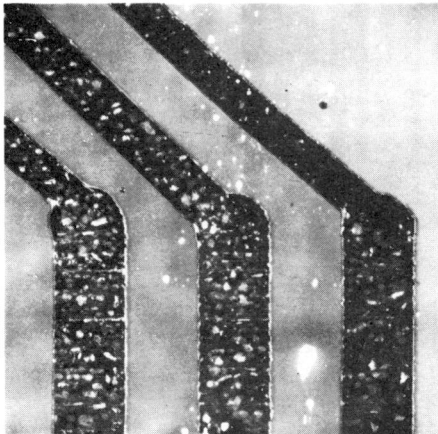

Fig. 3.13 This is a dense but weakly developed point corrosion. There is no marked formation of crystals, only a slight discolouration around each cell of corrosion. Test conditions are as given in Fig. 3.12. (Courtesy of the Danish Research Centre for Applied Electronics.)

Fig. 3.14 The picture shows a weak but unmistakable point corrosion resulting from pinholes. The test conditions are a hydrogen sulphide atmosphere holding a concentration of 10 p.p.m. H_2S, a relative humidity of 93%, and a temperature of 40°C, for 48 hours. (Courtesy of the Danish Research Centre for Applied Electronics.)

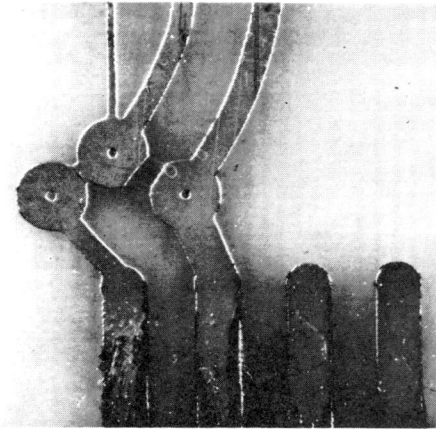

Fig. 3.15 An unmistakable corrosion of the edge of the contact fingers can be seen. The corrosion is due to the gold plating on the edges of the contact fingers being too thin. Test conditions are as given in Fig 3.14. (Courtesy of the Danish Research Centre for Applied Electronics.)

Fig. 3.16 Top left: Bare copper board before the test.
Bottom left: Same board after exposure to a hydrogen sulphide atmosphere. A very heavy and detrimental surface corrosion is evident. The surface is completely covered with a crystalline structure.
Right: This picture shows good gold plating with no traces of corrosion after the test. (Courtesy of the Danish Research Centre for Applied Electronics.)

38

i.e. copper plated, tin/lead plated and etched. During the gold-plating process, it is necessary to have an electrical connection to the contact fingers of the edge connector. The standard practice is to connect the individual contact fingers to a plating bar which is part of the pattern but outside the profile of the finished board, as shown in Fig. 3.17. When the board is profiled to shape, the plating bar is removed.

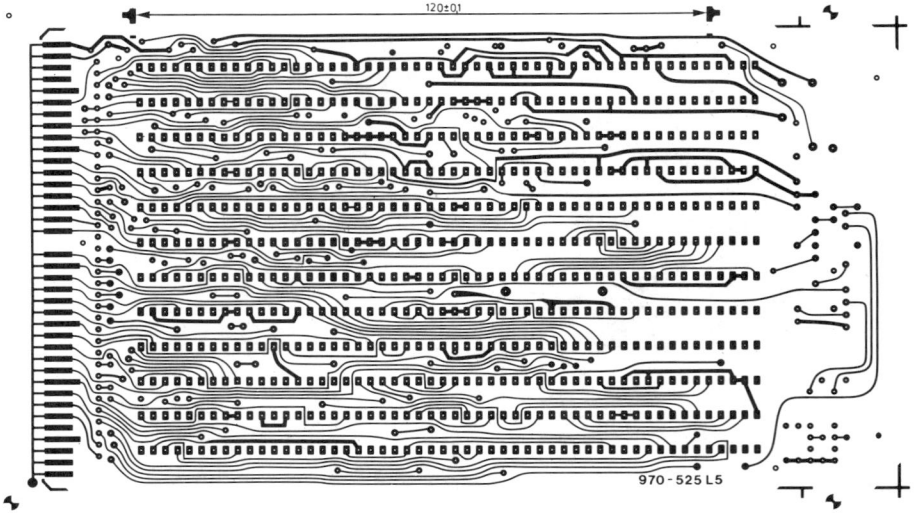

Fig. 3.17 Example of an artwork.

Very often the board will be tin/lead or tin plated, and the first step is to remove this plating in order to expose the copper. This involves the use of a special bath to dissolve the tin/lead, or the tin, without attacking the copper. The stripping process involves placing the board vertically in a shallow vat so that the stripping solution just covers the contact fingers. A sharp demarcation is achieved by covering the area just above the contact fingers with an etch-resistant tape. It is, of course, uneconomical to cover the total board with tape, so the operator has to be careful not to splash the stripping solution onto the rest of the board resulting in staining of the tin/lead or tin plating. When the stripping is complete, this tape is removed. Very often the bare copper area is polished mechanically, and mildly etched in order to ensure better adhesion of the subsequent plating. The mechanical polishing can be carried out in various ways, one of which is a fairly heavy-handed 'polishing' with fine steel wool. In this case the remarks in Section 3.7.3 concerning the smoothness of the underlying metal layer and the density of the gold deposit are extremely relevant, and the best way to ensure the specified quality level is to plate an underlying layer of nickel which will tend to remove scratches produced in the polishing process.

A second tape is now applied to the board to ensure a sharp demarcation of the gold-plated area. This tape is usually placed 0.5–1 mm (0.02–0.04 in) higher than before which means that there is a narrow transition zone between the edge of the tape and the copper exposed by the stripping process. The result is that gold plating takes place on top of the tin/lead plating in the transition zone giving improved protection against corrosion. If the tape has been positioned so that it covers part of the bare copper area, it is evident that a narrow zone of bare copper will be found between the tin/lead plating and the gold plating, with the attendant risk of corrosion. Some manufacturers find it cheaper to avoid this situation by omitting the second taping operation, leaving the first tape in place. The small unavoidable groove of unplated copper is then sealed by screen printing a strip of epoxy lacquer across the transition zone, for example as part of a solder resist or an insulating mask.

Gold is very expensive and should not be deposited where it is not needed. Therefore, it is usual to cover the plating bar and the connecting leads of the contact fingers with tape, before plating. As previously mentioned, the gold-plating process is preceded by nickel. During these processes, the board is placed vertically in a shallow vat with enough electrolyte to cover the contact fingers and part of the tape. An electrical connection is made to the plating bar and contact fingers. The contact fingers now act as cathodes, and the gold ions of the electrolyte will be deposited on the contact fingers as metallic gold. The anodes of the bath are either carbon anodes or platinum-clad titanium anodes.

There are several different types of gold baths in use, each with advantages and disadvantages. Briefly, there are alkaline bright cyanide, acid, and neutral gold baths. An advantage of the last two is that they do not chemically attack the base material which, to a certain extent, does happen with cyanide gold.

Some manufacturers avoid plating nickel under gold because of the risk of gold delamination. It is true that this may occur due to internal mechanical stresses in the deposited nickel layer or as a consequence of contamination of the nickel prior to gold plating. However, by careful control of the process conditions (constituents of the bath, temperature, pH values, current density, degree of contamination, and additives) it is possible to eliminate the risk of delamination. It is also sometimes stated that it is difficult to plate gold on top of nickel because the surface of nickel is rather passive. It is, however, possible to activate the nickel layer by means of a very mild etchant consisting of a mixture of dilute hydrochloric and sulphuric acids.

Occasionally, although all the contact fingers of the edge connector are present on a board, not all of them are used. Since gold is expensive, it would appear that a saving could be made by omitting these fingers from the artwork. However, it is sound policy to keep some spare contact fingers to be used in case of modifications. Alternatively, the customer could ask for the unused fingers not to be gold plated, but the cost of the additional selective taping operation is likely to be much higher than the value of the gold saved in this way. If the edge connector has a polarizing slot, the adjoining contact fingers should be left. In this way, it is easier for the customer's incoming goods inspection to check that the slot has been located correctly.

Finally, it should be noted that although gold is expensive the cost of processing may very well be an order of magnitude higher than the cost of the gold itself, depending on the gold-plated area. Take for example the board shown in Fig. 3.17, for which the processing cost of the nickel and gold plating is about £1.25 (July 1977 prices) whereas the price of the deposited gold is about 10 pence. As the handling cost can be regarded as independent of the board size, it is quite evident that a better price per gold-plated contact finger is obtained by increasing the number of contact fingers. This should be taken into consideration when determining the size and number of contact fingers which can be accommodated on a standard board.

3.7.5 Gold Plating of the Entire Pattern

Occasionally, there is a need to gold plate rather than tin/lead plate the entire board area. In thicknesses of 2 µm (0.00008 in) and above, gold acts as an excellent etch-resist, and furthermore, gold has low contact resistance and high conductivity which are of special interest in cases where boards are used in high-frequency applications. Also, gold retains good solderability even after long storage so the requirement for this finish is understandable. However, with heavier gold-plating thicknesses, for example, in excess of 3–4 µm (0.00012–0.00018 in), there is a higher risk of dry solder joints. During the soldering process the gold plating of a plated through hole is dissolved in the solder and forms an inter-metallic zone which, when the gold concentration reaches 20%, becomes very brittle and loses nearly all its strength. Due to the short soldering cycle, there is not sufficient time for the gold to diffuse completely through the solder, so it might very well occur that the critical concentration is produced in the transition zone between the solder and the gold plating. It should also be noted that solder joints on gold-plated solder pads assume a dark appearance which makes it difficult to differentiate them from dry solder joints. For the above reason, it is prescribed in various specifications, for example, the specifications of NASA (National Aeronautics and Space Administration), that the gold plating has to be removed prior to soldering. The conclusion must be, therefore, that gold plating of the entire board is dangerous and normally should be avoided.

If the board has contact areas to be used for mechanical contact devices, e.g. for a keyboard, often the entire board is gold plated. The reason is, usually, that selective gold plating of the contact areas only is considered too costly because several processes in the form of extra masking etc., are required. The price of gold is so high that in most cases it is cheaper to specify selective gold plating and to prepare the extra artwork. In this way, the above disadvantages concerning the soldering problems are avoided. In all cases, the reader is recommended to discuss the matter with the PCB manufacturer prior to finalizing his specification.

Very thin gold plating is sometimes specified for surface protection. The thickness can be 0.1–0.5 µm (0.000004–0.00002 in) in order to save gold, which in one way is justified by the fact that a thickness of 0.1 µm (0.000004 in) is sufficient to ensure the extended shelf life. This is, however, a rather doubtful measure because the very thin gold layer often turns out to be more or less porous, even if great care has been shown during the plating. Depending on the ways in which the boards are packed, despatched and stored, the underlying metal may oxidize due to the porosity of the gold layer. If the board is machine soldered, the thin gold layer is easily dissolved, but due to the short soldering time it is not possible for the applied flux to clean the surface of the underlying metal. The result can be serious dewetting: this means that the solder has pulled back from the surface prior to solidifying.

3.8 Machining

3.8.1 Introduction

Although the majority of the work in making a PCB relates to the image transfer, plating and etching, a significant amount of machining, including precision drilling of holes and profiling, is necessary. There may also be a need for special cut-outs and shapes, and, in the case of boards with edge connectors, chamfering of the edges and perhaps slotting.

The designer should select reasonable machining tolerances which take account of the characteristics of the specified material and the manufacturing tolerances of the technology and equipment used in manufacture.

Tolerances, as indicated on the master drawings, should never be reduced farther than absolutely necessary to meet the user's requirements. Very often the problem is that the designer does not appreciate these requirements and specifies in the normal way. Very close tolerances transfer the problem to the PCB manufacturer where they are reflected in the form of unnecessarily high prices, and, possibly in poor delivery performance because of high rejection rates at some of the production stages.

The object of this section is, firstly, to discuss the technology briefly, and secondly to familiarize the reader with the tolerances achieved in practice. The figures given must, of necessity, be general and it may be that particular manufacturers work to slightly different tolerances. The user should, therefore, consult his intended suppliers before finalizing his tolerance requirements.

3.8.2 Drilling

Holes which are to be plated through cannot be punched satisfactorily and are drilled as discussed in Section 3.4.2. Although punchable laminates are available for plated through hole applications, no leading PCB manufacturer uses them for quality applications at this time. The drilling operation can be performed in several ways.

3.8.2a Drilling by Direct Sight

The board is positioned manually under the point of the drill. In order to achieve good centring of the drill, the pad to be drilled should have a centre hole, preferably with a diameter of about 0.4 mm (0.016 in). The positional accuracy of the hole is unlikely to be better than ±0.25 mm (±0.01 in).

On some types of drilling machines, it is possible to mount a small optical device which projects a cross-hair target onto the board where the point of the drill is intended to make contact with the board when the drill is lowered. This device makes positioning of the board easier but the tolerance, again, is unlikely to be much better than ±0.25 mm (±0.01 in). Some drilling machines also feature a correlated adjustment of the drill feed and the drill speed, but even these machines are suitable only for drilling boards on a prototype or laboratory basis. Obviously, this method cannot be used for plated through boards where the holes must be drilled before the image is transferred to the board.

3.8.2b Drilling by Optical Sight

Considerable improvement can be gained by equipping the drilling machine with an optical sight as illustrated in Fig. 3.18. The drilling machine is designed such

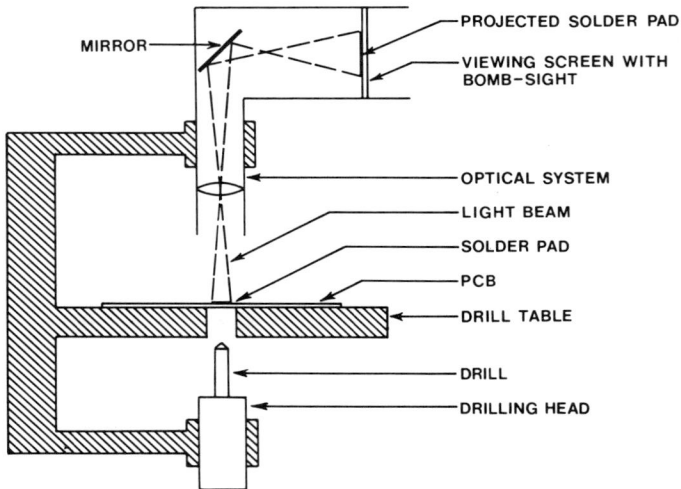

Fig. 3.18 Drilling with optical sight.

that the drill is located underneath the drilling table. The board is placed on the drilling table and the pad to be drilled is enlarged optically and displayed on the screen of the optical head. As the screen carries sight marks (cross-hairs and concentric circles) it is relatively easy to position the board correctly. By pressing a foot-switch, the board is clamped automatically and the drill is raised to drill the hole. The positional accuracy should be of the order of ±0.1 mm (0.004 in). One disadvantage of this method is that subdued lighting is necessary since the display on the screen is not bright.

3.8.2c Drilling with Jigs

The drilling machine described above can be modified to be semi-automatic by using a drilling jig and exchanging the optical system for a sensing device as shown in Fig. 3.19. The drilling jig is located on top of a stack of boards, using guide-pins or dowels. The drilling jig has guide-holes corresponding to the hole locations indicated on the master film. The guide-holes are conically shaped and the point of the sensor is easily located in them. When the point of the sensor has reached its lowest position, a clamping device is actuated clamping the jig and the

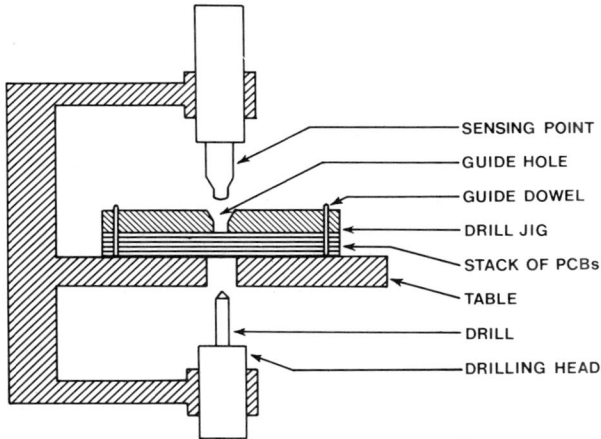

Fig. 3.19 Jig drilling.

stack of boards. The drill is raised, the hole drilled, and the drill is then automatically withdrawn and the clamp released. The operator then lifts the sensor point, moves the stack of boards so that the point catches the next guiding hole, and the process is repeated. The routes corresponding to the various drill sizes are indicated by lines of different colours thereby ensuring that the operator is fully aware of the numbers and sizes of the holes to be drilled. The drill jig is often made from 3-mm (1/8-in) thick perspex. The guide-holes can be drilled on the same machine by replacing the sensing device with an optical sight. A master film (scale 1:1) is fixed to the top of the perspex sheet, and the perspex is drilled in accordance with the pattern.

The positional accuracy of jig-drilling is of the order of ± 0.1 mm (± 0.004 in) which is equivalent to that which can be obtained by drilling with an optical sight. The advantages are reduced machining time and more uniform drilling quality.

3.8.2d NC Drilling

The development of semiconductors, and particularly integrated circuits, has resulted in an increased demand for printed circuit boards with high hole densities. For example, the board shown in Fig. 3.17 has about 900 holes and there are cases where larger boards with 5000 or more holes have been manufactured. Semi-automatic drilling machines are not practicable for these applications, partly because they are slow (30–50 hits per minute), partly because the operators suffer badly from fatigue resulting in reduced output and increased errors, and mainly because the machines are incapable of producing the accuracy necessary to meet the tolerance requirements. Detailed comments on high-density boards are made in Chapter 6, but it should be noted at this point that a hole with a finished diameter of 0.8 mm (0.031 in) will be drilled as 0.9 mm (0.035 in). The diameter of the solder pad may be as small as 1.4 mm (0.055 in) so the width of the annular ring is theoretically 0.25 mm (0.01 in). (The annular ring is the resulting circular strip of metal completely surrounding the holes.) Obviously there is not much room for error or inaccuracy and the methods of drilling described previously are unsuitable.

The numerically controlled (NC) drilling machine overcomes these disadvantages in the main, and, in spite of being very expensive, is being used more and more. An NC drilling machine can be programmed with a punched tape carrying all the necessary information on the positions and diameters of the holes. When the machine has received information on the position of a hole, the drilling table is moved very accurately to the desired position and the hole is drilled. The theoretical resolution is around 0.01 mm (0.0004 in) but, practically, drilling accuracy will be in the range ±0.025–±0.05 mm (±0.001–±0.002 in). Another advantage is the extremely high drilling capacity which, for the newer types, can be up to 200 hits per minute. If the machine is equipped with eight drilling heads, and there are three boards per stack, it is theoretically possible to drill 288,000 holes per hour. In practice, the capacity will be less due to lost time resulting from drill changes and loading and unloading the machine. New drilling machines have recently become available providing a facility for automatic drill changing thereby significantly reducing lost time on the machine. Other advanced machines with capacities up to 400 hits per minute are at an advanced stage of development at the time of writing.

The punched tape is derived from the master film in a digitizing machine which closely resembles the drilling machine. The drilling head, however, is replaced by an optical sight and, by moving the table on which the master film to be digitized is mounted it is possible to aim at the various pads, one by one. The coordinates of the holes are determined from the settings of the table.

3.8.3. Punching Holes

Punching is much cheaper than drilling because the punching die can pierce a larger number of holes in one operation. The quality of punched holes is, in general, such that the holes cannot be plated through because the hole walls will be too rough. In the case of plated through boards, non-plated through holes of a size or shape unsuitable for drilling, milling or routing may be punched. As mentioned previously, although punchable epoxy-glass laminate is available, the leading PCB manufacturers prefer to drill in order to ensure the quality of the plated holes.

On the other hand, phenolic-paper boards and to a certain degree, non-plated through epoxy-glass boards, are often punched. A pre-condition is relatively high batch quantities since the cost of the punching tool would be prohibitive for small quantities. Before choosing punching, the user should obtain alternative quotes for both punching and drilling. When the price of the tool and the drilling costs are known it should be easy to determine cost break even between punching and drilling.

3.8.4 Permanent Dimensional Changes After Drilling

The positional tolerances of the holes given above are valid with respect to the master film which determines the pattern. When the holes are drilled in the early phase of manufacture—which is always the case with plated through boards—there is a risk that distortion will occur in the later processes and the accuracy achieved at the drilling stage will be lost.

Epoxy-glass boards will probably suffer permanent dimensional change of approximately -0.05%. Fortunately, the relationship between the holes and the pattern is not affected in the case of plated through boards because the image transfer takes place immediately after drilling and before the processes involving heating and immersion in water, for example, plating and etching.

Phenolic-paper boards will probably suffer a dimensional change of approximately -0.12%. Normally the permanent dimensional change will be of minor practical importance to the customer since the play in the mounting holes will be sufficiently large to absorb the change. On the other hand, the designer must not lose sight of the above phenomena and specify unrealistic tolerances.

Problems might arise with very long boards if they are to be interconnected by means of connectors located at each end. It is much safer to move the connectors close together in an attempt to minimize the effects of dimensional distortion. Finally, problems may arise if automatic component insertion machines are used because of the requirement, in this case, for very tight positional tolerances.

3.8.5 Hole-diameter Tolerances

In general, the tolerance on the diameters of drilled holes is ± 0.05 mm (± 0.002 in) for diameters below or equal to 0.8 mm (0.031 in), and ± 0.10 mm (± 0.004 in) for diameters above 0.8 mm (0.031 in). These tolerances are accepted by nearly all PCB manufacturers and they should be achieved without the need for special attention. Price savings should, therefore, not be expected from specifications of drilled holes with tolerances on the diameters greater than those given above. It is possible to have a tolerance of approximately half the above values but usually the customer gains no practical advantage. It should be noted that keeping such tight tolerances can be a difficult matter which might reflect itself in a relatively high increase in price.

The tolerance on pierced or punched holes is normally ± 0.1 mm (± 0.004 in) regardless of the diameter. In order to avoid severe damage or fracture of the piercing punches, the rule of thumb is that the minimum diameter should be at least half the thickness of the board.

3.8.6 Cutting Base Material and Printed Circuit Boards

In order to utilize the production equipment (NC drilling machines, plating baths, etching machines, etc.) in the best possible way, the PCB manufacturer tries to avoid processing small boards individually as this requires too much handling. He prefers to manufacture a number of small boards on one production panel, which is processed as a unit until the individual boards are separated in one of the final production stages. By way of an example, the working film for the board shown in Fig. 3.17 will be prepared by a photographic 'step-and-repeat' process so that it contains three patterns. This is done for both sides of the board, and it is very important that the individual patterns are stepped very precisely, otherwise there will be a displacement or misregistration between the drilled holes and the pads.

3.8.6a Shearing

The base material is cut into convenient panel sizes using a guillotine having a shearing width of at least 1000 mm (40 in). Although epoxy glass laminates are much harder on the shear blades than phenolic paper, it is easier to shear the former satisfactorily because the glass cloth protects the edges against breakage or delamination. If a printed circuit board is to be clean-cut to its final shape, this can be done by shearing, and the accuracy depends almost wholly on the positioning of the sheet to the shearing blades. The board is aligned so that the corner marks are flush with the fixed shearing blade. Because of the unavoidable parallax, the tolerance cannot be expected to be much better than $\pm 0.3 - \pm 0.5$ mm ($\pm 0.012 - \pm 0.02$ in). For this reason, shearing is very seldom used for clean-cutting printed circuit boards to their final shape.

3.8.6b Sawing

Sawing is used exclusively for clean-cutting of edges, provided that the edges are straight and accessible for sawing. The saw is normally circular and is made from either good steel or is coated with diamond grains for longer life. The operator can either align the board to the corner marks, or he can use a sawing jig. In the latter case, the board is fixed to the jig by means of guide-pins or dowels which fit into guide-holes drilled within the contour of the board. The sawing jig is moved along a guide which is adjusted in relation to the saw, and a high degree of reproducibility is achieved.

The sawing jig reduces the tolerance to ± 0.2 mm (± 0.008 in). If the board is aligned by eye, the tolerance should be regarded as being the same as for shearing.

3.8.6c Blanking

Blanking is a process taken over from the metal industries and used for blanking both phenolic-paper and epoxy-glass laminates. If the batch quantities are high, it may be economical to adopt blanking of the contour. The more irregular the contour, i.e. with cut-outs, round corners or circular edges, the bigger is the advantage. The tooling costs must, of course, be measured against the potential savings. By way of example, a contour blanking tool for the board shown in Fig. 3.17 costs about £300–400 and the saving by changing from ordinary sawing to blanking is around 20 pence. A very coarse assessment, excluding the amortization of the tool, shows that the investment is justified if the quantity exceeds 2000 boards. For many electronics companies this is a very high quantity and some companies standardize the size of their boards, as far as possible, so that the blanking tool can be used for a range of models. The tolerance for blanking the contour of boards having a thickness of maximum 1.6 mm (1/16 in) is $\pm0.1–\pm0.2$ mm ($\pm0.004–\pm0.008$ in).

Very small boards are usually regarded as rather difficult to handle when the components are to be inserted or the boards are to be machine-soldered. In such cases, the blanking tool can be designed with a so-called 'return-to-the-blank' feature. This means that the board, after being blanked, is pressed back into the panel where it remains held by friction. The entire production panel is delivered to the customer who assembles and solders the panel as a unit and then pushes out the individual boards.

Blanking will, in general, produce rough edges, particularly when the tool becomes worn, but if the circuitry does not come too close to the edges, this should be of little practical importance.

3.8.6d Routing

Contouring by routing is performed by means of a machine called a pin router. With the aid of a routing jig the board is moved past a vertical side mill and routed. The board is provided with internal guide-holes (often named tooling holes) so that it can be attached to the jig. The routing jig, in turn, is guided in relation to the mill by holding it against a bushing concentric with the mill. In this way, it is possible to route the edges of the board with a tolerance of $\pm0.1–0.2$ mm ($\pm0.004–\pm0.008$ in), depending on the care shown by the operator. Pin routing gives a smooth edge but the process is fairly slow because a significant amount of material has to be removed. If the volume is high, and the edges are specified to be very smooth, the boards can be blanked slightly oversize and then pin routed. In this way, the amount of material to be removed is reduced and the routing speed can be increased.

Recently, numerically controlled routing machines for contouring printed circuit boards have appeared on the market. Some of these machines have four spindles and a rather complex electronic control system which ensures a constant working speed regardless of the shape of the board. It is claimed that an NC router can route approximately 1600 boards per hour of size 100×150 mm (4×6 in), whereas the capacity of simple routers is around 60 pieces per hour. In spite of the high price, NC routers have become fairly popular with the leading PCB manufacturers and it is to be expected that such machines will, in the long run, supersede the blanking method which requires expensive tooling.

In order to facilitate the insertion of printed circuit boards into matching contact receptacles, it is common practice to chamfer the connector edge on both sides. This is done using a small milling machine in which the board is moved at the desired angle past a small fast-running side mill.

The polarizing slot which serves to prevent incorrect insertion of the board in the connector and to ensure correct mating of the contacts, should preferably be pin or NC routed. Sometimes the slot is sawn, often by holding one of the edges of the board against a guide. If the overall contouring has been done by shearing or sawing without using a jig, the positional tolerance of the edge may be as much as $\pm 0.3 - \pm 0.5$ mm ($\pm 0.012 - \pm 0.02$ in). Therefore, the slot will be displaced by at least the same amount which, in many cases, is detrimental to the function of the edge connector. For this reason, the positional tolerance of the slot (not in relation to the edges but rather to the pattern itself) should be indicated on the master drawing. This leads to the introduction of a reference system which is discussed in Sections 4.14.1 and 8.9.3.

3.9 Sequence of the Total Manufacturing Process

In the previous sections the total manufacturing process has been broken down into the individual processes. We shall now collect them and describe the manufacturing sequence by means of flow charts. We shall discuss briefly the preparatory work (Fig. 3.20), fabrication of boards with non-plated through holes

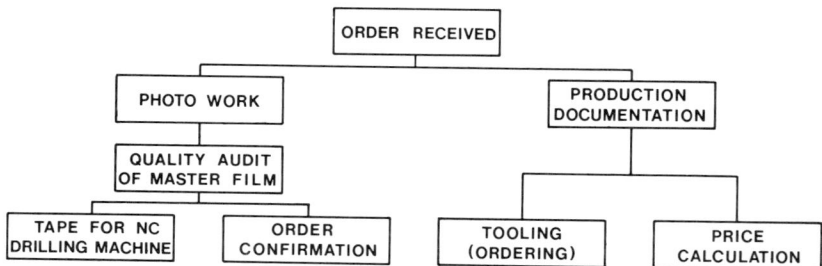

Fig. 3.20 The PCB manufacturer's preparatory work.

(Fig. 3.21), fabrication of boards with plated through holes (Fig. 3.22) and the fabrication of boards with plated conductors. The total manufacturing process is illustrated in a series of photographs (Figs. 3.23 – 3.41) provided by Bepi (Electronics) Limited, Galashiels, Scotland.

3.9.1 Preparatory Work

Although the flow chart in Fig. 3.20 is, on the whole, self-explanatory it may be worthwhile to add some comments.

The confirmation of the delivery time will, in most cases, be based on an assessment of the available capacity which is, usually, limited by traditional bottlenecks. These vary from manufacturer to manufacturer, but very often the most important bottleneck is the drilling process. If the order can be put through the department causing the bottleneck then there should not be any significant delays elsewhere in the factory.

Some PCB manufacturers do not wish to confirm the delivery before the artwork, and the master films derived from the artwork, have been quality audited and accepted. Amongst many other aspects, the registration of the pads from side to side is verified. This is a very important point since a misregistration of a pad on the component side, with respect to the corresponding pad on the solder side, might result in the drilled hole breaking the annular ring on one side of the board. Also, the manufacturer will check the dimensional tolerances of the board to ensure that they are within his capability, e.g. the diameter of the pads relative to the diameters of the drilled holes. It is quite evident that the smaller the pad is relative to the hole, the more important is the requirement for good registration between the two sides of the board. Finally, the conductor spacings and widths are checked. The most serious shortcomings in many artworks are misregistration and annular rings of insufficient width. One of the chief purposes of this book is, therefore, to highlight and discuss methods for the preparation of precision artwork suitable for the manufacture of high-quality printed circuit boards at reasonable cost.

The above quality audit of the documentation package delivered by the customer will, in many cases, delay the confirmation or acknowledgement. In return, the customer and the PCB manufacturer will benefit through improved quality and improved delivery performance.

Leading PCB manufacturers have many customers, each having his own specification and quality level. The production documentation must indicate clearly what is going to happen in the various production departments in order to produce the specified product. Usually, special instructions for non-standard processes and finishes are recorded at this stage.

If manufacture requires special tooling, for example, blanking tools, these will probably be purchased by the PCB manufacturer from a sub-contractor. The delivery time can vary from two weeks to three months depending upon the complexity of the tool.

The manufacture of printed circuit boards must be regarded as 'manufacture to customers' specification'. Only rarely are boards manufactured to be supplied 'off the shelf'. If an electronics company develops, for example, 100 new models of printed circuit boards each year, the purchasing manager would be overwhelmed were he to request and receive quotations for each model from a range of suppliers. The purchaser will normally have good business relations with a limited

```
        ┌─────────────────────┐
        │     CUTTING OF      │
        │ PRODUCTION  PANELS  │
        └─────────────────────┘
        ┌─────────────────────┐
        │    DRILLING OF      │
        │  LOCATION  HOLES    │
        └─────────────────────┘
        ┌─────────────────────┐
        │    CLEANING OF      │
        │   COPPER  FOIL      │
        └─────────────────────┘
┌──────────────────┐      ┌──────────────────┐
│ SCREEN  PRINTING │      │ LAMINATING  WITH │
│                  │      │    DRY FILM      │
└──────────────────┘      └──────────────────┘
                          ┌──────────────────┐
                          │    EXPOSURE      │
                          │  DEVELOPMENT     │
                          └──────────────────┘
        ┌─────────────────────┐
        │    INSPECTION       │
        │    TOUCH-UP         │
        └─────────────────────┘
        ┌─────────────────────┐
        │     ETCHING         │
        └─────────────────────┘
        ┌─────────────────────┐
        │  STRIPPING  OF      │
        │  ETCH-RESIST        │
        └─────────────────────┘
        ┌─────────────────────┐
        │  Ni/Au – PLATING    │
        │ (EDGE CONNECTOR)    │
        └─────────────────────┘
┌──────────────────┐      ┌──────────────────┐
│   DRILLING OF    │      │   PUNCHING OF    │
│      HOLES       │      │ HOLES & CONTOUR  │
└──────────────────┘      └──────────────────┘
┌──────────────────┐
│   TRIMMING OF    │
│    CONTOUR       │
└──────────────────┘
        ┌─────────────────────┐
        │  CHAMFER & SLOT     │
        │ (EDGE CONNECTOR)    │
        └─────────────────────┘
        ┌─────────────────────┐
        │  SOLDER  MASK       │
        │ (SCREEN  PRINTING)  │
        └─────────────────────┘
        ┌─────────────────────┐
        │    NOTATION         │
        │ (SCREEN  PRINTING)  │
        └─────────────────────┘
        ┌─────────────────────┐
        │   PROTECTIVE        │
        │    COATING          │
        └─────────────────────┘
        ┌─────────────────────┐
        │   INSPECTION        │
        │    DESPATCH         │
        └─────────────────────┘
```

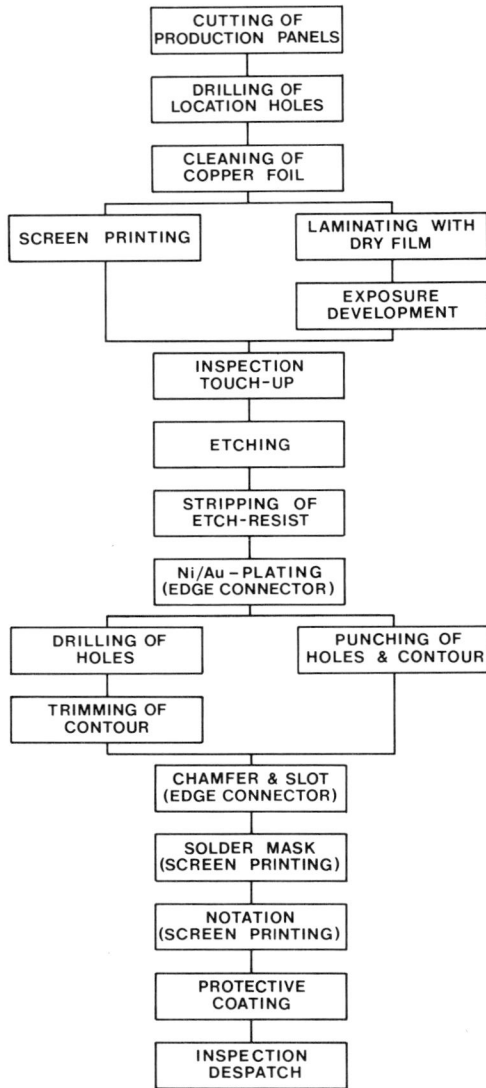

Fig. 3.21 Manufacture of non-plated through boards.

number of PCB manufacturers, and, in the course of time, he will establish a high level of confidence in the manufacturers' pricing, quality and delivery capability so that, on occasions, the order may be placed without quotation. Often it is urgent to start a trial production and, as it is a well-known fact that the documentation package for the printed circuit boards is always late, the quotation stage can easily delay the start by several weeks. It is, however, recommended that the

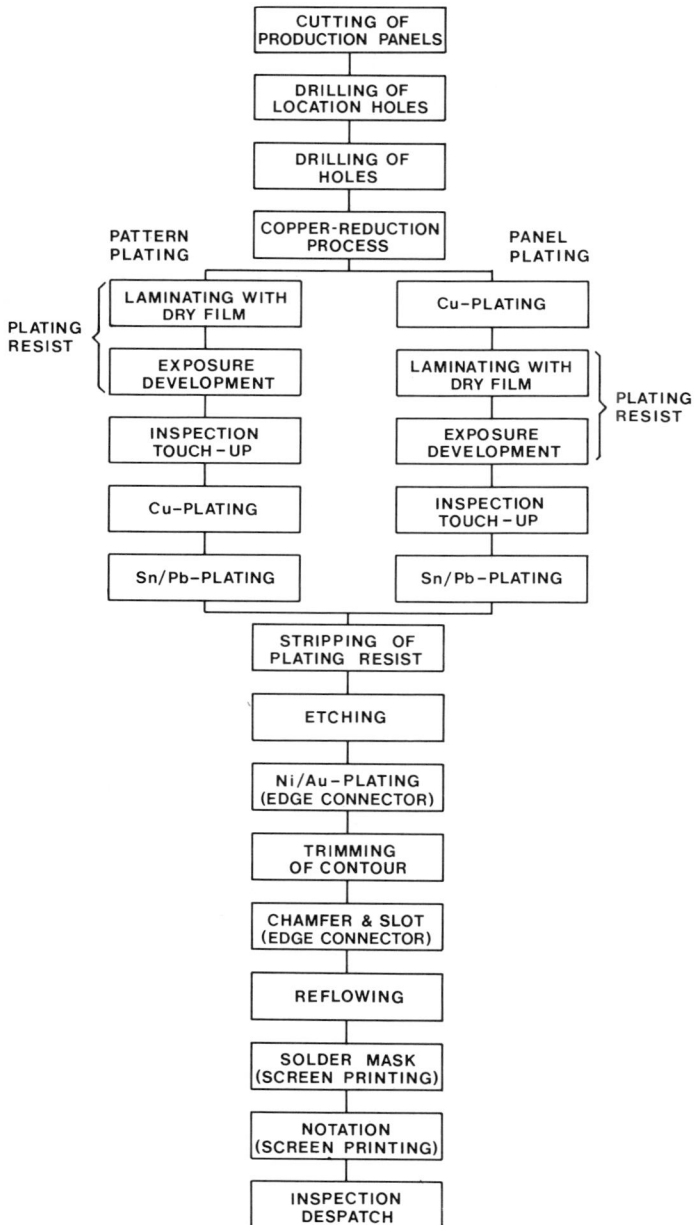

Fig. 3.22 Manufacture of plated through boards.

purchaser, from time to time, checks the market prices by requesting quotations from alternative sources.

The quality of the photographic work is critically important. It may, therefore, be practical to permit the PCB manufacturer to produce the photographic films. This way he will be responsible for the quality of the master films, a sensible step when one considers that he is, in all probability, best qualified to assess them.

If possible, the customer should make an arrangement whereby the PCB manufacturer automatically delivers a 1:1 positive copy of all films, whether they are for new models or up-issues of existing films. In this way, the customer ensures that his files are kept up-to-date, much appreciated by the incoming goods' inspector and also the the production assembly engineer. The cost of the extra film is usually included in the normal tooling costs by the PCB manufacturer and extra paperwork is avoided.

3.9.2 Manufacture of Boards with Non-plated Through Holes

The flow chart in Fig. 3.21 does not require much explanation since all part-processes except the application of the solder mask, the screen printing of the component identification and the surface protection, have been dealt with briefly in this chapter.

A solder mask—also called a solder resist—is a heat-resistant epoxy ink (very often a two-pack epoxy ink) which is applied to the board by a screen-printing process in such a way that it covers the pattern with the exception of the solder pads and the edge connector, if any. The main purpose of the solder mask is to prevent soldering faults which, in the case of automatic soldering (drag or wave soldering), occur as solder bridges between conductors and solder pads. Boards with narrow conductor spacings of 0.3–0.5 mm (0.012–0.02 in) often have a high susceptibility to solder bridging, even when the artwork is correctly produced. In the author's experience, the number of solder bridges in a fairly complex mother board can be as high as 15 but these can be avoided if a solder mask is applied before the soldering operation. The savings in the assembly department can far outweigh the additional cost of the solder mask—some 20 pence per board.

A component identification (component ident or notation) is printed on the component side and facilitates assembly work. The component notation, e.g. R101, will usually be covered when the component is inserted and will, therefore, be of little assistance at a later stage when repairs are necessary. Digressing for a moment, it should be noted that a drawing showing both the component notation and the component side of the board should be included in the service manual. This will make life much easier for the field service technician at the repair stage. The combined drawing is made by contact printing the films of the component notation and the component side on top of each other.

The protective coating serves to preserve the solderability of the solder pads which would otherwise be oxidized. The bare copper also oxidizes during manufacture and it is necessary to restore the solderability before the protective coating is applied. The last stage before the application of this protective coating is, therefore, a mild etch followed by a water rinse. The protective coating, which has to be applied immediately after the solderability has been restored, can be either a water dip, a colophony lacquer, or a tin/lead coating applied by roller tinning. In the last process, the board is passed between two rollers, the lower of which is running in molten solder, and transfers a thin layer of solder to the bare copper surface of the board.

3.9.3 Manufacture of Boards with Plated Through Holes

The flow chart given in Fig. 3.22 is based on image transfer using dry film. This process, called photomechanical printing, is used for fine-line boards having small conductor widths, narrow conductor spacings and high hole-to-pad ratios (where the diameter of the hole divided by the maximum dimension of the pad approaches unity). Screen printing, using wet inks, is used for less-demanding boards.

During the last few years there has been a noticeable trend towards complex printed circuit boards and, consequently, photomechanical printing has become more predominant. Recently, however, ultraviolet curing inks have become available which permit screen printing to closer tolerances than before. Therefore, there is now a trend towards using screen printing on some types of PCBs which were previously photomechanically printed.

3.9.4 Manufacture of Boards with Plated Conductors

In Section 3.3 it was mentioned briefly that it is possible to plate conductors and solder pads with tin/lead, and the method was indicated in Fig. 3.4. See also Section 3.4.1. This technique can be used for the manufacture of non-plated through boards and gives a protective coating of reflowed tin/lead plating to the conductors and pads. This protective coating is similar to that of plated through boards and is much better than the protective coating mentioned in Section 3.9.2 above.

Conductor plated boards are sometimes found in professional equipment where, for cost reasons, it is necessary to use a combination of non-plated through and plated through boards, but where the surface protection has to be of a very high quality in order to prevent corrosion in certain environments.

The flow chart shown in Fig. 3.22 also illustrates the above process if the copper-reduction stage is omitted. As the hole walls are non-conductive, neither copper nor tin/lead deposition will take place in the holes.

It is, of course, more expensive to produce a conductor plated board than a simple 'print-and-etch' board, but somewhat cheaper than a plated through board. The copper-reduction stage and the subsequent plating of the holes are among the most critical processes. It is necessary to perform a thorough inspection after these stages in order to avoid a high scrap rate in the later production stages. Therefore, the reduced price of conductor-plated boards compared with plated through boards is due to the saving of these processes and the corresponding inspection operations.

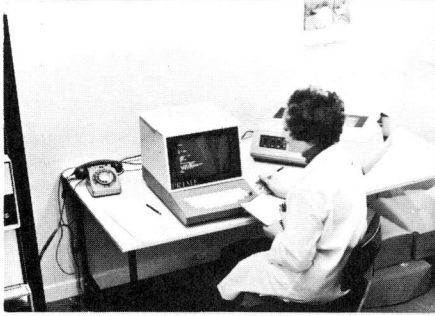

Fig. 3.23 Production planning and preparation of production documentation using a mini-computer.

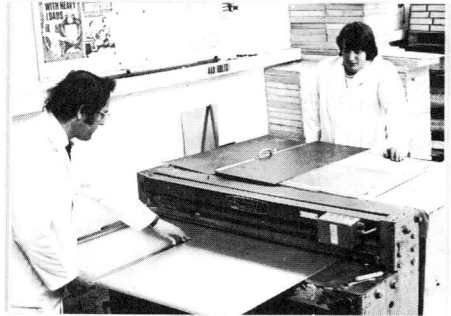

Fig. 3.26 Cutting production panels on a guillotine.

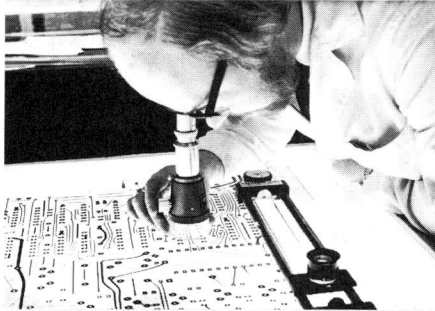

Fig. 3.24 Quality audit of artwork.

Fig. 3.27 NC drilling. The machine has four drilling heads and is sturdily constructed to ensure drilling accuracy. The electronic tape control unit is shown to the right of the drilling machine.

Fig. 3.25 Preparation of punched tape for the NC drilling machine. Note the image being projected on the viewing screen to the right of the operator's head.

Fig. 3.28 Manual drilling. The drill jig is placed on top of the stack of boards and is moved under the point of the sensing device. The drill head is mounted under the table.

Fig. 3.29 Lamination of dry film.

Fig. 3.32 Quality inspection of screen printing. Small faults are rectified by touch-up.

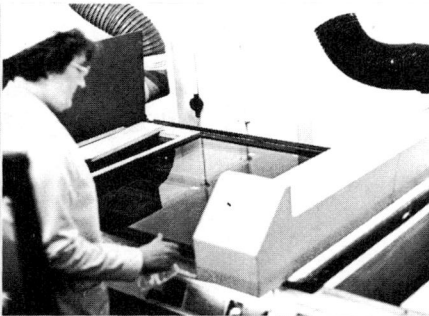

Fig. 3.30 Image transfer by means of dry film. The light source is located in a small 'carriage' which is passed along the top of the machine and exposes the board to light through the camerawork.

Fig. 3.33 A view down along the plating line shows the various vats. A computer-controlled travelling crane lifts a frame up from a vat and is about to move it to another vat.

Fig. 3.31 Screen printing. The screen printing machine is shown with screen-printed boards in the foreground.

Fig. 3.34 Etching machine. The boards are fed into one end of the machine and past the etching section.

56

Fig. 3.35 Gold plating of edge connectors. Note the shallow vats for the nickel and gold plating.

Fig. 3.36 Measurement of the thickness of the metal deposition by means of a betascope.

Fig. 3.37 Contouring using an NC router.

Fig. 3.38 Contouring by blanking.

Fig. 3.39 Chamfering of edge connector.

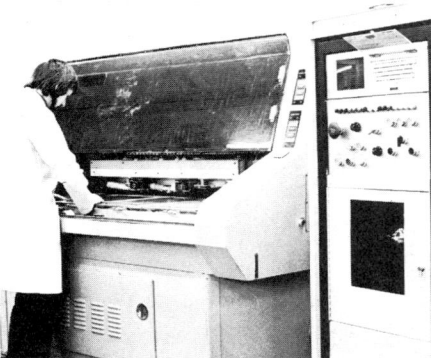

Fig. 3.40 Conveyorized reflowing, cooling, flux removal, and solder brightening.

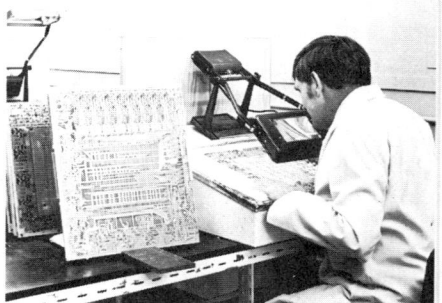

Fig. 3.41 Final inspection.

4

Purchase Specification

4.1 Introduction

A typical purchase specification is set out and discussed in this chapter. It must not be regarded as universally applicable in the form given here since not all customers have the same requirements. The reason for this discussion is to show how the information given in the preceding chapter can be translated into a series of instructions which, if followed, ensure that the boards manufactured will be to quality standards acceptable to the customer. Each clause of the specification is supported by explanatory comments of a practical nature for better understanding.

A general purchase specification serves two main objects. Firstly, it ensures that the customer has considered all the technical requirements of the boards which he is purchasing, and secondly it provides the basis of the agreement by the manufacturer to produce boards in accordance with these requirements. It is, therefore, important that the specification be clearly understood by both parties before an order is placed and accepted. Only by adhering to this procedure is there a good chance that the boards produced will be to the customer's quality standard.

The general purchase specification is supplementary to the detail requirements embodied in the artwork and drawings and enables general requirements for a whole range of boards to be omitted from the drawings and artwork of individual boards with greatly enhanced clarity in the presentation of the essential information.

4.2 General Conditions

4.2.1 Range of Validity

The purchase specification contains all the general technical requirements to be taken into consideration when ordering and manufacturing printed circuit boards. In the case of a conflict between the purchase specification and the documentation package (artwork and master drawing) the latter shall be taken as valid.

If the manufacturer's own quality standards or his manufacturing practices differ from the requirements given, and advantages in the form of cost savings or faster delivery can be obtained by following the manufacturer's own specification, the customer should be consulted.

When he has received the documentation package (see Section 4.3) the manufacturer is obliged to quality audit the contents and to object to requirements which, in his opinion, cannot be fulfilled in manufacture. He is also obliged to object if the quality of the artwork is unlikely to ensure an adequate yield of acceptable boards in production.

All modifications arising from the technical audit should either be confirmed in writing or up-dated issues of the artwork should be forwarded.

Comments

It will, from time to time, be found necessary to impose, in the case of individual boards, requirements which differ from those of the general specification. These requirements should be embodied in the drawings of the individual board where they will take precedence and the corresponding clauses of the general specification will be ignored.

In the author's experience, a customer is usually well advised not to force a manufacturer to follow methods other than those he already uses as manufacturing standards. To manufacture printed circuit boards requires a good understanding and close cooperation between the customer and the manufacturers, and it is usually worthwhile for the customer to give consideration to the views of an accredited manufacturer who is a specialist in his own field.

Regarding the quality audit of the artwork, the reader is referred to the beginning of Section 3.9.1.

4.2.2 Purchase Order

The printed circuit boards are identified in the purchase order by their part number and their drawing issue number. It is the printed circuit board manufacturer's responsibility to use the correct issue of the artwork for the solder and the component sides, the component notation and the solder mask as specified on the purchase order and master drawing.

Once the quality audit is complete and agreement has been reached with the customer on any changes and modifications, the manufacturer acknowledges the order and states a delivery lead time. If, during the production of the boards it proves impossible for the manufacturer to keep the promised time of delivery he should inform the customer immediately.

Comments

Although a quality audit of the artwork can delay the acceptance of the order it is advantageous to permit the manufacturer to make his assessment and to bring possible faults to light at this stage rather than later in the production process when resulting delays may be longer and more costly.

It is important that the manufacturer advises the customer of possible delays of delivery so that measures can be taken to minimize the resulting disorganization. The customer must, for his part, realize that it is a complicated matter to make printed circuit boards involving interdependent processes. Small things can upset the production plans of even the best and most reliable printed circuit board manufacturer.

4.2.3 Fulfilment of Order

Short delivery must not occur. An over-delivery of $+5\%$ or $+2$ boards, whichever is the greater, will be accepted. An order is not regarded as fulfilled unless all the items stated in the order are delivered in full.

Comments

Short delivery is completely unacceptable because of the resulting breakdown of the customer's production planning. By accepting a certain over-delivery the customer is in a better position to ask the manufacturer to start up with a larger safety margin than usual. Manufacturers do not like to be left with boards in excess of requirements as they become obsolete if the boards of a following order are up-issued.

4.2.4 Number of Batches

An order or part-delivery of an order should always be manufactured as a single batch. If this is not the case, e.g. because of failure in production, the manufacturer shall always inform the customer, and the boards originating from supplementary batches shall be delivered segregated in order to enable them to be inspected separately.

If the printed circuit board manufacturer wishes to supplement an order with excess boards of the same issue from an earlier order, he shall obtain approval if the boards are older than 10 weeks. Such boards shall be delivered segregated as prescribed above.

Comments

If the boards originate from different production batches, recurring faults may be found in one batch but not in others, for example, a plating fault due to a process being out of control. If the customer's Incoming Goods' Inspection is based on a certain sample plan, a mixture of acceptable and faulty batches can lead to an erroneous result.

By supplementing an order with surplus boards, the customer runs the risk that such boards have deteriorated and that the shelf life in the customer's stock is in consequence reduced, especially in respect of solderability.

4.2.5 Rejection of Boards

Boards which are rejected by the customer's Incoming Goods' Inspection shall be returned to the manufacturer. A test report shall also be sent immediately to the manufacturer, accompanied, where appropriate, by photographs indicating or highlighting the faults.

When the manufacturer accepts the reason for rejection, he is obliged to remake or rework the boards within three weeks, unless the customer accepts a longer replacement time. Alternatively, the customer is entitled to claim a credit note for the missing delivery.

The customer is also entitled to claim that irreparable boards be remade as per the latest issue, i.e. not necessarily the same issue as that of the rejected boards. In this case, additional tooling costs as well as a reasonably longer time of delivery will be accepted.

Comments

If the customer, at the planning stage, has incorporated a certain safety margin in the requested time of delivery, the manufacturer should be able to make replacement boards within this period of time. In this way, the customer can safeguard his production plans. In certain cases, however, the customer may be interested in cancelling the delivery, for example, if his own stock meets the demand. If it is only a matter of a few missing boards, this may also be an advantage to the manufacturer because it is very expensive to start up the whole production process again just to make a few extra boards.

Especially in the first phases of a project, when the issues are being up-dated at short intervals, and in cases where the demand can be met by means of deliveries from alternative suppliers, it is of no advantage to receive delayed replacement boards which are more or less obsolete.

4.2.6 Marking of Boards

All boards, where size permits, shall be marked with the manufacturer's name or company mark, and with the date of manufacture indicated as week number and year, e.g. 46–77.

All boards shall bear a final acceptance stamp.

The desired location of the company mark and the date marking will be indicated by a special frame on the artwork, usually on the component side. The company mark is to be included in the artwork by the printed circuit board manufacturer and the date marking stamped or inked on after the final acceptance of the board.

Markings shall be legible to the naked eye and shall in no way impair the electrical performance of the board.

Comments

On account of possible complaints about faults being found after the boards have passed the Incoming Goods' Inspection, e.g. a sporadically occuring reduction of solderability, it is expedient that the manufacturer and the date of delivery can be identified. When two or more suppliers are involved, it is particularly important that complaints of faulty work are directed to the correct company.

4.2.7 Packing of Boards

Boards over a certain size—as agreed on between the manufacturer and the customer—shall be packed individually in moisture-proof transparent poly-ethylene bags. The excess length of the bag shall be folded over and fixed with a piece of tape. The tape shall not seal the bag completely and its removal must not cause damage to the bag.

The stack of individually packed boards must not be held together by means of tape as the edges of the bags might be torn when removing the tape. Instead, the stack is to be packed in a large polyethylene bag and this may be closed by means of tape. The part number of the boards and the issue number to which the boards have been made shall be marked on the outer bag.

Small boards may be packed in moisture-proof polyethylene bags in lots of not more than 25 boards. The individual boards shall be separated by tissue paper, free from acid, sulphur, chloride, etc., which can impair the solderability of the boards. The part number and the issue number to which the boards have been made shall be marked on the outer bag.

If the boards are of the same type and are manufactured in two or more batches, these shall be identified by marking the batch number on the outer bag.

Comments

Opinions differ as to whether the bag should be sealed airtight or not. The author is of the opinion that the boards should be allowed to breathe since if water condenses in the bag it is probable that the solderability will be impaired.

It is important that the bags can be opened by the Incoming Goods' Inspection without being damaged. For long shelf life, the bags must be undamaged, and it would be wrong for the customer to have to re-pack the boards in new bags after the inspection, just because the tape is too sticky.

4.2.8 Test Reports

When requested in the purchase order, the manufacturer shall furnish a certified test report giving the results of tests performed on representative samples of the shipment to determine that the boards conform to requirements.

62

Comments

In the case of military and aerospace applications, the user requires a certificate of quality. For normal civil electronics' production this is not usually called for because of the significant cost of the testing involved.

4.3 Documentation Package

4.3.1 Master Drawing

The master drawing provides the following data, sufficient for the manufacture of the board:

(a) a dimensional drawing of the board giving the overall length and width together with size and dimensional location of special cut-outs, holes, notches and any other mechanical features
(b) a board specification where the individual data of the board is specified, e.g. board material, plating, finish, tolerances
(c) the part number and issue numbers of the artwork to be used.

Comments

It is very important to keep the various artwork issues in strict order. A numbering system, enabling these to be identified is discussed in Section 8.9.2 to which the reader is referred.

4.3.2 Artwork

A set of artwork comprises the following items:

(a) artwork for the solder side
(b) artwork for the component side
(c) artwork for the component notation
(d) artwork for the solder mask

The artwork is usually twice or four times full size. Exact copies of the original artwork, made by dry diazo contact printing on sensitized polyester films, are sent out as manufacturing information.

Comments

The preparation of artwork is time consuming and costly and the original is relatively fragile and cannot be folded or rolled for transport. For this reason, a contact print which can, if necessary, be rolled is the recommended method of sending this information to the manufacturer. Dry diazo contact printing has the additional advantages that it is easy to produce, very accurate, and inexpensive. The technique of making duplicate artwork is comprehensively described in Chapter 7.

4.3.3 Photomaster

Usually the manufacturer produces the photomaster (master pattern) by photo-reduction from duplicate artwork. He therefore has the responsibility for the quality of the photomaster, i.e. that it accurately reproduces all features of the duplicate artwork.

The manufacturer is requested to send, on a routine basis, positive films prepared from all the photomasters. This applies to all issues of the artwork. The charge for these films shall be included in the price of the camera work.

Comments

All inaccuracies of the photomaster will be repeated in the boards. In the author's experience it is better to place the responsibility for the conformity of the photomasters to the artwork supplied by the customer firmly with the PCB manufacturer.

Photographic film is, at its best, not an entirely stable material. Temperature and humidity changes and ageing will all produce stretching or shrinkage to some extent. In consequence, poor registration of the patterns on the opposite sides of the board may arise from these changes as well as from inaccurate image transfer.

In addition, photomasters are somewhat fragile and liable to be damaged in the production process. For this reason it is usual to have the photomasters prepared in the PCB factory where they can be replaced quickly should the need arise. The quality and age of the film used can be better controlled this way since the manufacturer makes day-to-day use of the material unlike his customer.

4.4 Copper-clad Laminates

4.4.1 General Requirements

The laminates shall be free from dirt, stains, corrosion products, finger-prints, and mechanical defects such as major scratches or bruises.

The laminate shall show no fractures, measling, crazing, blistering delamination, haloing, or weave texture and exposure. In non-critical areas a slight measling or crazing can be accepted.

There shall be no evidence of lifting of the copper.

Comments

Normally, the customer just specifies the quality of the end product, leaving the quality specification of the raw material to the manufacturer. On the other hand, there are good reasons for prescribing the use of a material which is known to produce satisfactory results.

Definitions of the defects listed are given below as extracts from *Acceptability of Printed Wiring Boards, IPC-A-600A* published by the Institute of Printed Circuits (IPC).

Measling: A condition existing in the base laminate in the form of discrete white spots or crosses below the surface of the base laminate, reflecting a separation of fibres in the glass cloth at the weave intersections.

Crazing: A condition existing in the base laminate in the form of connected white spots on or below the surface of the base laminate, reflecting the separation of fibres in the glass cloth and connecting weave intersections.

Blistering: A localized swelling and separation between any of the layers of the base laminate and/or between the laminate and the metal cladding.

Delamination: A separation between any of the layers of the base laminate and/or between the laminate and the metal cladding.

Haloing: A condition existing in the base laminate in the form of a light area around holes and/or other machined areas on or below the surface of the base laminate.

Weave Texture: A surface condition in which the unbroken fibres are completely covered with resin but still exhibit the definite weave pattern of the glass cloth.

Weave Exposure: A surface condition on which the unbroken woven glass cloth is not uniformly covered by resin.

The customer often calls for a high standard of appearance and finish of the boards and, consequently, boards which in all other respects are functional and excellent, are at times rejected by the customer's Incoming Goods' Inspection.

It is necessary to distinguish between functional and cosmetic requirements and the latter is of minor importance. It is quite obvious that fractures or delamination of the base material are unacceptable because the strength of the board will be impaired. Until now, measling has been a reason for rejection but a recent investigation carried out by IPC has shown that there is little justification for this. Measling is accompanied by an insignificant reduction in the strength of the material of approximately 10%. Localized crazing is usually harmless. It should, however, be noted that crazing around a hole implies that electrolyte can be absorbed during the plating process, which could have detrimental consequences. Blistering is caused by expansion of air trapped in small cavities near the surface. It can be regarded as a local delamination and, as long as it does not lift part of the pattern, can be considered as harmless.

4.4.2 Base Material

Unless otherwise stated in the master drawing, the base material shall be epoxy glass, NEMA type FR4, flame retardant.

Comments

The printed circuit board industry in its early days used exclusively phenolic-paper laminate as the base material and this, largely on account of cost, continues to be the practice in the manufacture of single-sided boards for entertainment products. The need for a more stable and heat-resistant material led to the use of epoxy glass for professional applications and with the dominance of plated through boards in this market, epoxy glass is now almost universally used for the professional product.

Flame retardant characteristics are obtained by adding compounds of bromine to the epoxy. This is satisfactory provided the quantity is not excessive in which case evaporation of the bromine can cause blow-holes in the solder joints.

4.4.3 Thickness and Tolerance of Base Material

Standard thickness	Tolerance on thickness
0.8 mm (0.031 in)	±0.165 mm (±0.0065 in)
1.6 mm (0.062 in)	±0.125 mm (±0.0050 in)
2.4 mm (0.094 in)	±0.230 mm (±0.0090 in)

The thickness most used, and normally used for plug-in boards, is 1.6 mm (0.062 in). Edge receptacles for this type of board are usually designed to accept board tolerances of ±0.18 mm (±0.007 in) and consequently are capable of accepting boards with some degree of warping.

4.4.4 Thickness and Tolerance of Copper Foil

Thickness	Tolerance on thickness
17.5 μm ($\frac{1}{2}$ oz/ft^2)	±5 μm (±0.0002 in)
35 μm (1 oz/ft^2)	+10 μm, −5 μm (+0.0004 in, −0.0002 in)
70 μm (2 oz/ft^2)	+18 μm, −8μm (+0.0007 in, −0.0003 in)

Comments

The thickness of the copper foil is not critical since the tolerance of the copper-plating thickness is in the region of 30 μm (0.0012 in).

A recent development is the introduction of a very thin copper foil of thickness 5 μm (0.0002 in) which has considerable advantages when precision fine-line work is required. It is not yet, however, readily available and some sources report that it is liable to have pinholes through which resin can flow to the outer surface during manufacture of the laminate.

For the present, 17.5 μm ($\frac{1}{2}$ oz/ft^2) copper foil is the thinnest commercially available, and is gaining in popularity because of its inherent etching advantages.

4.4.5 Warp and Twist

Thickness	Tolerance on warp and twist	
	Single-sided (mm/mm, in/in)	Double-sided (mm/mm, in/in)
0.8 mm (0.031 in)	0.015	0.010
1.6 mm (0.062 in)	0.010	0.007
2.4 mm (0.094 in)	0.008	0.005

Comments

Warping is checked by placing the finished board on a flat surface with the convex surface upwards and measuring the maximum vertical displacement.

Warp and twist depend on the distribution of the pattern over the board. Their importance should not be exaggerated since, in the completed equipment, the board is mechanically restrained in a flat position by screws to the chassis or by board guides. The flatness of an edge connector, however, is more important because it can be difficult to force a warped edge connector into the edge receptacle. For this reason, it is expedient to specify a tolerance for the flatness of the edge connector area. (See Section 4.9.5.)

4.5 Electroplating

4.5.1 General Requirements

Electroplating shall show no evidence of 'burning' unless this is quite unavoidable owing to the pattern configuration. Spurious deposits shall be removed from critical areas provided their removal does not detract from the appearance or performance of the board.

Exposed copper will be allowed at the edges of conductors and solder pads of non-reflowed boards.

Scratches in the tin/lead plating are permissible provided the underlying copper is not visible.

Requirements for gold-plated edge connectors are given in Section 4.5.5.

Comments

'Burning' of solder pads which are located far from other parts of the pattern can easily occur. The high current density to which these parts are subjected in the plating process results in a coarse and dark-coloured deposit usually described as 'burning'. Apart from the appearance, the effect is excessive thickness of the plating. The current density can be reduced in such areas by the addition to the pattern of 'idle' copper conductors—so-called 'robbers'—which divert some of the current and make the current density throughout the pattern more uniform. This technique is discussed in detail in Section 6.6.3.

Some faults, caused by pinholes in the plating resist, were mentioned in Sections 3.2.1 and 3.2.2. These result in unwanted areas or particles of copper on the board and if they are located between a pair of conductors so that the insulation resistance might be affected, they must be removed, for example, by scraping with a knife.

Scratches on the surface, due to insufficient care in handling the plated board should not be a reason for rejection of the board so long as the underlying copper is not exposed and protection against corrosion is still maintained. If the number of scratches is such that the board can be seen to have been grossly mishandled the board will, of course, be rejected. Faults in gold plating of edge connectors are illustrated in Fig. 3.12–3.16.

4.5.2 Copper Plating

The average thickness of the copper plating in holes shall be not less than 25 μm (0.001 in) and nowhere less than 15 μm (0.0006 in), excluding pinholes, voids and pits. (See also Section 4.6.4.)

The total thickness of the copper foil and the copper plating shall be not less than 35 μm (0.0014 in).

The purity of the copper shall be not less than 99.5%.

Comments

This specification of the plating thickness in a plated through hole is in close agreement with IPC–TC–500 (Specifications for Copper Plated Through Holes in Rigid, Two-sided Boards) and MIL-STD-275. It is generally accepted by printed circuit board manufacturers and also appears in the purchasing specifications of many electronics companies.

For satisfactory automatic soldering (drag or wave soldering) of a printed circuit board it is important that the heat transmission is sufficient to ensure that the molten solder wets the whole surface of the hole. One condition is that the copper layer in the hole is sufficiently thick; another is that the heat is not carried away from the solder pad on the top side, which would be the case if too many and/or too wide conductors are connected to the pad. If these conditions are not fulfilled, the temperature at the top of the hole will be insufficient to maintain the solder in a molten state resulting in failure to fill the hole with solder. A copper layer of 25 μm (0.001 in) has proved sufficient to ensure complete filling of the hole. Many PCB manufacturers define the solderability of holes assuming that no conductors are connected to the pad on the component side. In Section 6.5.7 we shall discuss such problems and indicate methods of improving the solderability by rather simple adjustments of the artwork.

It is superfluous to specify the minimum thickness of the copper plating of a conductor as the throwing power factor ensures at least as much copper on the

68

conductors as in the holes, cf. Section 3.5.1. However, in order to secure a reasonable total thickness, i.e. the thickness of the copper foil plus the thickness of the plating, it is usual to specify a minimum of 35 μm (0.0014 in).

4.5.3 Tin/Lead Plating

The specified thickness of the tin/lead plating in holes and on conductors and pads is at least 10 μm (0.0004 in), and not more than 20 μm (0.0008 in). The tin content of the tin/lead deposit shall be within the range 55–70%, the rest being lead.

Comments

In order to act as a reliable etch resist, the minimum thickness of the tin/lead plating is 6 μm (0.00024 in). Therefore it seems reasonable to specify a minimum of 10 μm (0.0004 in) to allow for variations in production. The maximum thickness of the tin/lead plating is specified for the following reasons: The thicker a tin/lead layer, the more heat is required to melt it in the soldering process since as well as bringing the tin/lead to soldering temperature, the latent heat of fusion has to be supplied. In automatic soldering fusion has to be complete in only 2–3 sec. If, therefore, the tin/lead layer on the hole wall is too thick, fusion and wetting could fail to cover the entire hole wall. Also, the reflowing process (see Section 4.5.6) causes the tin/lead layer to melt. If there is too much tin/lead on the solder pads and the attached conductors, the molten tin/lead tends to run into the holes reducing their diameter or even blocking them completely. Preferably, the tin/lead layer should not exceed 15 μm (0.00059 in), but it has proved necessary to choose a higher value. The commonly accepted plating tolerance of −0% and +100% would give a nominal thickness of only 7.5 μm (0.00030 in) which, in practice, is scarcely sufficient to provide an adequate etch resist.

There is, however, a further problem. The deposit on isolated areas of the pattern will be thicker than elsewhere, as described in the case of copper plating. Some improvement can be achieved by shielding the electrodes in the plating bath but the inclusion of 'robbers' in the pattern serves to overcome this trouble in both copper and tin/lead plating. This approach will be dealt with, in detail, in Section 6.6.3.

The composition of the tin/lead deposit is discussed in Section 3.5.4.

4.5.4 Tin Plating

When specified on the master drawing, or agreed between the manufacturer and the customer, pure tin can be substituted for tin/lead and the same thickness tolerances will apply.

Tin plating cannot be recommended since, like tin/lead, it is prone to shed slivers from the overhang of the plating (Section 3.5.5) and reflowing to eliminate these has to be done at the melting point of tin. At this high temperature there is considerable danger of damaging the board material. In addition, the organic additives used to accelerate the deposition of tin tend to cause dewetting in the reflow process. Nevertheless, pure tin plating was widely used in some countries, particularly in Germany. The new additive methods and the ultra-thin copper foils will reduce the slivering problem and it could be that, with these new techniques, reflowing will be superfluous. On the other hand, the reflowing process produces other advantages. (cf. Section 4.5.6.)

4.5.5 Gold Plating

Unless otherwise specified on the master drawing, edge connectors shall be gold plated as follows:

1. An underlying layer of low-stress nickel: minimum 5 μm (0.0002 in)
2. A top layer of hard gold: minimum 2.5 μm (0.0001 in)
    Vickers hardness: 140–160 kg/mm^2

Comments

The basis on which these standards have been set has been fully discussed in Sections 3.7.2–3.7.4.

4.5.6 Reflowing of Tin/Lead Plating

When specified on the master drawing, tin/lead plated boards shall be reflowed to remove any tin/lead overhang arising from the plating and etching processes.

The edges of conductors and pads shall be covered with tin/lead, and the tolerances on the diameters of the finished holes stated in Section 4.6.2 shall be met. The thickness of the tin/lead layer at the transition from the hole wall to the solder pad shall be not less than 0.5 μm (0.00002 in) after reflowing.

After reflow, there shall be no visual evidence of non-wetting, dewetting, solder webbing, incomplete fusion or slivers.

Comments

Reflowing is accepted as the only safe method of ensuring the absence of slivers from the finished board.

It is important that the tin/lead has been totally melted to ensure correct metallurgical recrystallization. Before reflowing, the tin/lead layer consists of areas rich in tin surrounding small discrete pockets rich in lead. After reflowing, the layer is recrystallized and appears as a homogeneous structure with a bright surface. The freedom from porosity in the reflowed layer ensures the important advantage of a greatly enhanced shelf life, typically from 6 to 12 months and, occasionally, depending on the storage conditions, even up to 24 months.

If a printed circuit board is viewed on edge, a reddish tinge will sometimes be seen. This tinge comes from the edges of the conductors which are not covered with tin/lead by the reflowing process. The component side, however, is not protected against corrosion of the conductor edges whereas the automatic soldering covers the conductor edges of the solder side of the board. In addition, there is a risk that a satisfactory wetting cannot be obtained even on the solder side owing to oxidization of the exposed copper.

Figure 4.1 shows a cross-section of a tin/lead plated conductor before and after reflowing. The conditions which determine whether or not the conductor edges

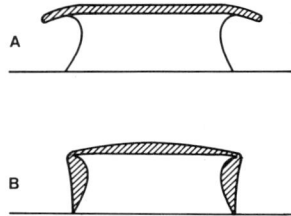

Fig. 4.1 Cross-section of a tin/lead plated conductor. A, Before reflowing; B, after reflowing.

will be covered with tin/lead after reflowing are complicated and will not be discussed here. Suffice to say that the profile of the conductors is a determining factor in achieving coverage of the edges.

In reflowing, the tin/lead at the edges of the holes becomes very thin since surface tension pulls the molten metal away from the sharp edge (see Fig. 4.2). In

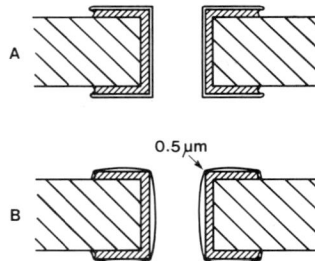

Fig. 4.2 Cross-section of a plated through hole (tin/lead plated). A, Before reflowing; B; after reflowing.

practice, a tin/lead thickness of 0.5 μm (0.00002 in) can be achieved. This does not usually cause any difficulty in the subsequent soldering. Where it is thought necessary (e.g. for space applications) to maintain a greater thickness, the reflow can be carried out at a lower temperature with, however, less certainty of complete recrystallization.

Reflowing reveals many of the defects which can occur in the tin/lead plating.

Non-wetting and dewetting, when they occur, indicate reduced solderability and, if extensive, justify rejection of the boards.

Definitions of some of the faults mentioned in the specification are given below:

Non-wetting: The occurrence of areas which have repelled the molten tin/lead in such a way that the underlying copper is visible.

Dewetting: The occurrence of areas where the molten tin/lead has pulled back from the surface, leaving a micro-thin layer of tin/lead. The underlying copper is not, in this case, visible.

Solder webbing: The occurrence of tin/lead which, after the process, displays a reticulated or netted surface.

4.6 Plated Through Holes

4.6.1 General Requirements

Plating of holes shall be smooth and uniform throughout the hole. The plating shall run continuously from the hole wall to the pad and the conductors connected to the pad. No epoxy smearing shall be found. Roughness shall not reduce either the plating thickness or the finished hole size below the minimum requirements.

The above requirements apply to all boards whether reflowed or not. All holes as drilled shall have a smooth wall without resin smear. No glass fibres shall be observed protruding from the hole walls after plating and reflowing.

Reflowed holes shall remain practically cylindrical and the reflowing shall extend throughout the hole.

Comments

A uniform and smooth surface of the plated through hole shows that the drilling and plating processes have been carried out correctly. Defects ranging from large voids to very large nodules indicate the likelihood of poor solder joints and are a reason for rejection.

Epoxy smearing, already mentioned in Section 3.4.2, forms a partition between the copper plating of the hole and the copper foil of the laminate, so that the connection from the hole wall to the copper foil goes via the plating on top of the pad as illustrated in Fig. 4.3, bottom right. Any tendency to cracking of the plating in the hole wall will be increased where this occurs. Epoxy smearing is only found by the customer when the holes are microsectioned. It is rarely found to be troublesome in plated through boards but is more serious in the case of multilayer boards where it can isolate a pad of one of the inner layers completely from the plated hole wall.

It is a fair requirement that all holes shall be clean, and have no glass fibres protruding from the hole wall. Holes filled with dust and sawing or milling chips, if they occur, indicate a lack of care in the latter stages of manufacture and poor Outgoing Goods' Inspection.

It is difficult to set up criteria for the quality of plated through holes, and especially for reflowed holes. It is, therefore, unavoidable that the general

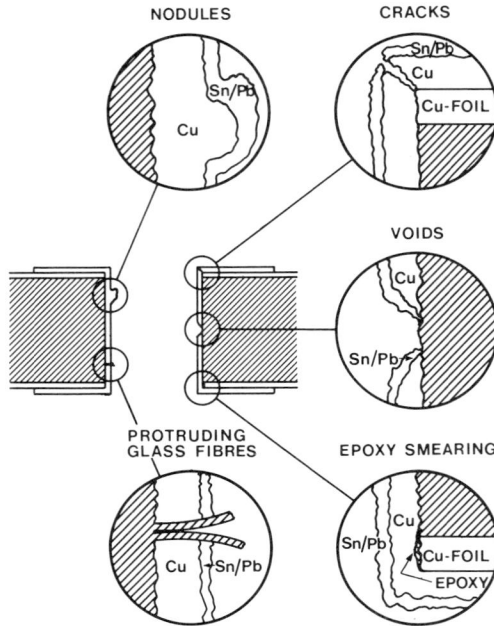

Fig. 4.3 Various faults in plated through holes.

requirements are less than precise. The fact that no official specification on the quality of reflowed holes exists is itself a comment on this difficulty. In practice, the customer is compelled to rely on the manufacturer's ability to produce good, solderable, plated through holes. He will soon become aware whether the manufacturer is able to control his processes so as to deliver, consistently, boards of satisfactory quality. Fig. 4.3 illustrates the various faults in plated through holes.

4.6.2 Diameter Tolerance for Finished Plated Through Holes

The tolerance on the diameters of finished plated through holes, whether or not reflowed, assuming an even distribution pattern, are as follows:

		Tolerance	
Nominal diameter d	*No reflowing*	*After reflowing*
$d \leqslant 0.8$ mm $(d \leqslant 0.031$ in$)$	± 0.10 mm $(\pm 0.004$ in$)$	$-0.1, +0.15$ mm $(-0.004$ in, $+0.006$ in$)$
$0.8 < d \leqslant 5$ mm	± 0.13 mm $(\pm 0.005$ in$)$	± 0.15 mm $(\pm 0.006$ in$)$
$(0.031 < d \leqslant 0.2$ in$)$		

Comments

In determining the drill size to be used, the manufacturer has to make allowance for the thickness of the plating such that the finished holes are within the diameter tolerances. The tolerances as given above are generally recognized as acceptable to most manufacturers. They do, however, make the assumption that no wide variation of the pattern density exists which would make control of plating thickness unduly difficult. No recognition of this appears in any official specification but in some cases it may be impossible to maintain the tolerances over the entire board. In such cases it may be necessary to permit waivers of the tolerance.

Reflowed holes are subject to the additional complication that the molten tin/lead tends to flow from the overhang on to the tracks and thence into the holes causing a reduction in diameter. This effect is more marked in less densely patterned areas of the board where the tin/lead deposit is heavier. For this reason, it is necessary to specify increased tolerances for reflowed holes. Since the minimum limit is determined by the needs of subsequent assembly, the increased tolerance has to be applied to the upper limit which permits the manufacturer to drill the holes slightly larger than normal. The larger number of variables which contribute to the finished diameter of a reflowed hole makes it expedient to have the widest possible tolerances consistent with subsequent component assembly. If there are some holes that require particularly close tolerances these can be specified on the master drawing.

Satisfactory soldering depends, among other things, on the relationship between the diameter of the solder hole and the diameter of the component lead. This is discussed in Section 6.5.4a.

4.6.3 Solder Pads of Plated Through Holes

The solder pads are designed with a nominal width of the annular ring (the ring limited by the solder hole and the circumference of the pad) of not less than 0.3 mm (0.012 in), or in the case of non-circular pads, with a nominal distance of not less than 0.3 mm (0.012 in) from the edge of the solder hole to the edge of the pad. On the finished board, the actual width or distance shall not be less than 0.05 mm (0.002 in). Any width less than this shall be the subject of special agreement between the customer and the manufacturer.

Isolated projections at the circumference of a pad will be accepted on condition that the size of the projection is no more than 0.2 mm (0.008 in), and at least 75% of the nominal spacing between the pad and any adjoining conductor or pad—as measured on the photomaster—is maintained. (See Fig. 4.4.)

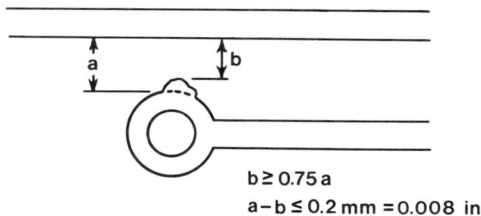

$$b \geq 0.75\,a$$
$$a - b \leq 0.2\,\text{mm} = 0.008\,\text{in}$$

Fig. 4.4 Isolated projection on a solder pad.

74

At most, two pinholes with a diameter not exceeding 0.15 mm (0.006 in) in any pad are acceptable provided that the pinholes do not occur at the junction of the pad and the conductor, or at the edge of the solder hole (Fig. 4.5).

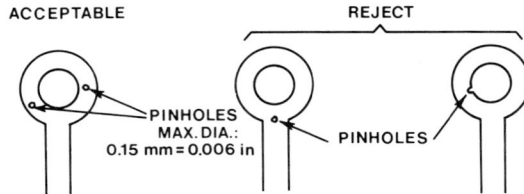

Fig. 4.5 Pinholes in solder pads.

Comments

Ideally, the photomasters for the two sides of a plated through board should have the pads in accurate register; in practice, slight variations will occur. In addition, there are manufacturing tolerances both in printing and in drilling. The sum of these variations must not exceed 0.25 mm (0.01 in) if the limit of 0.05 mm (0.002 in) for the minimum width of the pad is to be maintained. Occasionally, in manufacture, pads with broken annular rings will occur, and some latitude can be allowed in accepting these provided—

1. they do not occur where a conductor joins the pad;
2. the number occurring on any board is small;
3. only a small part of the circumference of the pad is broken.

Where the pattern is displaced relative to the drilling so that a large number of pads are broken this is a major manufacturing fault and the boards should be rejected. These faults are illustrated in Fig. 4.6.

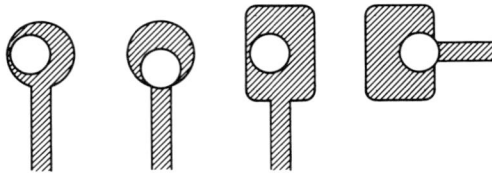

Fig. 4.6 Examples of misregistration.

The quality of the soldered joints is not significantly affected by breaks in the pads and in some cases boards have to be designed with broken pads (Fig. 4.7). For instance, if the hole diameters are increased to permit the use of automatic component insertion machines (see Section 10.3.5e) and conductors have to be run between pads on a 0.1-in grid, there is no alternative to accepting break-out of the holes.

In the author's experience the greatest benefit from demanding a minimum width of the annular ring is that this compels the manufacturer to show the necessary care in his processes to obtain a good registration between the two sides of the board.

Fig. 4.7 Example of 'extended' annular rings.

An unwanted projection at the circumference of a pad is most often due to poor etching or printing. This is not a reason for rejection of the boards as long as a sufficiently high insulation resistance is maintained, and the projections are few in number. A systematic occurrence would indicate that the manufacturing process is not fully under control.

The occurrence of pinholes at the transition between the pad and the conductor cannot, in any circumstances, be permitted as the result could be a discontinuity. Since the pinholes cause a reduction of the cross-section of a conductor, they must not be permitted to occur in any conductor where the current density is high, otherwise destructive local heating may take place.

4.6.4 Permissible Defects in Plated Through Holes

No more than 3 voids per hole are permitted. The total area of the voids shall not exceed 10% of the wall area. The largest dimension of a void must not exceed 25% of the circumference of the hole, or 25% of the thickness of the board (Fig. 4.8).

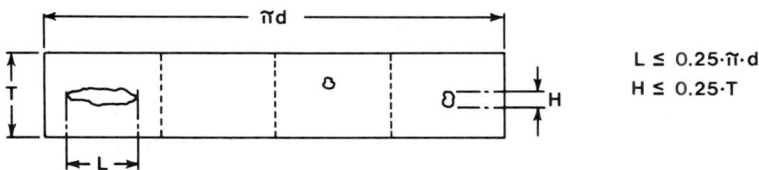

$$L \leq 0.25 \cdot \pi \cdot d$$
$$H \leq 0.25 \cdot T$$

Fig. 4.8 Fold-out of hole wall with 3 voids. T = Thickness of base material; d = diameter of hole.

No plated though hole shall have voids or cracks at the transition between the hole wall and the pad. The depth of the transition zone is 1.5 times the total copper thickness as measured on the surface of the board. (See Figs. 4.9 and 4.3, top right.)

Comments

The above requirement is as laid down in IPC-TC-500, *Specifications for Copper Plated Through Holes in Rigid, Two-sided Boards*. The maximum dimensions of the voids would appear to pose a problem for inspection either by the manufacturer or the customer. The fact is, however, that there will be either no voids (a perfect process control) or numerous and large voids. In practice, the borderline case, where it is difficult to decide whether a board is or is not acceptable, rarely occurs.

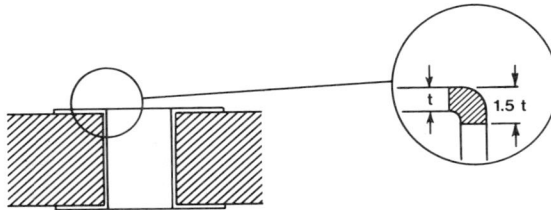

Fig. 4.9 Transition zone between solder pad and hole wall. t = Thickness of copper.

4.6.5 'Landless' Plated Through Holes

'Landless' plated through holes, i.e. holes with no pad on one side of the board, are used only in exceptional cases, and, if required, will be the subject of a special note on the master drawing.

Landless holes shall fulfil the requirements given in Sections 4.6.1–4.6.4. Furthermore, on the landless side of the board, the hole wall shall be flush with the board surface, i.e. no protruding collar can be permitted.

Comments

Landless plated through holes are occasionally used in single-sided boards in order to obtain stronger solder joints. In the plating process, growth occurs on the component side of the board raising a collar at each hole proud of the surface of the board. Such collars can break loose and remain on the component leads and may, in some cases, cause short circuits. To avoid this the holes can be tented with dry film on the component side but with the attendant disadvantages of poor circulation of the electrolyte in the holes (and consequently poor plating) and difficulty in complete removal of the electrolyte in the rinsing process.

In general, such a single-sided board costs nearly as much as a double-sided board and offers no real technical advantages.

In some cases, however, it may prove necessary to use a few landless holes in ordinary double-sided plated through boards (usually due to space limitations) or in order to keep the mutual capacitance as low as possible. If such holes are pointed out to the PCB manufacturer, he will remove protruding collars mechanically.

4.7 Non-plated Through Holes

4.7.1 General Requirements

Drilled holes shall be free from burrs and shall show no evidence of pads lifting from the base material.

Punched holes shall be used only if specially requested by the customer.

On plated through boards, holes marked 'NP' (non-plated through) shall not be plated through.

Comments

Burrs on the solder side can effect the solderability of the board.

Mounting holes for the board and for the assembly of more massive components often do not require to be plated through. These, and any other non-plated holes, can be drilled after the plating process in a second drilling operation. The expense of this operation can, however, be avoided by tenting these holes with the dry-film photo resist, thus preventing them being plated. The thin copper film from the copper reduction process is then removed during etching, leaving the holes unplated.

4.7.2 Diameter Tolerance on Non-plated Through Holes

Nominal diameter d	*Tolerance*
$d \leqslant 0.8$ mm ($d \leqslant 0.031$ in)	± 0.05 mm (± 0.002 in)
0.8 mm $< d \leqslant 5$ mm (0.031 in $< d \leqslant 0.2$ in)	± 0.10 mm (± 0.004 in)

Comments

These tolerances are a generally accepted standard which can be met by any good manufacturer. Closer tolerances than the above are available only at increased cost.

4.7.3 Solder Pads of Non-plated Through Holes

The minimum width of the annular ring—in the case of non-circular pads, the minimum distance from the solder hole to the edge of the pad—shall not be less than 0.2 mm (0.008 in) on the finished board unless otherwise agreed between the manufacturer and the customer.

Isolated projections at the circumference of a pad, and pinholes, shall fulfil the requirements for the solder pads of plated through holes as given in Section 4.6.3. A pinhole at the edge of the hole will be accepted.

The wider annular ring specified arises from the fact that the solder pad of a non-plated through hole has a somewhat poorer adhesion to the base material than that of a plated through hole. In order that the manufacturer may achieve a width of 0.2 mm (0.008 in) of the finished annular ring, the designer must specify an adequate size of pad. Also, if a component has to be removed during assembly or service a larger pad is less likely to be destroyed in the operation. As a rule of thumb, an area of 5 mm² (0.008 sq. in) is satisfactory.

4.8 Plated and Non-plated Conductors

4.8.1 General Requirements

The conductors of the board are normally designed with a nominal width of not less than 0.3 mm (0.012 in). All conductors shall be smooth and even, and there shall be no evidence of lifting of the conductors from the base material.

Comments

The minimum design width of the conductors is of particular importance as it determines the method of image transfer. Poorly defined conductor edges show that the manufacturing process is out of control.

4.8.2 Tolerance on Conductor Width

The tolerances are given with respect to the conductor widths as measured on the 1:1 photomaster. The values stated apply to 17.5 μm and 35 μm ($\frac{1}{2}$ oz and 1 oz) copper foil.

| | Tolerance on width | |
Nominal conductor width, w	*Plated boards*	*Non-plated boards* *Plated, reflowed boards*
0.3 mm < w ⩽ 0.5 mm (0.012 in < w ⩽ 0.020 in)	±0.10 mm (±0.004 in)	+0.05 mm, −0.10 mm (+0.002 in, −0.004 in)
0.5 mm < w (0.020 in < w)	+0.2 mm, −0.13 mm (+0.008 in, −0.005 in)	+0.10 mm, −0.13 mm (+0.004 in, −0.005 in)

In the case of 70 μm (2 oz) copper foil an additional tolerance of +0.07 mm (0.003 in) is permitted. It is assumed that the thickness of the laminate does not exceed 1.6 mm (0.062 in) and that the minimum diameter of possible plated through holes is 0.8 mm (0.031 in).

Comments

The tolerance on the width of conductors has to take account of factors such as the method of image transfer (screen printing or photo-resist printing), the function of the resist (etch resist or plating resist), the copper thickness and the etchant. The above specifications stipulate different tolerances for conductor widths below and above 0.5 mm (0.02 in), these tolerances applying to photo-resist and screen printing, respectively. It will always be the narrowest conductor width which will determine the method of image transfer.

The tolerances also take into account the ratio between the diameter of the smallest plated through hole and the thickness of the base material. The specifications are based on a minimum ratio of $0.8\,mm/1.6\,mm \sim 0.031\,in/0.062\,in = 0.5$. A smaller ratio would require a greater tolerance on the conductor width if the requirements regarding the minimum plating thickness in the holes are to be maintained. The smaller the hole, the longer the plating time, the thicker the deposit on the surface of the board, and the larger the growth in the conductor width.

The PCB designer's use of more than one tolerance on conductor widths is to be avoided since this would be confusing both to the manufacturer's and the customer's inspection. A general specification which can be met by well-established manufacturing techniques is clearly advantageous and any resulting loss of design flexibility is unlikely to be serious. This logic is basic to the rules of design as given in Chapter 6.

The tolerance on the conductor width follows the specification IPC-D-300E set by IPC, except for reflowed boards which are not specified. There is every justification for putting reflowed boards on the same footing as non-plated boards since the result of the reflowing process is to remove the overhang which represents the growth in width.

4.8.3 Undercut and Overhang

The undercut and the overhang which occur in plated boards, as shown in Fig. 3.11, shall be kept within the following limits:

Undercut

Nominal conductor width, w	Undercut per edge of conductor	
	$17.5\,\mu m\,(\frac{1}{2}\,oz)$	$35\,and\,70\,\mu m\,(1\,and\,2\,oz)$
$0.3\,mm \leqslant w \leqslant 0.5\,mm$ ($0.012\,in \leqslant w \leqslant 0.020\,in$)	$-0.04\,mm\,(-0.0016\,in)$	$-0.06\,mm\,(-0.0024\,in)$
$0.5\,mm < w\,(0.020\,in < w)$	$-0.06\,mm\,(-0.0024\,in)$	$-0.09\,mm\,(-0.0036\,in)$

Overhang

The overhang before reflowing shall be less than 0.05 mm (0.002 in) at either edge of the conductor. The total width of a conductor inclusive of the overhang shall meet the tolerance requirements given in Section 4.8.2. For reflowed boards, see Section 4.5.6.

Comments

The etch factor is assumed to be slightly above unity which normally should not present problems to the manufacturer. The overhang is of the same order of magnitude as the plating thickness which is 25 μm (0.001 in) copper multiplied by the throwing power factor, i.e. in total, 30–40 μm (0.012–0.0016 in) copper, to which should be added 10 μm (0.0004 in) tin/lead. If the plating tolerance is +100%, the total overhang amounts to approximately 0.10 mm (0.004 in) at each edge of the conductor. By using a relatively thick plating resist as shown in Fig. 3.11, the 'mushroom' effect can be reduced, and a total overhang of 0.05 mm (0.002 in) at each edge of the conductor can readily be achieved. Most of the overhang consists of tin/lead and it can, in any case, be removed by reflowing.

4.8.4 Conductor Edge Definition

The unevenness of the conductor edges shall be less than 0.1 mm (0.004 in) measured from peak to valley.

Comments

Uneven, and, in extreme cases, frayed edges of the conductors show that the manufacturing processes, especially image transfer and etching are not being satisfactorily controlled. The acceptable unevenness may seem insignificant in the case of wide conductors, but it is in fact 33% of a 0.3-mm conductor, i.e. not quite negligible.

4.8.5 Isolated Indentations along Conductor Edge

Isolated indentations of the conductors will be accepted provided that the conductor width is not reduced by more than 20%. The length of the indentation shall be shorter than the conductor width, or 5 mm (0.2 in), whichever is less (Fig. 4.10).

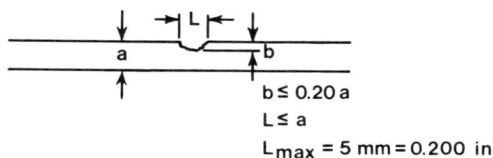

$b \leq 0.20\,a$

$L \leq a$

$L_{max} = 5\ mm = 0.200\ in$

Fig. 4.10 Isolated indentation of a conductor.

Comments

The width of conductors which carry heavy currents shall be so designed that excessive heating does not occur. No dangerous hot spots will occur if the indentations are kept within the stated limits.

4.8.6 Isolated Projections along Conductor Edge

Provided that 75% of the spacing, as measured on the 1:1 photomaster, is maintained, isolated projections of, at most, 0.2 mm (0.008 in) along the conductor edges are allowed (Fig. 4.11).

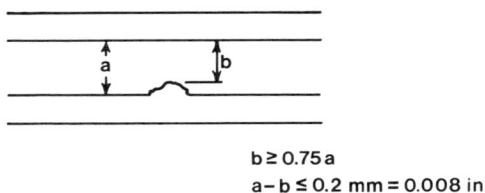

b ≥ 0.75 a
a − b ≤ 0.2 mm = 0.008 in
Fig. 4.11 Isolated projection on a conductor.

Comments

This requirement corresponds closely to that for isolated projections at the edges of solder pads (see Section 4.6.3).

4.8.7 Pinholes in Conductors

Pinholes up to 20% of the nominal conductor width will be accepted provided that the pinholes do not occur at the junction of the solder pad and the conductor (see Sections 4.6.3 and 4.7.3), and that the number of pinholes does not exceed 2 per 10 mm (0.4 in) of conductor length.

Comments

See the comments in Section 4.8.5

4.9 Edge Connectors

Three types of finishes of edge connectors are used: viz. tin/lead plating (also called solder plating), roller tinning and gold plating.

4.9.1 Tin/Lead Plating

When the board is tin/lead plated, and the edge connector has no other function than to make test connections during assembly testing, the edge connector shall be tin/lead plated with the rest of the board. The plating thickness is as specified in Section 4.5.3.

Comments

One of the cheapest ways of achieving test access to a board is to establish an edge connector which is tin/lead plated. This does not require additional operations if a polarizing slot is not required.

On account of rising gold prices, there is a certain tendency towards replacing gold plating of edge connectors by tin/lead plating. The long-term reliability of a tin/lead-plated contact system has not yet been fully established; some reservations are therefore recommended. The use of a tin/lead-plated edge connector for test purposes does not demand long-term reliability and is therefore fully acceptable.

4.9.2 Roller Tinning

Under the conditions described in Section 4.9.1, the edge connector can be roller tinned with the rest of the board. The thickness of the roller tinning is specified in Section 4.12.2. The roller tinning shall be even and smooth on the surface of the edge connector without formation of drops on the contact fingers.

Comments

Drops on the contact fingers will make insertion of the edge connector in the receptacle of the test equipment difficult and any material scraped off may, in the end, obstruct the receptacle.

4.9.3 Gold Plating

Gold plating shall be preceded by an underlayer of low-stress nickel. (see Section 4.5.5 concerning the required plating thickness.)

The finish shall be bright and smooth. The gold shall completely cover the surface of the contact fingers, including the edges. The only exception is at the end of the contact fingers where the board is machined (routed and chamfered). At the transition between the contact fingers and the conductors, the gold plating shall cover 0.5–1 mm (0.02–0.04 in) of the tin/lead plating in order to avoid any risk of exposed copper. No evidence of delamination of the nickel and the gold plating shall occur.

The following requirements regarding pinholes shall be satisfied:

maximum number of pinholes per contact finger: 2
maximum diameter of pinholes: 0.1 mm (0.004 in)

When stripping the tin/lead plating from the connector area prior to the nickel and gold plating, care shall be taken to preserve the tin/lead plating of adjacent circuits or annotations such as part numbers.

Comments

Delamination and failure to overlap occur not infrequently and it is therefore necessary to include these as reasons for rejection.

Fig. 4.12A shows a longitudinal section through a contact finger, and the overlap is clearly seen. As the procedure, which is described in detail in Section 3.7.4, requires two maskings with tape, there is an alternative and cheaper method employing only one masking. In this case, the nickel and gold deposits follow immediately after the stripping of the tin/lead layer. The small groove which might occur in the transition zone at the edges of the tape is then sealed by epoxy resin applied by screen printing (see Fig. 4.12B). Additional savings result if the sealing is applied as part of the solder mask.

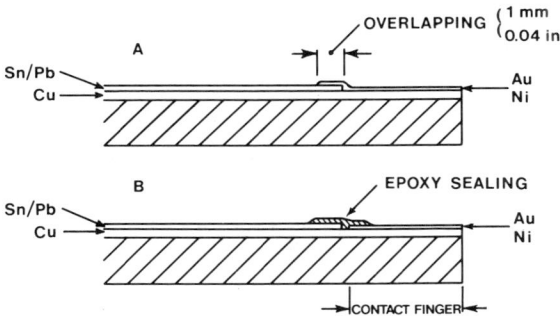

Fig. 4.12 Longitudinal section through a contact finger and the attached conductor.

Unwanted stripping of areas adjacent to the contact fingers can occur, especially if the board is provided with two edge connectors, the space between them being used for conductors or various inscriptions. Careful masking of these areas is required.

4.9.4 Chamfering

Both sides of the edge connector shall be chamfered at an angle of 45° (Fig. 4.13).

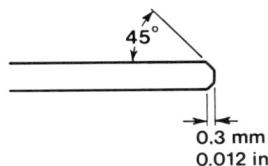

Fig. 4.13 Chamfering of an edge connector.

84

Comments

The purpose of the chamfering is to protect the receptacle against wear of the gold plating and to facilitate insertion in the receptacle.

4.9.5 Bow Along Edge Connectors

To facilitate insertion of the edge connector into the receptacle, the maximum bow (warp) along the edge connector area shall not exceed 0.25 mm (0.01 in) overall or 0.005 mm/mm (in/in), whichever is the greater.

Comments

The reader is referred to Section 4.4.5. It should be noted that heating in the reflow process can cause serious bending of the board.

4.10 Component Notation

A component notation and location marking shall be printed on the boards by screen printing. The adhesion shall be good so that the quality is not impaired during the course of production, i.e. assembly, soldering and testing.

The colour of the printing ink shall be in contrast to the base and the image shall be free of smears and have good definition. No printing ink shall be found in the holes of the board. The printing ink shall in no way impair the electrical performance of the board, especially the insulation resistance.

In any cases where a component notation is printed on top of the solder mask, the ink shall not be damaged by soldering at 260°C for 5 secs.

Double-sided boards which are tin/lead plated and reflowed or roller tinned shall have these processes completed before the component notation is printed.

In the case of bare-copper boards, the component notation shall be printed and cured before the protective coating is applied (see Section 4.12).

Comments

The colour of the printing ink shall show good contrast with the base material, the colour of which is nearly always greenish. Furthermore, it shall show good contrast with bright metallic areas, for example, ground planes. Yellow affords a better contrast than other colours, including white and black.

If the component notation is printed directly on the tin/lead-plated surface, reflowing causes serious distortion of the lettering.

4.11 Solder Mask and Insulation Mask

In practically all cases, a solder mask is specified. As soon as the technology makes it possible, solder masks for complex high-density boards shall be applied as dry-film solder masks with very small clearances around the solder pads.

For the present, a screen-printed solder mask will be specified. The printing ink shall be of a type agreed between the manufacturer and the customer. Normally the colour shall be green, and the solder mask must be able to withstand contact with solder at 260°C for at least 5 sec without damage. A slight wrinkling is acceptable where the solder mask is printed over large tin/lead-plated areas. No solder webbing shall be found after the soldering of the board.

The solder mask shall not significantly reduce the insulation resistance between conductors on the board.

A solder mask shall protect against the formation of solder bridging during automatic soldering of the board. This requires the same care and precision in printing the solder mask as in printing the conductor pattern.

The artwork for the solder mask shall not—without the consent of the customer—be modified by the printed circuit board manufacturer in order to facilitate the application of the mask.

Screen-printed solder masks shall be designed with a nominal clearance of 0.3 mm (0.012 in) around the pads. Conductors within 0.42 mm (0.017 in) of the solder pads must be fully covered by the solder mask. A slight degree of over-printing of the solder pads (maximum 0.15 mm or 0.006 in) will be accepted.

After the introduction of dry-film solder masks, the clearance around the pads will be nominally zero, and it will be accepted that the dry-film solder masks cover the outer 0.1 mm (0.004 in) of the solder pads.

In the case of tin/lead-plated boards, the solder mask shall not be applied until reflowing has been performed. On roller-tinned boards or boards having a protective lacquer, the solder mask shall be applied prior to the roller tinning or to the application of the protective lacquer. Adequate measures shall be taken to retain the solderability after printing and curing the solder mask (see Section 4.12).

Insulation masks on the component side of the board, are screen printed using the same ink as that used for the solder mask. All the requirements for the solder masks apply also to the insulation masks, with the exception that a larger clearance around the pads can be permitted.

Comments

It is difficult to screen print a solder mask with the accuracy necessary to prevent solder bridging over the whole area of the board. When a dry-film process for solder masks becomes available, however, it will be possible to apply a solder mask with an accuracy consistent with the conductor pattern, eliminating solder bridging completely. Unfortunately, dry-film solder masks are not commercially available in Europe at present; screen printing has, therefore, to be used. This can be easily understood since photomechanical printing of the conductor pattern is preferred to screen printing because of its greater precision. It follows that the less-precise process cannot be expected to produce a mask in perfect register with the conductor pattern and that very careful attention must be given to the screen-printing process if satisfactory results are to be achieved.

The insulation mask serves to prevent short-circuits on the component side, for example, between a conductor and a component. It is not unusual that the insulating lacquer of a component body is too thin to prevent a short-circuit to a conductor which lies under the component. Since it is not necessary on the component side to have such small clearances around the pads as in the case of solder masks, the artwork should be designed with larger clearances to facilitate printing and thereby reduce costs.

4.12 Protective Coatings

4.12.1 General Requirements

Bare-copper boards shall have a protective coating in order to preserve solderability. The copper surface shall be clean and fully solderable immediately before the protective coating is applied, and the preparation of the surface shall not impair the solderability. Abrasives such as pumice shall not be used as there is a risk of embedding particles of the abrasive in the copper thereby incurring a risk of dewetting. A mild etching of the copper surface before the application of the protective coating is preferred.

Roller tinning and the application of a protective lacquer are the methods used for protecting the solderability of bare-copper boards.

Comments

It is well established that mechanical cleaning by means of abrasives causes serious dewetting and must on no account be used. Roller tinning superficially covers a copper surface which is contaminated with particles of abrasives but subsequent soldering will display dewetting. A mild etching by which some few microns of copper are etched away is the acceptable method of cleaning and will leave a surface with good solderability.

4.12.2 Roller Tinning

Roller tinning shall be performed with 60/40 solder. The thickness of the applied layer shall be not less than 5 μm (0.0002 in) for conductor widths up to 2 mm (0.08 in). On wider areas, a reduced thickness is acceptable.

Holes will have been drilled prior to the roller tinning, and any which become closed in the process shall be re-opened. Solder drops on the contact fingers of edge connectors will not be permitted (see Section 4.9.2).

Comments

A common method of preventing closure of holes is to roller tin the boards before drilling the holes. This impairs the solderability as a result of oxidization of the copper rims which are exposed by the drilling.

4.12.3 Protective Lacquers

The type of protective lacquer chosen shall be compatible with the type of flux used in the customer's soldering process. Usually an undercoating of a water-dip lacquer shall be used in order to prevent oxidation during the drying process after etching. The water-dip lacquer shall be followed by a matching top-coating lacquer.

Comments

In the experience of the author, roller-tinned boards are less solderable than boards having a protective lacquer provided that the copper surface has been adequately cleaned prior to the application of the protective lacquer, very often by means of two or three mild etchings. It is essential not to air-dry the copper surface as this inevitably causes oxidation of the copper, so a water-dip lacquer stage is preferred. Experience has shown that water-dip lacquers are generally detrimental to satisfactory soldering because, to some extent, they age and become insoluble, but if the boards are transferred without drying from the vat with the water-dip lacquer to the vat with the matching top-coating lacquer, near-perfect solderability should be achieved.

4.13 Soldering

4.13.1 General Requirements

When printed circuit boards are soldered by automatic soldering methods (wave or drag soldering) the solder shall completely wet the pattern, including the plated through holes, if any, an other surfaces such as ground planes which have to be soldered.

The soldering shall produce a bright, adherent coating, free of dark or granular areas, and there shall be no evidence of non-wetting, dewetting or solder webbing.

Comments

Solderability is a prime requirement of printed circuit boards. Non-wetting, dewetting and solder webbing are defined in Section 4.5.6. Non-wetting and dewetting are due to contamination of the surface, whereas solder webbing is due to a base material (or solder mask) not being completely cured, or an incompatibility between the protective lacquer and the solder flux.

4.13.2 Blowholes

No blowholes shall be found after soldering plated through boards.

88

Comments

Blowholes, if any, are usually found on the solder side of the board, and appear as craters in the solder joints. They are due, very often, to water entrapped in the base material or in small pores of the plating. During soldering, the water evaporates and—depending on the degree of soldification of the solder—causes an open crater or a concealed pocket in the solder joint. Such blowholes can be very detrimental to the reliability of the solder joint and corrective action must be taken whenever they are observed. Although there may be other reasons for blowholes, for example, too high a content of organic brighteners in the plating solution or poor plating, moisture seems to be the most common cause. An effective remedy is to bake the boards for one hour at 105°C prior to soldering.

4.13.3 Soldering Time and Temperature

The boards shall be able to stand the following soldering conditions:

Soldering temperature: 260°C ±5°C
Soldering time: not less than 5 sec.

Comments

These requirements are commonly accepted by printed circuit board manufacturers.

4.13.4 Solderability after Storage

The boards are stored under the following conditions:

Temperature: 15–35°C
Humidity: 45–75% relative humidity

Full solderability is required after storage for the minimum periods indicated below:

Bare-copper boards with surface protection: 3 months
Tin/lead-plated boards: 6 months
Reflowed tin/lead-plated boards: 12 months

Comments

A long storage time is always desirable, and quite essential for many electronics companies. The storage times given above are accepted by most printed circuit board manufacturers. The 12-month storage time for reflowed tin/lead boards is fairly conservative. The author has found that boards 30 months old exhibited excellent soldering characteristics. It is, however, essential that the boards are stored in polyethylene bags, and, furthermore, that rubber goods are not stored in the same room.

4.14 Machining

4.14.1 General Requirements

Acceptable contouring methods are sawing, routing or blanking. The machining must not cause delamination, crazing, fractures or other damage to the board, e.g. scratching of the surface.

A reference system is required to ensure the correct relationship between the pattern and the outline of the board. Particularly important is the location of the polarizing slot of an edge connector with respect to the contact fingers. Two tooling holes which are used to locate the board for machining are introduced within the contour of the board. The coordinates of these holes are given with respect to the reference system. When digitizing the board, it is assumed that the datum point of the reference system is taken as the reference point, that the tooling holes are introduced by their normal coordinates as given in the master drawing and that the tooling holes are drilled in the same operation as all the other holes of the board. This implies that internal scraching of plated through tooling holes is accepted.

Comments

In many cases, it is not necessary to obtain a very accurate correspondence between the pattern and the contour, so the profiling can very well be based on the corner marks of the artwork. The problem usually arises in the case of plugable boards with edge connectors. Frequently, it is imperative to achieve a well-defined relation between the contact fingers, the polarizing slot, and the contour of the board. In order to facilitate the machining, the customer should in his own interest, introduce internal tooling holes suitable as fixing points for the jigs or fixtures used when profiling the boards. Without such a reference system there is a much higher risk of the polarizing slot being displaced, in extreme cases, so much that the reliability of the external connection is subject to doubt.

4.14.2 Dimensional Tolerances

Unless otherwise indicated on the master drawing, the tolerance of the overall dimensions is ± 0.3 mm (± 0.012 in).

Comments

Narrow dimensional tolerances normally add to the cost of the board, so it is desirable to specify tolerances as wide as possible. This does not necessarily mean that the manufacturer will saw a board instead of using the NC router, if a tolerance of ± 0.3 mm is specified. For cost reasons he will probably choose the NC router.

The tolerances associated with the various methods of machining have been discussed in Section 3.8.6.

4.14.3 Positional Tolerance for Polarizing Slots

Polarizing slots shall be either punched simultaneously with the blanking of the entire board or routed. Sawing of the slot is not permitted. As an aid to obtaining the tolerances given below, tooling holes are introduced as described in Section 4.14.1.

Width of slot:	±0.1 mm (±0.004 in)
Location of slot:	±0.1 mm (±0.004 in)

The location is measured with respect to the adjacent contact fingers of the edge connector, the centre line of the slot being taken as reference.

Comments

The specification allows a worst-case displacement of the edge of the polarizing slot of maximum 0.15 mm (0.006 in), 0.1 mm (0.004 in) resulting from the location tolerance, and $\frac{1}{2} \times 0.1$ mm ($\frac{1}{2} \times 0.004$ in) from the width tolerance.

4.14.4 Methods for Locating Holes

Normally all the component or solder holes of a board are determined by the position of the solder pads on the photomaster derived from the artwork. Tooling holes are located by their coordinates in the reference system as described in Section 4.14.1.

Comments

The holes can be located in three fundamentally different ways:

1. The location is taken directly from the photomaster and is digitized by means of an optical digitizer (cf. Section 3.8.2d). This method normally gives the lowest production costs and the best registration between the solder pads and the solder holes and is preferred by most manufacturers.
2. The location is determined by coordinates given for all holes. This implies that the location of any pads on the photomaster agrees closely with the coordinates given in the master drawing. If the artwork is manually prepared, there is a long tolerance chain, the weakest links being the absolute accuracy of the grid used for the taping of the pads, and the positional tolerance of the pads themselves. Therefore, the registration between the solder pads and the solder holes will, inevitably, be poorer than that obtained by the first method.
Note: A tooling hole, which is located with respect to a reference system, does not depend on the accuracy of the artwork but is determined directly by its coordinates.
3. The location is based on a modular system, and the NC tape is made to agree with the modular steps indicated, i.e. from nominal dimensions and not from an optical measurement of the artwork as described in the first method.

As described in the second method, the location of the pads must comply perfectly with the modular system, which usually is a precision grid. As a matter of fact, a high degree of accuracy is difficult to achieve, cf. Section 6.4.1, where the positional accuracy of a pad is discussed. To this is added the inaccuracy of the grid itself, cf. the remarks in Sections 6.3.3 and 8.2.3.

4.14.5 Positional Tolerances for Holes

The tolerances given below assume that the holes are drilled with a numerically controlled drilling machine and that the NC tape information is derived from the 1:1 photomaster.

The radial deviation of any drilled hole from its true location on the photomaster, or from its nominal coordinates with respect to the reference system, shall be less than 0.08 mm (0.0031 in) for board lengths less than 200 mm (8 in) and less than 0.10 mm (0.004 in) for board lengths above this value.

Comments

Where boards are assembled manually, the above specifications are usually adequate. If, however, automatic component insertion machines are used, the above tolerances might lead to some difficulty. Increasing the diameter of the solder holes can make automatic component insertion much easier. This is discussed in Chapter 10.

5

Layout of the Board

5.1 General Considerations

The producibility, price, performance, reliability and serviceability of a board depend on the quality of the board layout. The chief aim of this chapter is therefore to consider the general principles of board layout including determining the optimum size, locating the components with due regard to their internal connection, and determining the board's external connections.

It should be noted that the practical taping of the board pattern is determined by the layout. This means that the PCB designer should take into account the probable problems of taping when laying-out the board. It is of vital importance that the PCB designer understands completely the 'rules' for taping; for example, the allowable conductor widths, the minimum conductor spacing and the preferred diameters of the pads. Only in this way can he produce a layout which can be realized as a correctly taped artwork without a succession of modifications or rectifications. A procedure for design of the artwork should be formulated and, to facilitate this, a number of Design Rules are stated in Chapter 6 which taken together ensure that no important factor is omitted. A discussion of the theoretical conditions which underlie these rules is included.

Below we shall discuss some of the conditions of production which the designer can influence by his layout.

The electronic circuitry of the finished boards displays a higher degree of uniformity than was the case with the old point-to-point wiring method, in which the components were soldered individually to terminals and the interconnections made with wires. The wiring was very operator-dependent and could vary a great deal from series to series, causing unwanted couplings or showing recurrent mounting faults in a series. The PCB designer should keep in mind that easy and reliable assembly depends on a board design offering easy location and insertion of the components.

Another advantage of printed circuit boards is the possibility of automatic soldering which is not practicable with point-to-point wiring. An obvious precondition for automatic soldering is that the board itself is solderable. For example, the PCB designer should ensure that the clearance between component leads and the hole walls is such that capillary attraction of the solder up into the holes is achieved. Again, the PCB designer (as well as the draughtsman) can invite solder bridging during automatic soldering unless he routes the conductors in certain ways to avoid this happening.

In order to simplify automatic testing of the assembled boards, the designer should establish suitable input/output connections. Even in the case of plug-in boards, i.e. boards provided with connectors, it will be necessary to have access to certain test points of the circuitry. In some cases, these points can be brought out to surplus contacts in the board connector, but in others it may be necessary to introduce a special connector, the cheapest form being an edge connector which is tin/lead plated together with the remainder of the board. The distribution of the input/output terminals of the working connector is also important as the interface for the test equipment can be simplified. The power terminals for example should always have the same location.

When all the circuitry, including the power supply unit, is assembled on a single board (full integration) the board is often large, e.g. 250×300 mm (10×12 in) or more. In this case, it is advantageous if the PCB designer takes care to build in the possibility of convenient separation of some of the circuits; he should at least aim to separate the power supply from the rest of the circuitry. The introduction of such measures requires good cooperation between the engineering groups involved: the circuit and PCB groups in the development department, and the test equipment and processing engineering groups in the production department. A rational testing procedure should be prepared at this stage when all the possibilities are still open.

5.2 Board Size

In many types of equipment the size of the board is determined by the requirements of the cabinet and little freedom is left to the PCB designer. It is quite a different matter if the above restriction does not exist, and the equipment is a large digital system comprising many integrated circuits (ICs). In such cases, the packaging engineer faces the problem of determining the optimum size of the printed circuit board itself. The selection depends on many factors and will inevitably finish up as a compromise depending upon the company's existing or future policies with regard to development, production and after-sales service. We shall now discuss the various factors which affect the layout and size of a board but, obviously, variations in company costs, overheads and the complexities of the circuitry, restrict the observations to those of a general nature.

5.2.1 System-dependent Factors

From the viewpoint of circuitry or system, it is expedient to divide the system into logical or functional sub-units. In order to effect rational testing procedures, sub-units should always be placed as complete entities on boards. Also, two or more sub-units, belonging more or less together, can be placed on the same board. It will be the largest sub-units, therefore, which determine the required board area and the overriding factor will usually be the number of ICs within these subunits. Thereafter, the smaller sub-units are combined rationally, taking into account the need for their interconnections in such a way that the total size approximates to the size of the larger sub-units in terms of, say, the number of ICs used. This can require that the boards become rather large, comprising 50, 100 or 200 ICs, depending on the type of system.

An advantage of this approach is a quicker and more effective test procedure since the sub-units, together with their interconnections, can be tested as sub-systems. Allocating each of the sub-units to an individual smaller board means that each sub-unit can only be tested separately. Consequently, any faults originating from unfortunate combinations of boards, or from marginal coupling effects caused by the interconnections, cannot be found until later, when all the boards have been plugged into the equipment.

Other factors, however, militate against very large boards. The larger the number of circuits on a board, the disproportionately larger is the board area required for interconnections between circuits. Consequently, the degree of utilization of the board area is reduced. Every competent PCB designer will endeavour to shorten the interconnections by intelligent grouping of the circuits. He will also aim at placing circuits with external connections as close to the connector as possible. However, some circuits must, due to lack of space, be placed at the far end of the board, for which reason a fairly large part of the board area may be lost in establishing external connections.

A well-arranged and expedient test procedure for a very large board normally requires that the system can be divided into isolated logical or functional sub-units using suitable circuit interruption devices such as small plugs. In this way, the testing and troubleshooting of separate sub-units can be undertaken independently of each other. Such measures, however, take up some space on the board. In this context it should be noted that some automatic test systems will stop when the first fault is detected. After the fault has been rectified the test procedure will start again from the beginning, possibly limited to a check of the main data, until the stopping point is reached. The reason for this test procedure is that it is impossible to ensure that adjustments already made are correct, or alternatively, that the rectification of the fault has not had some side-effects on circuits already tested and adjusted. The larger the board, the greater will be the risk of faults, which in turn implies a larger number of re-starts of the test routine. This disadvantage can be overcome by dividing the circuitry into several independent circuits. This corresponds, in a way, to cutting the board physically into pieces, but this would cause a considerable increase in the total handling time, i.e. the time used for connecting and disconnecting the boards. For test reasons, a medium-sized board, having the possibility of separating the electronic circuits on it, would seem to be most advantageous.

The inherent complexity of the very large board restricts the possibilities of modification, and it is well known that changes in the circuitry will from time to time be called for, very often in the form of extensions. If the sub-unit which is to be modified, is located on a separate small board, the modification will, of course, only affect this board, but, if the sub-unit is part of a large board it is very likely that the same modification will affect the adjacent board areas. In extreme cases the modification cannot be effected without a complete re-design of the board. It is a useful precaution in the case of a fairly large board to provide a spare area corresponding to 5% of the total number of ICs which have to be accommodated on the board. Despite such foresight, there is a latent risk that changes in the circuitry of the very large board cannot be introduced without major re-design of the board and in rare cases of the complete equipment.

The division of a large board into small boards facilitates troubleshooting and repair, especially in the field or when performed by a less-qualified repairer. Very often the service policy will be aimed at the localization of a fault to a specific board which can then be exchanged and returned to the manufacturer for repair. In this way improved flexibility and better utilization of the equipment and reduction in the stock-holding of spare boards can be achieved.

Finally, it should be noted that partitioning in the form of standardized circuits placed on separate boards, e.g. a digital-to-analog converter, or a micro-computer, enables such boards to be used in more than one type of equipment as ready-developed sub-units and possibly also carried in stock. Obviously, this is impossible if the circuits are physically embodied in the circuitry of a large board.

As was mentioned in the introduction, many factors are involved in the determination of the optimum board size. If the designer, after hard and prolonged consideration, finally reaches a suitable and convenient board size, he will very often be urged to standardize on this size. He must, however, always recognize that circuits are becoming increasingly complex which implies an increasing demand for board area. This demand for board area is off-set to some extent by the fact that ICs which accommodate for a growing number of functions, are constantly being developed. Nevertheless, the risk in standardizing on a small- or medium-sized board is evident.

5.2.2 Constructional Factors

Until now we have considered the size of the board from a purely electronic viewpoint. In the following items we shall discuss certain constructional factors such as mechanical strength, cooling and connectors.

5.2.2a Resonant Frequency

Small boards of, say, 100×150 mm (4×6 in) present little, if any, problems in terms of mechanical stability. The situation is different when the board size is increased, as there is the risk that the resonant frequency of the board falls within the frequency range of the mechanical forces to which the equipment is subjected, both in transit from the electronics manufacturer to the customer, and in operation. The PCB designer should bear in mind that components mounted on a board may be subjected to considerably higher accelerations than the equipment as a whole. The amplification depends *inter alia* on the ratios between the frequency of the external force and the resonant frequencies of the various elements transmitting the force to the components in question (cabinet–chassis–board–component). It is, perhaps, not well known that amplifications of up to 5 or 10 times can occur, so the problem can be significant.

In the case of large boards of 300 × 300 mm (12 × 12 in) or more, the centre of the board can be subject to relatively large deflections caused by external forces. Unless spacers are introduced, there is a risk that components having non-insulated terminals or electrically live housings, can short-circuit intermittently to the solder side of an adjacent board. One of the more serious problems resulting from such large resonant deflections, however, is the risk that some of the solder joints or the terminal leads of the components fail due to metal fatigue. Fig 5.1

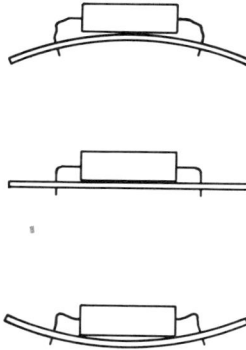

Fig. 5.1 Board subjected to resonant vibration.

shows a somewhat exaggerated drawing of the centre area of a board which is subject to a resonant vibration. The terminal leads move backwards and forwards, especially when they have been bent through 90°. Small cuts or scars caused by the mechanical lead-forming operation will often be the sites for initiation of lead fracture.

The resonant frequency can be changed fairly simply by the introduction of stiffening devices such as metal angles, or by the use of additional mounting holes and/or supporting studs. The resonant conditions can, of course, be calculated. Practical information obtained from vibration tests of prototype equipment will in most cases be more satisfactory.

5.2.2b Cooling

Let us assume that 200 ICs have been mounted within a volume of one litre, achieved by having a component density corresponding to that of the board shown in Fig. 3.17, and a board spacing of 10 mm (0.4 in). If the estimated average power dissipation per IC is 50 mW, a total quantity of heat equivalent to 10 W is generated. Even in the case of a lower packaging density, the quantity of heat to be removed is so large, that cooling, by free convection, radiation or conduction is not sufficient. It therefore becomes necessary in such cases to use forced convection which requires a built-in fan or blower. This should be designed to deliver the volume of air necessary to restrict the temperature rise of the equipment within the permissible limit. It is important to achieve an even distribution by establishing, for example, special air-ducts, and to secure a sufficiently high velocity of air in order to keep the surface temperature of the hot components reasonably low. In order to prevent dust entering the equipment, a filter should be used. This, clearly, can become choked and a thermal circuit-breaker should always be included in the design to disconnect the equipment from the mains in the event of excessive temperature rise from this cause.

Temperature-rise is a factor which limits the allowable component density, and, in turn affects the board size. The problem of heat distribution should in all cases be carefully analysed since thermal design should not be undertaken on a cut-and-try basis. A fuller treatment of the criteria affecting thermal design is beyond the scope of this book and the reader is referred to the technical literature. At the very least, the heat distribution should be tested in a mock-up as there is a risk of local 'hot-spots' even when full thermal analysis has been carried out. Figures 5.2 and 5.3 show equipment cooled by forced convection. In Fig. 5.2 the air flow is divided so that part of it is used for cooling a power supply and the remainder for cooling the boards. In Fig. 5.3 a deflector is used to guide the air flow.

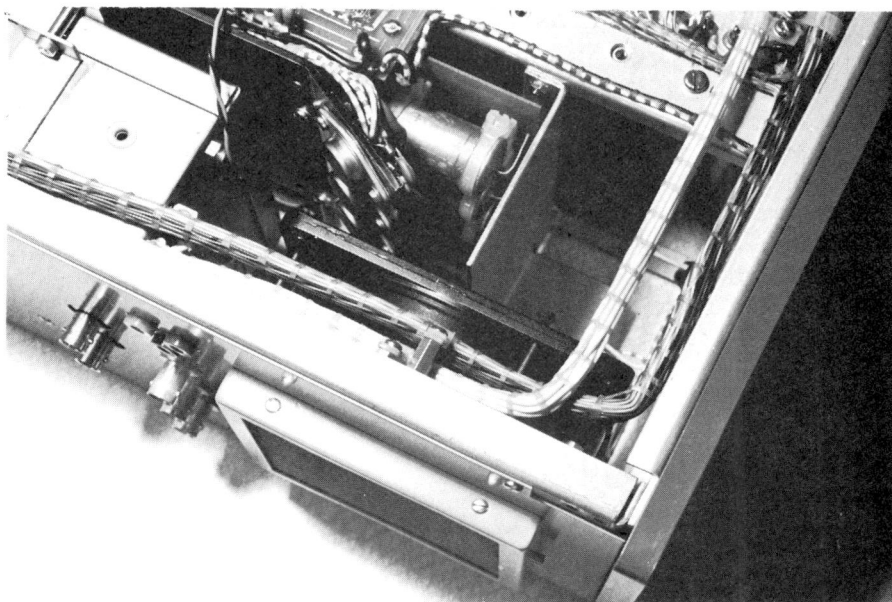

Fig. 5.2 Forced convection cooling. Note the partition that divides the air flow.

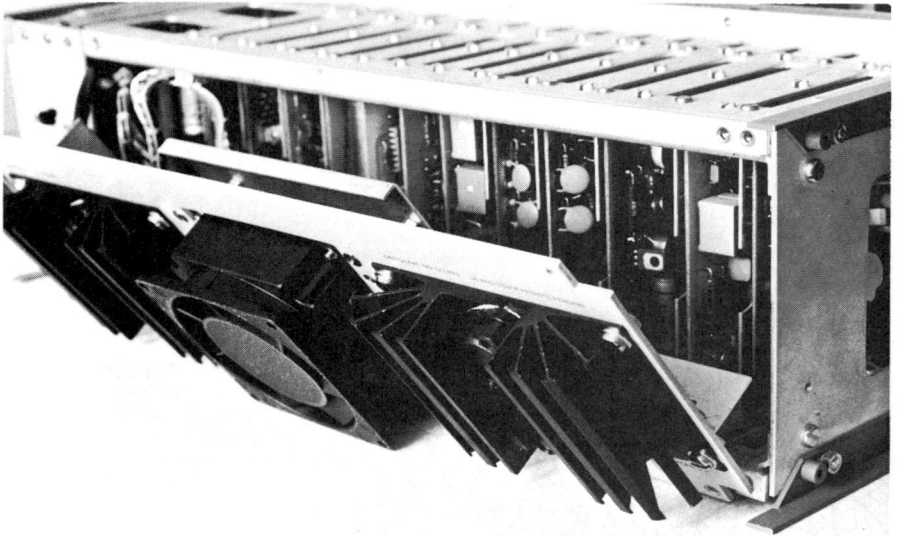

Fig. 5.3 Forced convection cooling. Note the deflector just behind the fan.

5.2.2c Connectors

The overall connector size obviously depends on the number of terminations which is determined by the circuitry. Occasionally, a standard connector is chosen for reasons of design or rationalization. In these cases, the connectors may have unused contacts. Large boards often require a large number of contacts, and, in order to secure a reliable connection in the case of low-level signals, it is necessary to specify a rather high contact pressure of perhaps 50–100 g (approximately 2–4 oz). This can produce a mating force of several kilograms in extreme cases so that the connector becomes the limiting factor when determining the board size. A high mating force reduces the sensitivity during insertion to such a degree that the operator may be unable to 'feel' when the male contact misses the female and is, therefore, damaged. Zero-force connectors exist but they are rather expensive and also fairly large.

5.2.3 Handling-dependent Factors

The maximum board size is often affected by handling limitations—an obvious point which is occasionally overlooked.

5.2.3a Artwork Size

How big an artwork is it possible to prepare on the PCB drawing room's light tables when a reasonably large scale is desired? Although in special cases it may be necessary to work on a scale of 2:1, a scale of 4:1 is recommended. The basis for this is given in Section 6.4. Commercially available light tables can measure up to 500×1000 mm (20×40 in) covering board formats up to 125×250 mm (5×10 in) on a 4:1 scale, whereas home-made light tables can have dimensions of 1000×1250 mm (40×50 in) or 500×2000 mm (20×80 in), covering board formats up to 250×300 mm (10×12 in) and motherboards* up to 125×500 mm (5×20 in) excluding allowances for drawing margins.

What is the capacity of the artwork file? Artworks should be stored flat, preferably horizontally, and drawer files are preferable. As far as the author knows, the largest commercially available files have a drawer size of 700×1040 mm (28×41 in) covering board formats up to 170×250 mm ($6\frac{1}{2} \times 10$ in) in a scale of 4:1.

5.2.3b Camera Size

What is the artwork capacity of the photographic equipment? It should be borne in mind that the risk of non-linearity (distortion) increases with increasing size. The upper limit often lies around 1500×1200 mm (60×48 in) which gives a board format of 375×300 mm (15×12 in), if the artwork is made on a 4:1 scale. Larger artwork, for example, 400×1600 mm (16×64 in) for a motherboard of 100×400 mm (4×16 in) can be accommodated with a little ingenuity. The first step is to make two 4:1 overlapping reductions which can be combined to cover the entire board. The next step is to make a contact print of the combined negatives. Obviously this is a complicated method which demands careful processing, not least a very accurate registration. It is nevertheless practicable.

5.2.3c Manufacturer's Equipment

What is the maximum board size which the PCB manufacturer can produce economically? The answer depends upon the capacity of the equipment he uses, the degree of utilization of that equipment, and the raw materials. Thus it is a matter of how well the format matches the jigs used in the plating vats to obtain full utilization, or how well it matches the size of the laminate sheets to avoid excessive waste. Another example is the image transfer; if the board is screen printed, the inaccuracies will be greater, the greater the size, because of the unavoidable stretching of the screen. If the board is photo-resist printed, the limit could possibly be determined by the width of the laminator or by the exposure equipment. Likewise, the NC drilling machine can be a restriction. In the case of very big formats, only one-half of the drilling heads can be used and the capacity of this very expensive piece of machinery is halved. This will, of course, be reflected in the price of the board.

---

* A motherboard is a printed circuit board provided with fixed receptacles (connectors). This board serves to establish all the interconnections between the various plug-in (daughter) boards, which make up the complete system. Automatic soldering gives better soldering quality than manual soldering, so it is important that the motherboard can be passed through the soldering machine.

Clearly, a good appreciation of manufacturers' limitations is required by the designer and each manufacturer is limited by different criteria. As a general rule of thumb, the manufacturer should be consulted when the projected board size is in excess of 300 × 400 mm (12 × 16 in).

5.2.3d Customer's Equipment

What are the limitations of the customer's own production department? A typical example is the drag soldering machine. One model restricts the width and the length of the boards to 280 mm (11 in) and 400 mm (16 in), respectively. A larger model allows a width and a length of 400 mm (16 in) and 700 mm (28 in) but very few companies need such capacity. A commonly used wave-soldering machine has a working width of 380 mm (15 in) and without, in principle, restrictions on the length.

The soldering machine used is critical if, for example, a 19-in rack is used. In this case the motherboard can easily be 400 mm (16 in) long and as it usually carries a large number of female connectors each having many contacts, the area of the bath required in a drag soldering machine is, obviously, large.

Again, the physical size of the assembly tables, the physical capacity of the terminal lead-cutting machine, the limitations of the assembly jigs or the capacity of the automatic component insertion machine are factors which must be taken into account.

5.2.4 Price versus Board Size

The price per unit of area can be expressed as a function of the board size which can give some indication of the optimum board size. The considerations given below assume constant area and constant component density which means that there will be the same number of holes independent of whether the circuit is placed on few large boards or many small boards.

It is assumed that the PCB manufacturer's equipment can be best suited for medium-sized boards which can, therefore, be regarded as the optimum board size. Larger boards are usually higher priced per unit of area than medium-sized boards. The larger boards certainly require less handling per unit of area, but waste of raw material and the production limitations discussed in Sections 5.2.3b and 5.2.3c are important considerations.

The scrap rate is another important matter. Let us assume that the failure rate per unit of area is independent of the board size and, therefore, can be considered fairly constant. The scrap rate expressed in number of boards, will increase disproportionately with increasing board size.

Very small boards exhibit a slightly higher price per unit of area than medium-sized boards. By a step-and-repeat photographic process, a reasonably high number of boards can be manufactured on one production panel. Therefore, although the board requirement may be large, the number of production panels will be small, so that the production costs per panel, and consequently per board, will be disproportionately high.

In order to demonstrate the points made above, the author has investigated the cost of various changes to the board in Fig. 3.17. The modifications are shown in Fig. 5.4, where A is the original board, B is half of A, C is twice A, and D and E

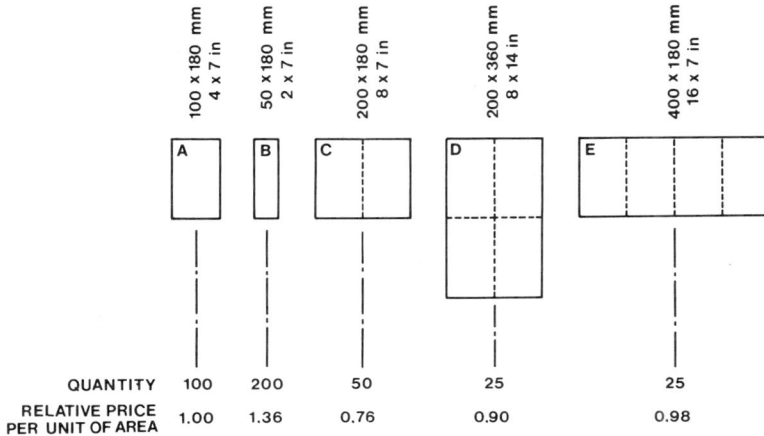

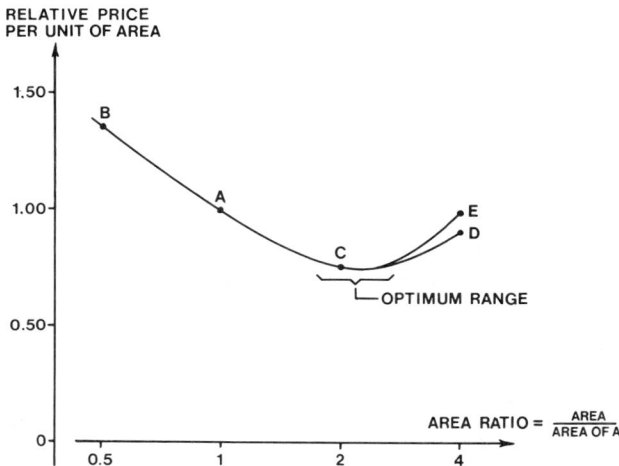

QUANTITY	100	200	50	25	25
RELATIVE PRICE PER UNIT OF AREA	1.00	1.36	0.76	0.90	0.98

Fig. 5.4 PCB price as a function of the area ratio.

are four times A. The quantity of boards of each size is such that the same total area is obtained in all cases. The price per unit is calculated relative to A. The curve shows the relative prices per unit of area as a function of the ratios of the area of the boards relative to A. In order to widen the curve, the X-axis has been made logarithmic. The curve has a minimum at an area ratio of 2.5 corresponding to a board format of approximately 180×250 mm (7×10 in). The PCB

manufacturer in question agrees that this size is his optimum. The results correspond fairly well with other investigations where a board format of 200 × 250 mm
(8 × 10 in) is said to be the optimum size.

5.2.5 Standardized Boards

The German standard, DIN 41494, describes the standardization of board sizes
corresponding to the commonly used 19-in rack system. These sizes are shown in
Table 5.1.

Table 5.1 Standardized PCB sizes. Extracted from DIN 41494 by permission of
Deutsches Institut für Normung e.V.

	Front panel		PCB size				
Size	Height		Height		Length		
	mm	inch	mm	inch			
C	132.5	5.22	100	4	100 mm	160 mm	220 mm
D	177	6.97	144.5	5.69	(4 in)	(6.30 in)	(8.66 in)
E	221.4	8.72	188.9	7.44			
F	265.9	10.47	233.4	9.19			

The author would now like to describe a standard board which has proved very
versatile and has been used in many applications. The board measures
100 × 160 mm (4 in × 6.3 in) and contains six double rows of solder holes in a
module of 2.54 mm (0.1 in). Each row accommodates six ICs, and the entire
board 36 ICs. This gives an area utilization of 445 mm² (0.69 sq. in) per IC, including the area of the edge connector. This area utilization corresponds fairly
well to values given in technical literature where a conservative value is 645 mm²
(1 sq. in) per IC or, more rigorously, 516 mm² (0.8 sq. in) per IC. In the case of
computer-aided designs (CAD), it is often stated that optimum design costs are
obtained with an area of 645 mm² (1 sq. in) per IC.

Figure 3.17 shows a modified standard board which is obtained by extending
the normal standard board by 20 mm (0.75 in). The extension gives room for a
special connector placed opposite the edge connector. These two standard boards
have found extensive use, and Figs 5.5 and 5.6 show the internal construction of
two recently introduced instruments into which they are incorporated. Although
the standard boards are, perhaps, not quite optimum in terms of price, the size has
proved to be an excellent compromise between reasonably large sub-assemblies,
convenient testing and repair, high component density and reasonable cabinet
dimensions. Figure 5.7 shows an instrument based on a single large board.
Despite high component density, the packaging density (volume utilization) is
rather low, and furthermore, larger modifications of the circuitry are difficult to
put into effect. This design was necessary since for practical reasons the instrument had to be shallow.

Fig. 5.5 Location of standard boards in a modern instrument. Note the shielding of the three boards at the left. Note also that two of the boards are provided with an output connector.

Fig. 5.6 Location of standard boards in another modern instrument.

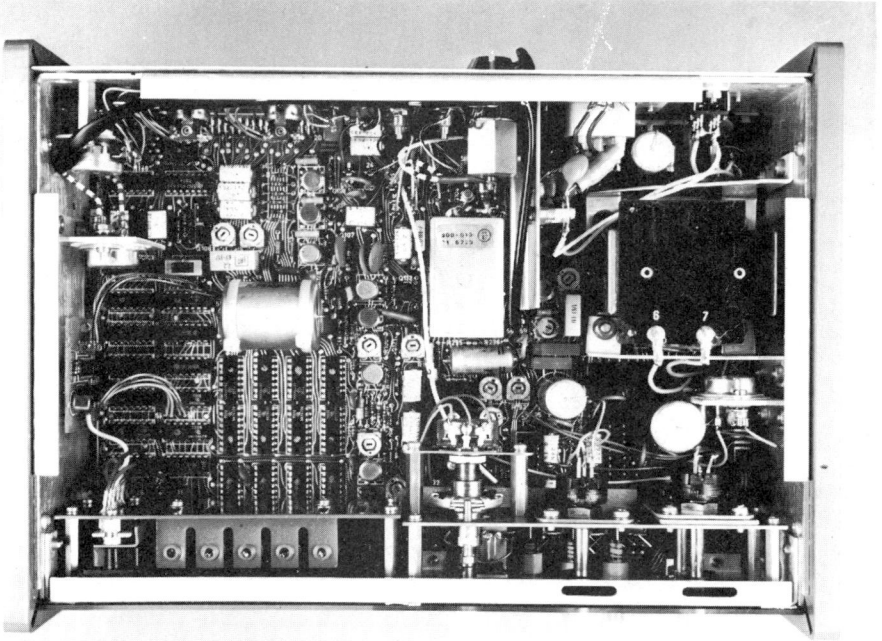

Fig. 5.7 Instrument based on a single, large board.

5.3 Practical Layout Procedure

Now we shall deal with the practical layout procedure. The following description can only be a treatment of the general principles as every PCB designer has his own method of approach and, furthermore, the variations in printed circuit boards are almost infinite. In order to ensure clarity, the descriptions will be limited to the manual layout of single-sided and double-sided boards, at the exclusion of computer-aided designs and multilayer boards. Computer-aided designs, however, are described in detail in Chapter 9.

In the author's opinion it is, for several reasons, worth the extra trouble of preparing a careful layout before proceeding to the taping stage. An artwork must be correct in all its details, and furthermore, be clean and sharp in order to ensure a good photomaster which forms the basis of the manufacture of the board. It is almost inevitable that, if the layout and taping stages are combined, changes to taped parts of the pattern will become necessary, particularly if the circuitry is complicated and the space limited. Consequently, the layout becomes more dominant than the taping and, as a result, the accuracy and correctness of the taping will be compromised. Also, modifications to the taping will inevitably lower the quality; the drafting film will become dingy, some of the adhesive will be left on the surface, the conductors will become frayed at the edge, and if a solder pad is damaged, the replacement may be a little displaced. In short, one finishes up with a poor-quality artwork. One other point: a special procedure for taping double-sided boards assumes that the hole pattern is already determined and taped. After that, two identical diazo contact prints made on stabilized polyester film are taped with the conductors of the solder side and the component side, respectively. Finally, many PCB departments are divided into designers and draughtsmen who have special expertise. If the layout and taping stages are combined, then this job allocation becomes inapplicable. The result of destroying this arrangement will be under-utilization of expertise and a lower product quality. The author considers that the best and cheapest results are achieved when a careful layout precedes the taping stage. The latter can then be carried out without modifications and problems. To the experienced designer, much of the following description will be obvious, but it may be that a re-assessment of his own approach could produce improvements in his organization and procedures.

5.4 Schematic Diagram Considerations

The schematic diagram forms the basis of the layout work and it is, therefore, important that it is complete and brought up to date. By a complete schematic diagram, the author means a well-ordered, ready-drawn diagram and not a collection of small pieces of paper filled with corrections, and corrections to corrections, and possibly, covered with various calculations. The layout and drawing of schematic diagrams is often part of the printed circuit board department's duties, so the first task is to ready-draw the schematic diagram, which should then be approved by the circuit designer.

The circuit designer, who may have been working on the design for a long time and is probably aware of the pitfalls which should be avoided during the layout stage to prevent spurious signals, unwanted couplings, etc., can certainly assist the PCB department to produce a good working layout with minimal problems. Therefore, an exchange of information and ideas before the layout work is started will pay dividends later.

5.5 Circuit Considerations

5.5.1 Introduction

The PCB designer must minimize the size and the influence of the parasitic circuit elements such as the resistance and inductance of conductors, and the mutual capacitance and inductance between pairs of conductors, in order to ensure that the design of the pattern does not adversely affect the operation of the circuit. Also, he should avoid capacitive and inductive interactions with adjacent circuits.

We shall now deal with the factors taken into account by the PCB designer to achieve the above objective. Only seldom are the effects pre-calculated since this is very time consuming and unlikely to be wholly accurate. However, once the PCB designer has acquired good working habits and has sufficient experience, the results will normally also be adequate or good. The desirability for an experienced

PCB designer is emphasized when one considers that, in many cases, it is impossible to improve significantly an inadequate layout. The only practical solution is to start again.

5.5.2 Conductor Length

Since the parasitic inductances and capacitances are proportional to conductor length, all conductors should be made as short as possible. This implies, in practice, that the electronic components should, as far as possible, be so located that the interconnections are limited to a length corresponding to the physical dimensions of the components. High impedance points are particularly sensitive to parasitic capacities resulting from the physical size of the conductors. In the case of conductors connecting low impedances, the resistance and inductance are most significant.

5.5.3 Conductor Width

Narrow conductors are used for high-impedance signal lines when the stray capacitance must be kept small and the series resistance and inductance are of little importance.

Wide conductors are used for low-impedance signal lines when low series resistance and inductance are essential and the stray capacitance is unimportant.

If the width, only, of a conductor is changed, the product of inductance per unit length and capacitance per unit length is constant. For high-frequency applications, it is possible to design the conductor width to achieve a prescribed characteristic impedance.

5.5.4 Ground Conductors

In order to achieve correct performance of the circuitry, it may be necessary to split the ground connection into several independent systems. A common ground conductor may be undesirable as voltage drops along the conductor may introduce false signals. Under unfavourable conditions these signals may, for example, occur in series with the input signal to a gate and, if of sufficient magnitude, may trigger it. It may be necessary, therefore, to introduce separate ground conductors, e.g. a digital ground conductor, an analog ground conductor, and a reference ground conductor. The PCB designer must try to avoid a common ground conductor or, alternatively, try to reduce the length or to increase the width of the ground conductor as much as possible.

Short and wide conductors may also prove important. Where this is so, the circuit designer will inform the PCB designer.

5.5.5 Ground Planes

Particularly in the case of high-frequency circuits, it may be necessary to introduce a ground plane, both to achieve a shielding effect (to avoid capacitive couplings), and to ensure a high-frequency return line of low inductance.

Mutual capacitances can be reduced if the signal line is placed close to the ground plane or, alternatively, is inlaid in the ground plane. This reduces mutual capacitances at the expense of increased capacitance to the ground plane which is usually less harmful.

Reducing the distance between the signal line and the matching return line or ground plane reduces both the stray inductance and the mutual inductance to other current loops. To summarize, small conductor spacings and wide ground planes ensure lower impedance and less interactive wiring, and with that, a reduced risk of stability problems.

A special design is the so-called 'stripline', which is a type of transmission line consisting of a narrow conductor of accurately controlled width on one side of the board and a ground plane on the other. A careful calculation of the geometrical relationship should already have been done by the circuit designer, but when taping the artwork it is necessary to compensate for the changes in conductor widths caused by the manufacturing processes.

5.5.6 High-impedance and Sensitive Circuits

Direct current circuits, having high-impedance points, may have to be shielded by means of a ground plane, possibly supplemented by a metal shielding box attached to the other side of the board. In amplifiers of nanoampere-range sensitivity it is necessary to protect the input against leakage currents by means of a guard-ring. If the insulation resistance of the base material due to its hygroscopic (water-absorbing) properties is so low that the input impedance is reduced significantly, it becomes necessary to keep the sensitive part of the circuit free of the board. This can be achieved by drilling a free hole in the board for the input terminal of the amplifier, and by raising the circuit on Teflon pillars or the like.

5.5.7 Mutual Location of Components and Conductors

Components and conductors should be so located with respect to each other that the inevitable mutual capacitances occur at places where their effects on the performance of the circuit are minimized.

Certain high-impedance, sensitive circuit points must be guarded against undesirable and detrimental cross-talk from other parts of the circuitry with higher signal levels. This is best achieved by making the areas of the receiving and the transmitting elements (conductors, solder pads, and components) as small as possible, by maintaining adequate space between these elements and by using the interlying area for neutral and insensitive parts of the circuitry. Decoupling capacitors can be utilized as shielding elements. Several types of capacitors are so constructed that the outer-foil electrode acts as a shield, if grounded.

Figure 5.8 shows how the PCB designer can affect the parasitic conditions by modifying the layout. The circuit is shown in Fig. 5.8A, and the corresponding equivalent diagram is shown in Fig. 5.8B, where the capacitive and inductive parasitics are shown: $C_{yn} = C_{y1}$, C_{y2}, and C_{y3} are the stray capacitances to an adjacent signal source Y. $C_n = C_1$, C_2 and C_3 are stray capacitances to ground, whereas $L_n = L_1$, L_2, and L_3 and $L_{0n} = L_{01}$, L_{02}, and L_{03} are the stray inductances in the signal and the ground lines, respectively. If the layout is 'open', as shown in Fig 5.8C, the dominating parasitics will be L_n, L_{0n}, and C_{yn}. A 'closed' and less sensitive layout is shown in Fig 5.8D. In this case, the capacitances C_n are the dominating parasitics, whereas C_{yn}, L_n, and L_{0n} are considerably reduced.

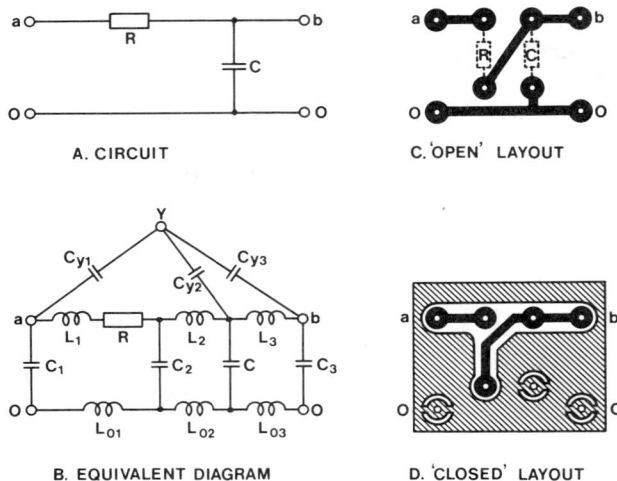

A. CIRCUIT   C. 'OPEN' LAYOUT

B. EQUIVALENT DIAGRAM   D. 'CLOSED' LAYOUT

Fig. 5.8 Example of the way in which the parasitic elements can be neutralized.

5.5.8 Mutual Location of Current-carrying Conductors

By careful adjustment of the lengths, widths and spacings of the conductors, it is possible to reduce either the stray inductances or the stray capacitances by a significant amount—possibly by a factor of 10.

As the circulating signal currents run in closed loops, the PCB designer should always aim at concentrating these loops in such a way that signal currents originating from different parts of the circuitry circulate in separate loops with no common ground conductors whose impedance would cause unwanted couplings.

The desired signal separation is achieved by observing the following rules:

1. The circuit is laid out so that signal levels increase progressively from one end of the board to the other.
2. The power supply is connected to the end of the circuitry having the highest signal level.
3. Decoupling capacitors are placed to force the signal currents in certain loops in order to guard other parts of the circuitry against spurious signals and, vice versa, to attenuate spurious signals transmitted from other parts of the circuitry.

5.6 Thermal Considerations

The PCB designer should know each of the heat-sensitive and/or heat-developing components so that he can locate them advantageously with respect to one another. Normally, heat-developing components are raised above the board to avoid discoloration or damage both to themselves and to the board. The sizes of any necessary heat sinks should also be made available to the PCB designer.

5.7 Definition of Components

Semiconductors are usually indicated in the schematic diagram by their type designations; the physical sizes can easily be found from data sheets. Other components are usually identified by their electrical values only, sometimes with a stated tolerance.

The physical size of a component depends almost entirely on the type selected. In 90% of cases, the type of component to be used is known by the PCB designer. The remainder will probably be special and the circuit designer should identify and define these to the PCB department. In any case, the circuit designer must be responsible for the ultimate choice of components and, therefore, for the parts' list.

When all the components have been defined, the PCB designer has to procure data sheets on all unknown components. It is more practical if he establishes a 'library' of his own, covering the more-or-less special components. Next time the information is needed, it will be at hand. Furthermore, it would be expedient if the PCB designer were to procure a sample of each component and, in this way, build up a collection of samples. Sometimes samples can show details which do not appear sufficiently clearly in the dimensional outlines given in the data sheets. On the other hand, it must be a rule never to settle for measurement of the physical dimensions of a sample component. It could be an advanced laboratory sample subject to certain modifications, so it is always better to rely on data sheets until such time as the component is officially released.

5.8 Mechanical Requirements

Another important task for the PCB designer is to obtain reliable information regarding the physical size of the board. Normally, this requires very close contact with the mechanical design department. Initially, it is a question of determining the outline dimensions and the mounting holes, if any, of the board. All height restrictions should be stated so that the PCB designer can take these into account when locating large components. In order to locate heat-dissipating and/or heat-sensitive components in the best possible way, the PCB designer should be informed about the construction of the equipment, including the location of the board (horizontal or vertical) and the method of cooling (free or forced convection).

Certain components mounted directly on the board require specific positions, e.g. front-panel operated components such as push-button switches or variable resistors. Also, a connector may have to match up with an aperture at the rear of the equipment or with a fixed receptacle.

Very often a board is not rectangular and various board cut-outs must be determined. Here the PCB designer should endeavour to specify wide tolerances. A common problem of PCB manufacturers is the specification of cut-outs with sharp angles: they prefer rounded corners (see Fig. 5.9). If the application does

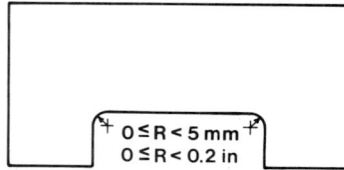

Fig. 5.9 Cut-outs in PCBs.

not require sharp corners it is better to give the PCB manufacturer a free hand, for example, by specifying $0 \leqslant R < 5$ mm (0.2 in), where R indicates the radius of the rounding. This implies that the profiling can be done by pin or NC routing, using a 3.2-mm (0.125-in) mill.

The PCB designer has to have all other mechanical parts defined, including mounting elements such as brackets, clamps, clips, shielding boxes and heat sinks. It is not sufficient to know the location of the mounting holes; the PCB designer must also know the area occupied by the component in question in order not to run conductors across this area. In this connection the PCB designer must bear in mind that screw heads and nuts require a certain space! A classical fault is that the screw head short-circuits some conductors and the classical remedy is to place an insulating washer under the screw head.

The above information should make it possible for the PCB designer to draw and to indicate the dimensions of the profile of the board and the mounting holes, and to highlight the 'forbidden' areas. This drawing is the basis for the subsequent layout work and the draughtsman must have access to it as all the mounting holes must be placed precisely. Similarly, it is important to know the limitations of the 'forbidden' areas so that the draughtsman will be able to run the conductors with the prescribed spacing.

5.9 Methods of Component Mounting

In order to make a layout which results in a professional and easy-to-assemble board, the PCB designer must know the normally used mounting methods and general layout practice. In this context, it is very important that the PCB designer is familiar with the limitations set by the production equipment used by the assembly department. It is no use the PCB designer specifying that the terminal leads of a resistor should be bent to a four-module span equal to 4×0.1 in (4×2.54 mm) (lead centre-to-centre distance), if the lead-forming machine cannot bend to a span shorter than five modules. For their common good, the assembly department should always keep the PCB department informed of all method and equipment changes.

Below, we shall start with a few ground rules, which are mandatory on the PCB designer.

(a) A component lead should always be inserted in a solder hole and soldered from the opposite side.

Note: The exceptions are flat-packs and similar devices, and also where it proves necessary to attach some extra components to solder pads on the solder side in order to cope with some technical problems occurring in a production series.

(b) Only one component lead should be inserted per solder hole.

(c) A solder joint should never be mechanically over-stressed. A rule of thumb is: Non-plated through boards, 10 g (0.35 oz) per solder joint; plated through boards, 25 g (~1 oz) per solder joint.

(d) Components of the same type should be bent to the same span of modules.

(e) The components should, as far as possible, be placed parallel to one another.

5.10 Axial Lead Components

This is, perhaps, the most common type and comprises chiefly, resistors, capacitors, and some diodes. The leads are bent through approximately 90°, inserted in the solder holes and soldered (see Fig. 5.10). From practical considerations, the number of different component spans should be limited because the lead forming (bending and cutting the leads) is automatic. It is advantageous to lay down a standard span for the same type of component, for example five modules = 5×0.1 in for a 0.25 W resistor. Usually, the PCB layout is based on modules of 0.1 in (2.54 mm) but in some cases it is necessary to use a finer module of 0.05 in (1.27 mm) and in rare cases 0.025 in (0.635 mm).

The span selected depends first and foremost on the length, k, of the component body (see Fig. 5.10). In order not to damage the joints between the component body and the leads, the 90° bend must not be placed too close to the component body. A rule states that the minimum distance, a, shall be at least 0.8 mm (0.031 in) + $1.5 \times$ lead diameter (see Fig 5.10). The lead diameter is very often 0.4–0.5 mm (0.016–0.020 in) so the bending should not be closer than 1.5 mm

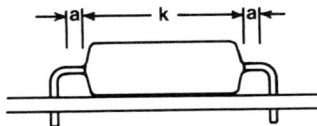

Fig. 5.10 Lead forming for mounting parallel to the board.

(0.06 in) to the component body. If the lead is welded outside the component body, the point of bending should be measured from the weld in order to avoid stress. The span becomes $k + 2a$, which is rounded off to the nearest larger module.

Often the lead-forming machine has a clamping device which grips the component lead just outside the component body in order to ensure that the bending force is not transferred to the joints. In certain machines, the clamping device requires 2.5 mm (0.1 in) which makes it impossible to obtain the above free distance of 1.5 mm (0.06 in), cf. the introductory remarks about the communication necessary between the PCB department and the assembly department.

In the case of fragile components, especially glass-encapsulated diodes, the leads, or at least one of them, should be formed with a stress-relief loop as shown in Fig. 5.11. In this way the lead(s) will be able to absorb the mechanical stresses

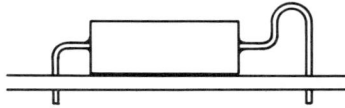

Fig. 5.11 Lead forming with a stress relief loop.

to which the component body can be exposed, for example, during a vibration of the board (cf. Fig 5.1), or due to heating caused by the soldering machine.

The methods of mounting shown in Figs 5.10 and 5.11 assume that the component body rests against the board. If the board is double sided, conductors may run beneath the component body and thus the insulation of the component should be checked to ensure that it is adequate to prevent voltage breakdown. This, however, is normally no problem due to the low voltages used in semiconductor circuits. Nevertheless, a very inconvenient type of fault due to poor layout practice is sometimes encountered: An interfacial hole, also called a via hole, is placed just beneath the component body. During automatic soldering, the solder is sucked up through the hole and damages the insulation of the body, which is very often an insulating lacquer, resulting in a short-circuit. Especially in the case of digital boards which are drilled in a standard hole pattern, the PCB designer is tempted or forced to place a component across a via hole belonging to the hole pattern. If it is not possible to move either the component or the via hole due to limitations of space, or because the PCB manufacturer has already a stock of semi-manufactured boards, the component should be raised from the board. This can be achieved by using stand-off lead forming, snap-in lead forming or, simply, by threading small glass beads on the leads.

Snap-in forming means that the leads are bent in a zigzag shape which serves to lock the leads in the solder holes in order to prevent the components falling out during transport to the soldering machine (see Fig. 5.12). To digress, it should be borne in mind that the snap-in method, by its nature, demands uniform reflowed (fused) plated through holes. For this reason, it is of the utmost importance that the PCB designer and draughtsman take adequate measures to enable the PCB manufacturer to maintain the diameter tolerances (cf. Sections 4.5.3 and 6.6.3). Practice has shown that after balancing the board pattern (by introducing robbers) it is possible to obtain a board assembly using snap-in leads with practically no problems. Without robbers it is difficult to use this method due to variations in hole diameters.

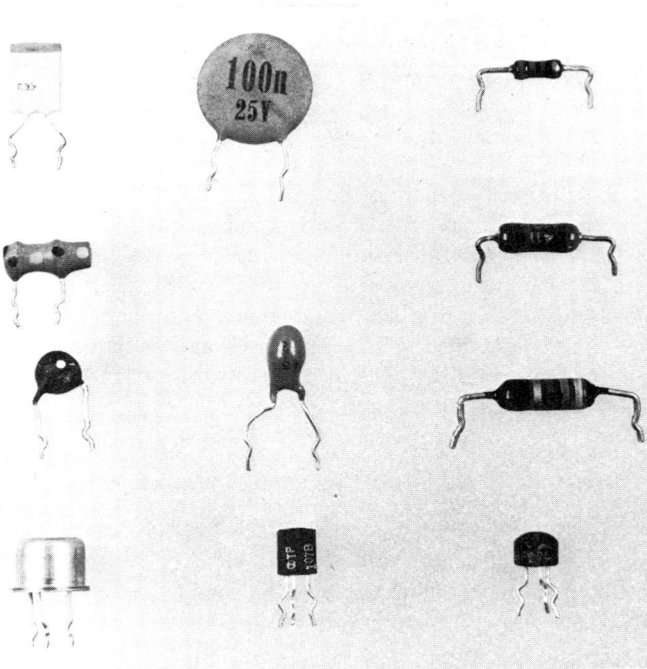

Fig. 5.12 Examples of snap-in leads for various components.

Where space is limited, it is possible to mount axial lead components in a vertical position, as shown in Fig. 5.13, where one lead is bent as a hairpin. Thus it is possible to achieve a 'span' of 0.1 in (2.54 mm), but it should be noted that the method might mean that the hairpin must be insulated using sleeving.

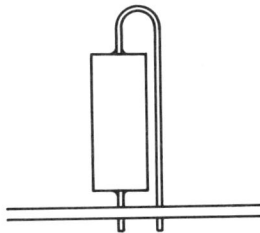

Fig. 5.13 Lead forming for vertical mounting.

Heavy components, as defined in Section 5.9c, should be secured mechanically in order to safeguard the solder joints. Clamps and clips may be used (Fig. 5.14) but in all cases it is a condition that the component must be secured to the board

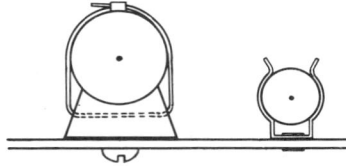

Fig. 5.14 Examples of component clamps and clips.

prior to soldering, otherwise there is the risk of stressing the solder joints when the component is secured. In the case of non-plated through holes, there might even be lifting of the solder pads. Rotation of clamps or clips can be prevented using two fasteners or one fastener plus a non-turn device inserted in a hole of the board. If the clamp is made of metal, adequate distance from adjacent conductors and pads must be maintained. The insertion of an insulating washer under the clamp is not recommended as pockets of moisture, which could affect the circuitry, might occur under the washer.

5.11 Radial Lead Components

This group comprises mainly capacitors, certain resistor types, variable resistors (also called potentiometers), semiconductors such as integrated circuits (ICs), most transistor types, mechanical components such as switches and transformers intended for PCB mounting, and connectors.

Usually ceramic capacitors have been dipped in lacquer for surface protection, and some of this lacquer can easily overflow onto the leads. These capacitors should not be soldered as shown in Fig. 5.15A because the gases produced by the lacquer during soldering are detrimental to the solder joints. The capacitor should be raised from the board either using glass beads or spacers (Fig. 5.15B) or by a suitable lead forming, e.g. an offset mounting, as shown in Fig. 5.15C.

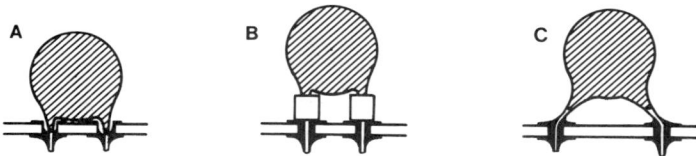

Fig. 5.15 Mounting of ceramic capacitors.

Many electrolytic capacitors are mechanically secured by means of a mounting ring with feet. The PCB designer must realize that the flaps which are used for tightening the ring around the capacitor body, require some space, and furthermore he must ensure that the screw used for tightening the ring is accessible. It is an important rule that mechanical clamping devices must not be part of the electrical circuit. When designing single-sided boards, it can be a temptation to use, for example, a standing clamp to bridge some conductors, but for several reasons the PCB designer must be dissuaded from this. One reason is that the surface treatment of a clamp precludes good solderability and results in poor electrical connections. Also, the solder joints become dry through mechanical movements or vibrations. The author recollects a case of a push-button switch which was seated on a single-sided board and secured by soldering the small mounting tabs of the common switch frame. The frame was used to bring a ground conductor from one switch section to the next and so on, but after use for a short period of time, excessive electrical noise occurred. The reason was that the small impacts caused by the release actions of the push-buttons via the frame were transferred to the solder joints which, in turn, fractured and became dry.

Most semiconductors (ICs and transistors) are radial lead components. Transistors are usually soldered to pads arranged in a cluster corresponding to the botton view configuration of the transistor, i.e. the leads are brought straight through from the body of the transistor to the solder holes of the board. In many types, however, the lead circle diameter is so small that straight-through mounting is not possible and offset mounting must be adopted. The leads are then formed to fit into solder holes arranged in a cluster having a significantly larger diameter in order to give sufficient space for larger solder pads.

ICs are available in three basic types: TO-70, flat-packs and dual-in-line. They are described below.

The TO-70 Type

An IC of this type is mounted in a can corresponding to the original and well-known three-lead TO-5 transistor, but with 8, 10, or 12 leads. The lead circle usually has a diameter of 0.2 in (5.08 mm), but lead-circle diameters in the range 0.1 in (2.54 mm) through 0.141 in (3.58 mm) to 0.23 in (5.84 mm) also exist. Within the types TO-70, 71, 73, 74, 75, 76 there are significant differences between the lead circles, the numbers of leads, and the angles between the leads, and a mistake can make it impossible to insert the IC directly. The PCB designer must ascertain the maximum pad size he can obtain if straight through mounting is to be used. In many cases, through mounting is not feasible and in others the space available is such that the pads would be so small that, after drilling, little or no annular ring would be left. The maximum diameter of the pad would, in such cases, be 1 mm (0.04 in), and with a lead diameter of 0.5 mm (0.02 in) the finished solder hole should not be smaller than 0.7 mm (0.028 in). This means that the hole should be drilled with a 0.8-mm (0.031-in) drill, leaving a theoretical annular ring of 0.1 mm (0.004 in). Manufacturing tolerances will absorb more than 0.1 mm and it is very unlikely that a PCB manufacturer would be prepared to undertake the job unless hole break-out was permitted. Added to this is the problem of running the necessary conductors in a practical manner. The distance between two solder pads under the above circumstances is so small that it is impossible to

116

route a conductor of reasonable width between the pads. This must mean that it is only possible to obtain access to the pads from the outside of the cluster. The tendency is, therefore, to offset the leads, either by preforming them or by using a spacer which spreads them, thus producing a larger lead-circle diameter. Perhaps more commonly, the leads are preformed to match a normal dual-in-line pattern. The latter method gives some layout advantages which are mentioned later in the discussion of dual-in-line ICs.

Flat-packs

Because flat-packs are small and have a small lead distance—0.05 in (1.27 mm)—they have certain uses. Flat-packs can be mounted in two fundamentally different ways:

1. Directly on top of non-drilled pads as a surface mounting either by welding, or more commonly, by reflowing (Fig. 5.16). The process of reflowing involves soldering the component using only the small amount of solder already present on the pad and the lead, i.e. additional solder is not applied.

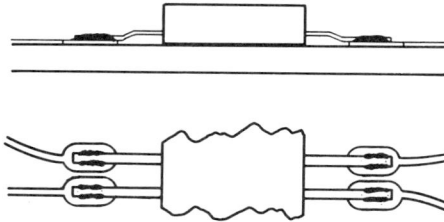

Fig. 5.16 Mounting of flat-packs.

2. By preforming the leads to correspond to an in-line pattern, followed by normal soldering. The very small distance—0.05 in (1.27 mm)—between the leads, is unfortunate but an improvement can be obtained by off-setting adjacent pads and leads on the same side of the flat-pack, as shown in Fig. 5.17.

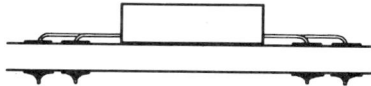

Fig. 5.17 Mounting of flat-packs with offset terminal leads.

The first method has the great advantage that it is possible to run conductors on the opposite side of the board, under the solder pads. Thus it is possible to achieve a high component density and to save the drilling of a significant number of holes, which must be performed in the second method. On the other hand, the number of via holes will be increased in order to connect the conductors of the opposite sides with IC pads.

Flat-packs can be mounted on both sides of the board, using the first method, and a still higher component density can be achieved. Unfortunately, the welding or the soldering produces fairly serious technological problems, of which the handling and the accurate positioning and fixing during the welding or soldering operations seem to be the most severe. These disadvantages have limited the use of flat-packs and have led to the development of the dual-in-line package.

Dual-in-line packages (DIPs)

This type of integrated circuit has become the most popular type for several reasons: The package is fairly sturdy, is easily inserted on a board and soldered or can be mounted in sockets. The lead configuration makes the dual-in-line package suitable for even a very complicated layout and a large number of conductors can be routed lengthwise through the dual-in-line cluster. It is also possible to run a conductor between two adjacent solder pads.

The testing of mounted dual-in-line ICs can be achieved relatively easily by means of a logic test probe which is clipped to the IC leads. The PCB designer, however, must not locate the dual-in-line ICs too close to one another as it will be impossible to open the jaws of the test probe and locate them on top of the ICs. For this reason, it is practical to leave a certain amount of free space between the ICs as shown in Fig. 5.18. In the author's experience, a $1\frac{1}{2}$ module equal to a 0.15-

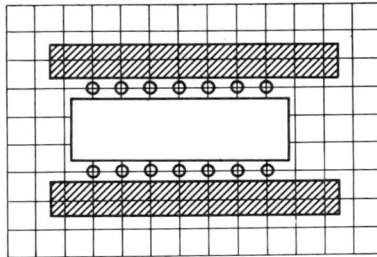

Fig. 5.18 Free space (cross-hatched) along a dual-in-line IC.

in (3.8-mm) wide space on both sides of the IC is sufficient. No other components should be located within the free space; only via holes and jumper wires are allowed. The length of free space for a 14-lead IC should be the same as that for a 16-lead, because of the test probe, and located assymmetrically as shown in Fig. 5.18. In practice, this means that there will always be room for 16-lead ICs on the board. Figure 5.19 shows the maximum achievable density of dual-in-line ICs, if reasonable consideration is to be given to conductors running lengthwise through the cluster.

When laying out dual-in-line ICs on non-plated through boards, the PCB designer must ensure that the solder pads are sufficiently large to prevent them lifting from the board. Even when the solder pads are relatively small, lifted pads will not normally occur during or after soldering. The problem arises when a board is defective and has to be repaired. If the operator is not very careful, the desoldering of a dual-in-line IC can result in some lifted pads. There is a rule of thumb which states that the minimum area of non-plated through solder pads shall be 5 mm² (0.0078 sq. in), but this comes into conflict with the need to run

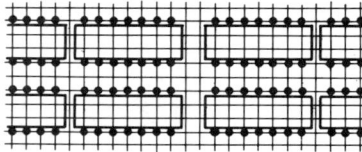

Fig. 5.19 Maximum packaging density for dual-in-line ICs.

conductors lengthwise and transversely in relation to the IC. This need is even more important in the case of single-sided, non-plated through boards because of the limited possibilities for routing the conductors. Such boards are, in fact, not particularly practicable even where a limited number of dual-in-line ICs are to be mounted. For reliability, the PCB designer should never overlook the above rule of 5 mm² in his struggle for a component density corresponding to that of double-sided, plated through boards.

A detailed discussion of the size and shape of the solder pads is given in Sections 6.5.4 and 6.5.5.

In addition to the 14- and 16-lead ICs, other ICs with still more leads, for example 24 leads and 40 leads, are in use. The above points must be adapted to meet the actual conditions and, clearly, this is a problem for the PCB designer to solve.

Jumper wires

Especially when laying out single-sided boards it is often necessary, here and there, to bridge across transverse conductors. This is done by means of jumper wires which, in their most simple form, consist of a piece of insulated (sleeved) wire formed in the same way as a radial lead component. Clearly, standard lengths should be used, thereby restricting the number of wire lengths required, and the lengths should always be whole numbers of modules so that an automatic lead-forming machine can be used.

A seated and soldered jumper wire is a fairly cheap device, so the difference in price between a single-sided, non-plated through board and a double-sided, plated through board can pay for many jumper wires. The author recollects a case where about 100 jumper wires were used in order to realize the board as single-sided, non-plated through. In the long run this proved to be uneconomical due to increasing wages and a reduction in the price differential between the two types of board.

Remaining Components

The remaining radial lead components are adjustable resistors, trimmers, transformers, switches and connectors. These components, with the exception of the trimmers, are mounted in the same way as other radial lead components with preformed leads. Excepting trimmers which are normally very light, all the other components are usually provided with clamping devices, very often in the form of small lugs. If this is not the case, other measures should be taken in order to secure the component properly. Figure 5.20 shows how a flat power transformer is secured by means of a clamp made specially for the purpose.

A heavy transformer for use on a PCB should always be mounted and secured

Fig. 5.20 Fixing a power transformer by means of a metal clamp.

before the soldering of the board takes place. This is the only practical way to avoid mechanical stressing of the solder joints. Because of various impacts or resonant vibrations, the board cannot and should not carry the transformer without mechanical support. A simple supporting system uses two brackets as shown in Fig. 5.21.

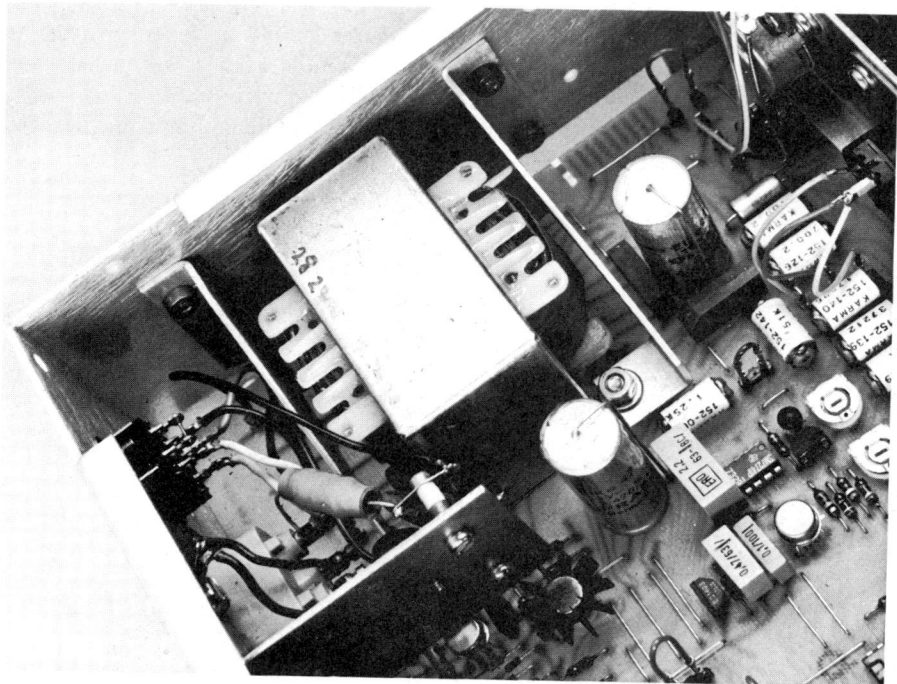

Fig. 5.21 Mounting a power transformer on a board. The mechanical support is provided by means of two brackets.

In conclusion, let us consider a couple of examples which show how important it is that the PCB designer visualizes the components mounted on the board. The first example concerns trimmers: In spite of a careful study of the dimensional sketches, the PCB designer can easily mislocate the trimmers by positioning them so close side by side, that the rotary contacts practically touch each other when turned to the most unfavourable position. A practical check using a piece of perforated board with a 0.1 in (2.54 mm) module would have demonstrated this classical fault. Another type of fault, with some safety aspects, is found in the case of transformers for PCB mounting. Although the transformer terminals meet the requirements for creepage distance at the point where they emerge from the bobbin, the transformer manufacturer sometimes overlooks the fact that the creepage distance available on the board itself is reduced by the pads, particularly in the case of non-plated through boards which require fairly large pads.

5.12 Determination of Layout Scale

The next problem with which the PCB designer is confronted, is the layout scale, and here the author would assert that the best layout scale is 4:1, unless the sketch becomes so large that it cannot be accommodated on the drawing board. Some PCB designers prefer to work in a 1:1 scale because the layout corresponds exactly to the actual size of the components, but it is only a matter of custom to work in a 4:1 scale. Why should the PCB designer choose a 4:1 scale instead of a 1:1 scale, or in an emergency a 2:1 scale? The principal reason is the level of accuracy necessary in the taping of the artwork. This will be dealt with in Chapter 6. As the draughtsman has to work on 4:1 it is, obviously, an advantage to produce the layout sketch to the same scale. Even if the layout sketch is not quite accurate in that the components are not located precisely, the draughtsman can use the sketch as a good indication of the location of the pads. The only condition is that the pads are placed in accordance with an underlying grid which serves as a guide. If the layout sketch comes in another scale, the PCB designer has to scale-up all dimensions which really means re-drawing the layout sketch. This involves the risk of introducing errors and, furthermore, it is more difficult and time consuming to check the artwork—or to see how far the taping has progressed—against a layout sketch made in another scale.

Once the PCB designer becomes familiar with a 4:1 scale, he will soon realize that he can work much more quickly and easily with the generous space which it affords.

Use of the recommended scale of 4:1 means that his eyes must adapt to see everything enlarged four times. A good aid is to draw component sketches in a 4:1 scale and a very convenient method of sketching the board layout on the basis of such component sketches is described in Sections 5.22.3 and 5.22.4.

5.13 Location by Means of a Grid

Generally, the components are located according to a grid. As was mentioned in Section 5.10, distances are usually expressed in terms of modules, where a module corresponds to 0.1 in (2.54 mm). Therefore, the grid is ruled into squares having side lengths of 0.1 in. For a layout in the 4:1 scale, it is practical to use a grid where every fourth grid line is accentuated. If the intervening grid lines are included, the PCB designer has the facility to operate with quarters of a module.

It is common practice to position the pads at intersections of the grid lines. In the vast majority of boards, increments of whole modules (1:1) are sufficient, but, if necessary, they can be supplemented with half-module increments. In complex boards with high component densities, on the other hand, it may be necessary to operate with quarters of a module in order to utilize the space as effectively as possible. Therefore, to facilitate the layout work, a grid system constructed with grid lines accentuated and non-accentuated in accordance with the chosen scale is recommended.

5.14 Rough Layout

The PCB designer normally starts the layout by planning and sketching a rough location of the components. If the format of the board is rectangular, the highest packaging density is obtained if the components are parallel to one edge of the board. Alternatively, some components may be located parallel to one edge and others parallel to the other edge. With this approach, apart from higher component density and good cosmetic appearance of the assembled board, there are significant advantages in terms of ease of assembly and fault location.

As has been mentioned earlier, components or groups of components demanding external connections, e.g. input or output amplifiers, are placed as close as possible to the board connector. The PCB designer should, of course, be aware of undesirable couplings such as cross-talk between the input and output and should not locate the terminals close together. The remaining components are now located so that they follow the flow in the schematic diagram in order to achieve the shortest possible interconnections (see Sections 5.5.7 and 5.5.8). It is practical to locate, first, those components requiring the largest board areas and then to fill the remaining areas with smaller components. During this phase the PCB designer must bear in mind maintenance considerations. The minimum requirement is that every component can be replaced without having to desolder other components. In order to meet more than the minimum requirement, the PCB designer should think three-dimensionally and avoid location of small components in the shadow of large components as shown in Fig. 5.22A. For the same reason, the space beneath cylindrical components should be left free as shown in Fig. 5.22B.

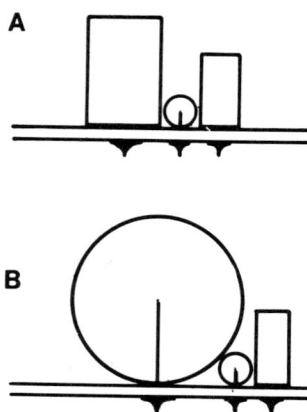

Fig. 5.22 Inadvisable location of components.

How closely packed can the components be? Components have dimensional tolerances and assembly departments have practical limitations. The clearance between heat-dissipating components should be kept as large as possible but for normal 'cold' components the rule given below, and illustrated in Fig 5.23, should be followed:

$$A = \tfrac{1}{2}\,(d_1 + d_2) + 0.3 \text{ to } 0.5 \text{ mm } (0.012 \text{ to } 0.020 \text{ in})$$

where A is rounded-up to the nearest whole or half module (scale 1:1), and under limited conditions to the nearest quarter module. The PCB designer should attempt to achieve a fairly uniform packaging density. If, in the event of one of the boards in a system being only half-filled with components there is, of course, not

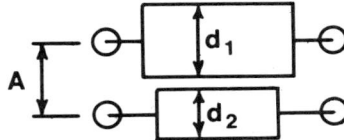

Fig. 5.23 Clearance between components.

much point in spreading the components over the total board area. In such cases, it is better to maintain the same packaging density as before and leave the unused part intact for later modifications or extensions. In order to facilitate later modifications to assembled boards, it is prudent to place surplus pads (for example, located in dual-in-line patterns) on such areas, and also to run some conductors down to the free terminals of the connector.

5.15 Terminal Allocation

The location of the connector has already been discussed. Regarding the terminal allocation, i.e. the distribution of certain terminal numbers to certain functions, there are two alternatives: a standardized and a random allocation. The former restricts the layout of plug-in boards but facilitates, on the other hand, the layout of the motherboard which serves as an interconnecting means for the individual boards of the system. The latter alternative facilitates the layout of all the plug-in boards and complicates the layout of just one board, namely the motherboard. In certain cases, the layout of the motherboard can be so difficult that it is impossible unless the motherboard is designed as a multilayer with the consequences of higher prices and longer delivery times. In the author's experience, a standardized terminal allocation is to be preferred. As stated above, the random terminal allocation complicates the layout of the motherboard which is already difficult due to the fixed location of the connectors and the small clearances between them. For this reason, it may be particularly difficult to achieve a reasonable layout, especially if the conductors travel from one side of the motherboard to the other. The practical approach is to route the conductors on the motherboard straight through the board with a minimum of deviations and to concentrate the design effort on each of the plug-in boards and on the sequence in which they are located on the motherboard. The PCB designer should always try to locate the plug-in boards in such a way that the interconnections are as short as possible.

5.16 Special Layout Considerations for Digital Boards

The best way to locate dual-in-line ICs on digital boards with many ICs is in straight rows, either longitudinally or transversely. The spacings between the rows depend on the area necessary for the signal, voltage and ground conductors. The voltage and ground conductors must usually be of a prescribed width and they will have a major effect on the row spacing. In order to minimize electrical noise, the voltage and ground conductors should be located side by side, or, if convenient, on the two sides of the board, above one another. Often it is advantageous to place these conductors along the edge of the board. Section 5.19 describes an X–Y coordinate wiring system in which conductors in the X-direction (usually lengthwise) are kept on one side, mostly the solder side, and conductors in the Y-direction (crosswise) are kept on the other. By placing longitudinal voltage and ground conductors along the edge of the board, conductors running crosswise are not blocked although the above rule is broken.

Normally all the dual-in-line ICs are turned the same way although in certain cases it is possible to gain some advantage by turning every second one through 180°. In particular, it is possible to simplify the conductor routing when the voltage and ground conductors are connected to opposite corners (lead numbers 7 and 14). The method has some unfortunate points which militate against its use. More assembly faults in the form of ICs being mounted the wrong way will occur even when the 'polarity' is indicated by the screen-printed component notation. Also, the repair work will be more difficult because the operator must remember to count the leads alternatively from the opposite corner in order to localize the correct leads.

As the reader may realize, many dual-in-line ICs contain four separate circuits, e.g. gates which, in the schematic diagram, are drawn independent of their physical location in the various ICs. Therefore, part of the rough layout is to combine these circuits into complete ICs with the shortest possible interconnections both internally and externally to other ICs.

Unused gates should not have open inputs because of the risk of noise and they should be connected to suitable potentials.

Very often it is worthwhile to use a standard pattern for digital boards. A standard board, where the dual-in-line ICs are placed with a spacing of three modules equal to 0.3 in, was described in Section 5.2.5 (Fig. 3.17). The edge connector has 37 contact fingers on each side, insulated from each other so, in total, 74 contact fingers are available. As it is not always possible to determine beforehand on which side of the board a contact finger is to be used, a row of via holes is located adjacent to the contact fingers. Originally, the standardized pattern comprised some via holes located here and there between the rows. These via holes, however, turned out to be more harmful than useful for the conductor routing, as they were rarely located where they were needed and usually obstructed the conductors. For this reason, these via holes were removed, and today the PCB designer uses via holes only where they are needed.

A distinct advantage of the standardized pattern is that contact prints of the pattern, including the contour, registration marks, edge connector, etc., are ready for use and a significant amount of work is avoided. Also, the PCB designer simply starts his layout on the basis of the standardized pattern.

5.17 Bus System

In many cases, the voltage-distribution system restricts a practical layout. The voltage conductor, for example +5V, and the ground conductor, obstruct transverse signal conductors, or the current is so heavy that the conductors must be very wide, thereby reducing the packaging density. A convenient solution is to use specially designed busbars which are seated on the board and soldered like any other component. An obvious advantage is that the PCB designer can work three-dimensionally and allow the signal conductors to cross beneath the bus. Further, as the bus is placed on its edge, it occupies little space. The most simple, and by far the cheapest, form is a bar provided with small solder lugs as illustrated in Fig. 5.24.

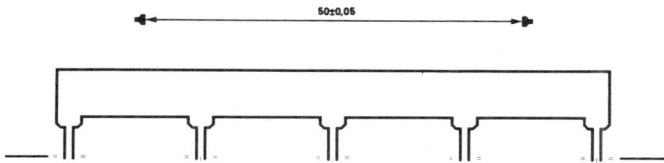

Fig. 5.24 Busbar produced by chemical milling.

If the quantity required is insufficient to justify the use of a blanking tool, the bus can be made by chemical milling. The material is often 0.5-mm (0.02-in) copper sheet, and the finished buses can be tin-plated in order to ensure good solderability. If the etch tolerance is known, it is possible to prepare the necessary artwork on a suitable scale. The etcher normally uses a multiple photomaster prepared from the original artwork by a step-and-repeat process. In practice, the contour line of the bus should be drawn sufficiently wide so that, after being printed on the copper sheet as a negative pattern, it disappears during etching. In this way the bus is released from the copper sheet, but, in order to prevent its falling to the bottom of the etching vat, the etcher interrupts the contour in several places. The bus is produced, therefore, with a number of small lugs which secure it to the sheet. If the lugs are correctly designed, taking into account the etching tolerance, it is possible for the etcher to break-out the buses from the sheet. Commercially made busbar of the type just described are available. In certain applications sandwich types, with built-in capacitors, are used but, of course, the prices are significantly higher.

5.18 Checklist for the Rough Layout

It can be seen that there are many factors to take account of when preparing the rough layout and it would not be surprising if one of these was overlooked. A list of these factors such as shown in Checklist Number 1 (see below) enables the designer to verify the completeness of his work. Separately, the points of the checklist seem obvious and banal, but even the most experienced PCB designer will, undoubtedly, be able to recollect cases where he omitted to take into account all the points made.

Checklist Number 1. Mechanical Considerations

1. Is the board size compatible with the manufacturing equipment?
2. Has the board the correct mechanical dimensions?
3. Has the assembled and populated board sufficient clearance when inserted in place?
4. Is it possible to plug the assembled board into place without fouling other components or equipment?
5. Are the input and output connectors and the board-mounted controls located correctly with respect to the cabinet or the chassis?
6. Can the internal adjustable components be adjusted easily?
7. Are 'forbidden' areas taken into account?
8. Are the components located with adequate clearances?
9. Are the component leads bent to cover spans of multiples of 0.1 in (2.54 mm) modules, and not bent too close to the component body?
10. Have leads of components of the same type been bent to the same span?
11. Are the components accessible for replacement?
12. Is it possible to clip logic probes over the IC packages?
13. Is the packaging density fairly homogeneous across the board?
14. Does the board hold spare areas for possible extensions or circuit modifications?
15. Are temperature-sensitive components located reasonably with respect to heat-developing components?
16. Are heat sinks located to fulfil their cooling function?
17. Have the overall cooling conditions been considered?
18. Are heavy components mechanically secured?
19. Is the board sufficiently supported at places where heavy components such as transformers are located?
20. Will the assembled board be able to withstand vibrations?

5.19 Fine Layout

After the PCB designer has finished the rough layout, the next step is the detailed routing of the conductors. Here we shall deal only with reasonably large double-sided plated through boards. There are several courses of action open but in order to limit the content of this section, we shall confine ourselves to generalities.

In order to achieve a reasonable solution to the problem of conductor routing, the PCB designer must have a good appreciation of the rules and requirements associated with the taping of the artwork (see Section 5.1). These rules and requirements will be dealt with in detail in Chapter 6 where we shall consider the accuracy requirements of the artwork. Below we shall assume that the PCB designer carefully observes all the rules so it is possible for the draughtsman to produce good artwork from the layout.

The special sketching technique used by the PCB designer was tacitly implied when the rough layout was described. Again, in this section the same assumption will be made. The various techniques of sketching are considered in Section 5.22.

5.19.1 Procedures

The inexperienced PCB designer usually starts by laying as many conductors as possible on one side of the board. Generally, he places the most critical conductors, such as the voltage or ground conductors, then the less critical conductors and so on, until the whole side of the board is filled to such a degree that he has to continue the layout on the other side, often by introducing a considerable number of via holes. This procedure is not particularly rational and the result suffers from several obvious shortcomings:

1. There is no guarantee that the conductors on the solder side take into account the requirements set by automatic soldering.
2. Modifications are often very difficult.
3. Many conductors can be unnecessarily long.

One result of the above procedure is that there may not be sufficient space on the other side either and a third layer is introduced—a multilayer. The way is now open to introduce a fourth layer and so on. It may be rather controversial but, in some cases, where there are no special requirements for the voltage and ground planes, it is difficult to understand why the designer has used expensive multilayers.

In the author's experience it is possible to produce quite large circuits comprising, for example 75 dual-in-line ICs, as double-sided plated through boards as shown in Fig 5.25, by adopting a more systematic layout procedure.

The preferred procedure is based on the use of a directionally orientated conductor routing system, which means that all longitudinally directed conductors are kept on one side of the board, while all transversely-directed conductors are kept on the other. At the transition point the conductor is transferred from one side to the other by means of a via hole. There are, of course, exceptions to this procedure. For example, a conductor which runs along a row of solder pads may have to be bent a little in order to 'land' on one of the pads. On the whole, the conductors are routed in accordance with the X- and Y-axes of a coordinate system. Assuming reasonable board area, no conductor needs to be more than 40% longer than the direct line of connection between the end points, and, in most cases, increases of only 10% to 20% are necessary.

In CAD boards (CAD, computer-aided design) a conductor may 'circle' around the board. This is caused by the fact that the computer cannot survey the

Fig. 5.25 Example of a board with many dual-in-line ICs.

situation in the same way as the PCB designer, nor can it take the initiative and move components in order to find more direct routes.

In the vast majority of cases, the voltage and ground conductors will be defined, so the first step is to lay these conductors. Further details are given in Section 5.19.2. If a terminal allocation has been adopted, all the conductors from the connector to the respective points of the circuitry are laid down in accordance with the X- and Y-axes. It will often be advantageous to reserve certain longitudinally located 'channels' for conductors which are to be routed to the far end of the board. When these conductors have been laid, the next step is the systematic layout of the remainder of the pattern. Frequently it is preferable to start with the dual-in-line ICs closest to the connector and from these the PCB designer should work his way to the far end of the board. Wherever possible, conductors should be laid in areas which are already populated and the clear, or unpopulated areas, should be kept free. Experience will show that the free areas will be useful for laying down the final conductors when space has become paramount.

When routing the conductors, the PCB designer has to be as far-sighted as possible and he should intuitively feel whether the conductor which has just been laid is likely to clash with conductors to be laid down later. Occasionally, it is an advantage if special groups of circuits are dealt with separately, e.g. the chain of counter–latch–decoder/driver–LED display packages making up a complete digital display. This sub-layout can then be standardized and used in other later applications.

5.19.2 Connection of Supply Voltage and Ground

All the dual-in-line ICs on digital boards have to be connected, in the most practical way, to the voltage supply and to the ground. Therefore, a distribution system consisting of a voltage and a ground main conductor with branches can be introduced. As the ICs are normally located in clusters, the distribution system can form a particular pattern which is repeated all over the board. Local variations occur, however, depending on the supply voltage and the ground pinning of the ICs (lead numbers 7 and 14, 4 and 10, or 4 and 11). The layout and execution of the distribution system depends on the location of the ICs which, in turn, depends on the format of the board and the location of the connector. Under all circumstances, it is very important that the distribution system be so placed that the routing of the signal conductors does not become too tortuous or, indeed, impossible. Hence the voltage and ground conductors should not be wider than absolutely necessary. In order to reduce coupling to the signal conductors, the voltage and ground conductors should run close together or on different sides of the board, but opposite one another. Various layouts of voltage and ground distribution systems are shown in Fig. 5.26, including an example of the method of turning every second IC through 180°, as discussed in Section 5.16. (Refer also to the busbar system mentioned in Section 5.17.)

5.19.3 Decoupling of Supply Voltage

The supply voltage is normally decoupled on the board by a ceramic capacitor. The schematic diagram very often shows a number of $0.1\,\mu F$ decoupling capacitors corresponding to one for every five or six ICs. The onus is on the PCB designer to place these decoupling capacitors in the most practical way to achieve a fairly equal distribution. Frequently the decoupling capacitors are located at the main trunk of the distribution system at the starting points of the branches which supply the five to six ICs. These capacitors deliver the current peaks which occur when the circuits are triggered. Due to the unavoidable inductance of the conductors it is important to place the capacitors close to the ICs within, say, 75–100 mm (3–4 in).

5.20 Cleaning-up the Layout

The PCB designer must realize that it is very seldom that he will achieve a perfect layout first time and many rectifications and modifications are likely to be necessary during the course of the work. When the basic layout has been finished, it is cleaned up and simplified. For example, it may happen that the number of transverse routings can be reduced or conductor routings can be changed so that

130

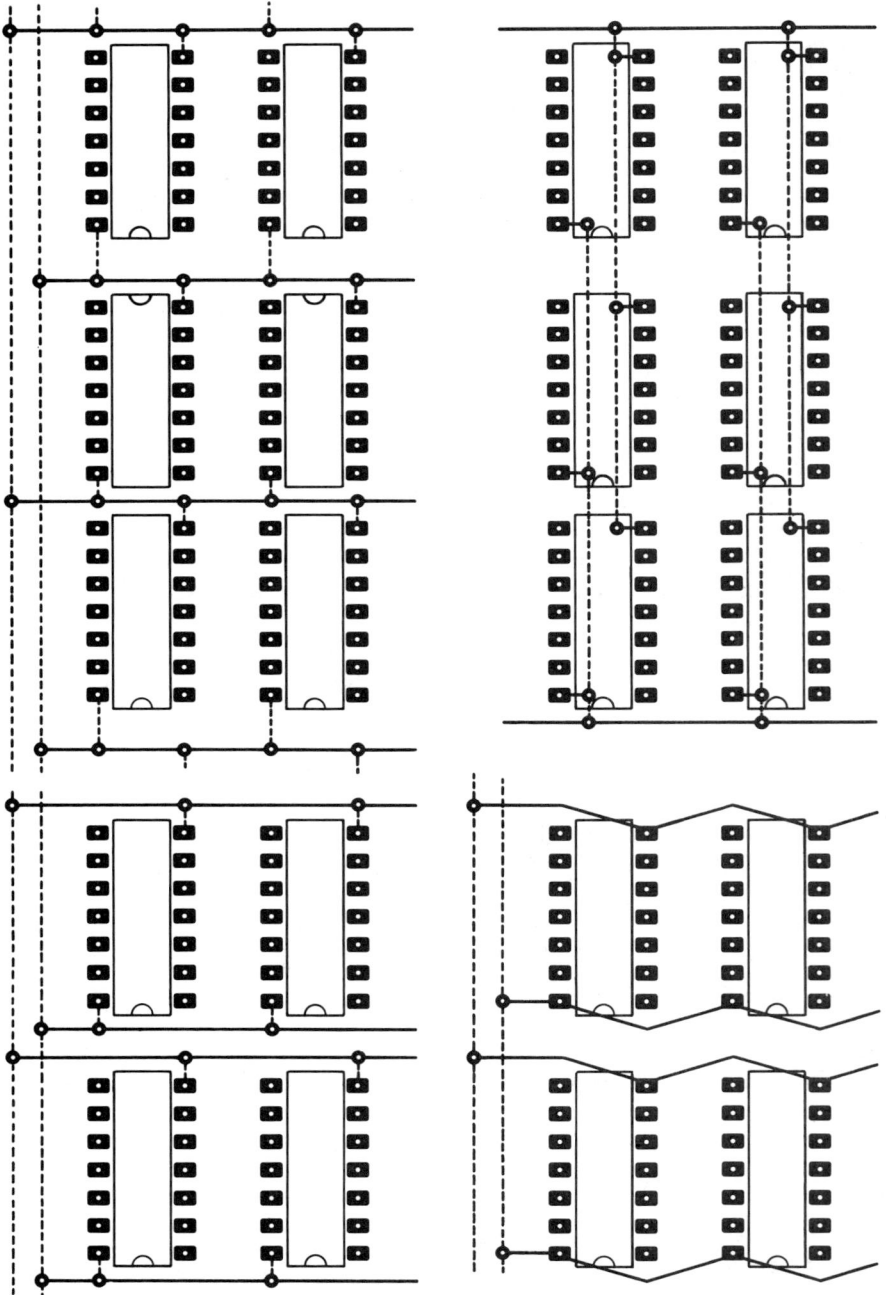

Fig. 5.26 Examples of voltage and ground distribution systems.

the conductors do not pass between pairs of pads. An example of a clean-up is shown in Fig. 5.27.

The PCB designer must make allowance for the fact that taping consumes space and he should not attempt to run through narrow gaps more conductors than it is possible for the draughtsman to tape without compromising the required minimum values of conductor spacing or conductor width.

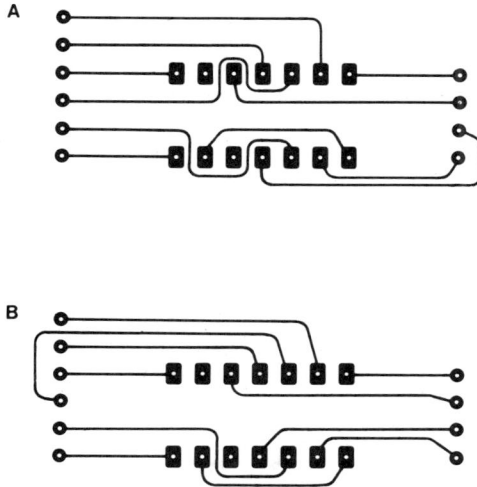

Fig. 5.27 Simplification of a layout. A, the first layout; B, the simplified layout.

Further, the layout should be overhauled and the pattern 'opened-out' as much as possible. This means that parallel conductors should not run closely side-by-side over long distances if there is sufficient space for the conductor spacing to be increased. Closely-spaced conductors are a potential source of fault to the PCB manufacturer as it is difficult to remove the plating resist completely from between the tracks. (See Section 3.6.)

In computer-aided designs, parallel conductors are often closely spaced over long distances and conductors often pass very near a row of solder pads although there may be adequate space to permit larger spacings. These shortcomings are due to the fact that the computer routines are arranged to ensure only that the minimum spacing requirements are fulfilled. Usually there are no routines to ensure that the conductors are arranged to use the space available.

5.21 Checking the Layout

The layout must be checked before the taping of the artwork is begun. If possible, modifications to the artwork itself should be avoided. The fewer the rectifications, the better the quality of the artwork (cf. Section 5.3). Therefore, the PCB designer must check the layout against the schematic diagram in a systematic way, point by point. All connections which are checked and found to be correct are struck out on both the schematic diagram and a blueprint of the layout sketch, preferably with a crayon. Instead of a blueprint, a sheet of clear film can be placed on top of the layout sketch and the checked conductors marked with a felt pen. When the designer has completed his check all connections on both the schematic diagram and the layout sketch should have been struck out. Discrepancies indicate faults which, of course, must be rectified before the taping is started. In this way, the PCB designer has proof that the layout is correct and in accordance with the schematic diagram, but, as previously noted, the schematic diagram cannot give all the necessary information. Therefore, he should finish this stage with Checklist Number 2, which covers all the electrical considerations.

Checklist Number 2. Electrical Considerations

1. Do the component locations give a logical signal flow?
2. Have undesirable couplings been avoided, e.g. from the output to the input terminals?
3. Is the grounding system divided into separate ground conductors?
4. Have common conductors, causing false signals, been taken into account?
5. Have inductive loops been avoided or reduced?
6. Have conductors with a critical resistance, capacity, inductance, or insulation been considered?
7. Have critical requirements affecting insulation been specified?
8. Have the temperature conditions of current-carrying conductors been examined?
9. Have the power-supply conductors been designed with adequate width to withstand a possible short-circuit in the system?
10. Have safety precautions (sufficient creepage distance) been taken where mains voltage is present on the board?
11. Are adjustable components connected to give a logically correct direction of rotation?
12. Have the unused gates of ICs been connected correctly?
13. Has the number of jumpers been kept as low as possible, and are the jumpers bent in modules?
14. Are all the components unequivocally defined?
15. Have precautions been taken to ensure that alternative components with slightly different mountings and component locations can be used?
16. Has a terminal allocation to the motherboard been considered?
17. Is it possible to separate the individual parts of the circuitry in order to facilitate the testing of the board?
18. Is it easy to connect the board to the test equipment?

5.22 Layout Sketch

Now we shall consider the PCB designer's layout sketch and the sketching methods involved. The sketch must be so made that it forms the basis of subsequent taping of the artwork. For practical reasons, it is advantageous to do the sketching as seen from the component side. All the solder pads and the conductors should be located in accordance with the rules valid for the taping of the artwork. Very precise location is not necessary since the draughtsman will place the solder pads at the intersections of the grid lines and will also ensure that the minimum spacing requirements are met. The scale ratio has been mentioned in Section 5.12. It should be realized that boards having a high packaging density very often demand minimum spacings of the order of magnitude of 0.35–0.40 mm (0.014–0.016 in), and it can be very difficult to ensure that the taped conductors maintain a minimum spacing of 0.8 mm (0.031 in) when the artwork is made in a 2:1 scale. Even in the 4:1 scale it can be difficult to maintain a minimum distance of 1.6 mm (0.063 in).

We shall now discuss the fundamental methods of sketching, followed by a consideration of the information necessary for the draughtsman to tape the artwork.

5.22.1 A Simple Method of Sketching

The most simple sketching method is to use a piece of grid paper, often Bristol board printed with a standard grid pattern. The contour of the board, mounting holes, 'forbidden' areas, edge connectors, etc., can be indicated by pencil sketches. In order to avoid confusion, this sketch should always show the component side of the board. The components are then located as previously described, by pencilling their contour and solder holes, the latter normally being located at the intersections of the grid lines. Some PCB designers prefer to lay down the conductors concurrently with the components, while others prefer to locate all the components before laying down the conductors. Whether he follows the first method or the second, the PCB designer will normally have to make frequent recourse to his eraser and the sketch will become grubby rather quickly.

Incidentally, if the layout is made in a 4:1 scale, it may be difficult to obtain grid paper in the size necessary for large boards.

5.22.2 An Improved Method of Sketching

A better method of sketching is to use a polyester film printed with a precision grid as a permanent underlay on the drafting table. Such films can be obtained in sizes up to at least 36 × 48 in (900 × 1200 mm). A sheet of drafting film is placed over the underlay and the contour is pencilled as described in Section 5.22.1. Simple components are drawn directly but, in the case of more complex components, models can be placed under the drafting film in order to facilitate the pencilling of the outlines. The remainder of the procedure is as described above, but because the surface of the film is considerably stronger than that of the grid paper, the sketch is relatively undamaged by rectifications. Both procedures are easy to follow and do not call for large investment. An obvious disadvantage is, however, that rectifications and changes can be rather difficult to undertake and, usually, the clarity of the sketch is not all that it might be, especially if the board is filled with components and conductors.

5.22.3 Sketching by Means of Puppets

The procedure now described overcomes the disadvantages mentioned above. The method has been used for some years with very good results in terms of quality of layout and time saving. It is estimated that the latter can be as much as 25%. Also, there is a further advantage when the component notation drawing is produced.

The system is developed by Bishop Graphics Inc., and is called the Puppets Printed Circuit Design System. The system consists simply of a collection of layout patterns showing the outlines and terminal locations of commonly used components (see Fig. 5.28). The Puppets (trade mark) adhere so positively—without the use of glue—to any glossy drafting film that it is impossible to displace them unless they are peeled off by lifting a corner. This is a convenient way to undertake the layout work since the puppets can be located very quickly and can be re-arranged easily until the optimum component distribution is achieved.

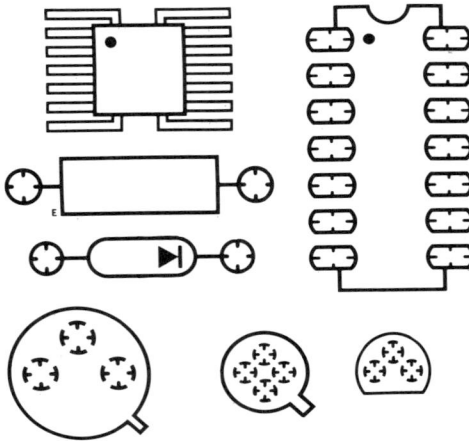

Fig. 5.28 Examples of puppets.

In this procedure, a drafting film is placed over the permanent precision grid. The drafting film must be glossy (clear) on the top side to make it possible for the puppets to adhere to the surface. If the other side of the drafting film is matt, the contour of the board, mounting holes, etc., can be pencilled on that side. When pencilling, the board has to be seen from the solder side in order to show the component side when the drafting film is turned with the glossy surface upwards. If the PCB designer prefers to sketch the contour as seen from the component side, this means that he must work on the glossy surface. Since drawing ink does not adhere well, a felt pen should be used. Alternatively, the contour or profile edge can be taped, or indicated by means of corner marks. Edge connectors can be indicated most easily using a connector pattern which is taped to the drafting film.

Before starting the rough layout, the PCB designer should collect all the necessary puppets. The Puppets' System also contains blank material which enables the designer to create his own 'custom' component pattern by drawing the outline using a special permanent black marking pen. Therefore, he can make-up those puppets which are not available in the kit. The next step is to mark the individual puppets with their reference numbers as indicated in the schematic diagram, e.g. R101. This can be done either by using small index labels or by making small index puppets which adhere to the larger puppets.

The rough layout procedure follows the instructions given in Sections 5.14 to 5.16. The puppets are located on the drafting film and adhere easily and, because it is easy to make alterations, the PCB designer can produce an optimal layout. With experience, he should be able to locate the puppets, making due allowance for the space required by the conductors.

When the component location appears satisfactory, a layer of drafting film is placed over the puppets. This drafting film has to be matt on both sides: matt on the under-side to prevent the puppets adhering, and matt on the upper-side to facilitate pencilling. The drafting film is taped along the upper edge so that it can be lifted easily thereby permitting access to the puppet layer. This is important since it may be necessary to introduce some minor changes to the component locations.

The conductors are pencilled on the overlay in accordance with the instructions given in Sections 5.19 and 5.20. The puppet system has the great advantage that the designer can work on the conductor layout, sketching and erasing as necessary, without affecting the puppet layer. If the board is double sided, the conductors of the second side are drawn on the same sheet of drafting film using a colour code. For example, a black pencil can be used for the component side, and a green crayon for the solder side. Another possibility is to use dotted lines for the conductors on one of the sides but this is not so legible.

Division of the ground connection into separate grounding systems as described in Section 5.5.4 can be indicated by colour coding. For the draughtsman's benefit these critical conductors are very often pencilled in full width, i.e. double line, and the various ground systems can be indicated by using different colours for the filling or cross-hatching of the conductor area. When the layout is complete and checked, the taping of the artwork can be started. Often the PCB designer wishes to re-use his puppets as soon as possible, and makes a contact print of the puppet layer. Although an ordinary copying machine can be used if a matt protecting cover sheet is placed over the puppet layer, the contact printing system described in Chapter 7 is particularly suitable. The puppet layer is first used when taping the pad master, as all terminal locations are indicated, and later when the component notation drawing is to be made, as the contour of the components is also indicated.

The Puppets' System meets the PCB designer's requirements for an easy and fast method of laying-out the board and can be recommended also on the basis of ease of rectification.

5.22.4 Sketching by Means of Semi-sticky Layout Symbols

If the drawing office has contact printing equipment, as described in Chapter 7, at its disposal, it is easy to make semi-sticky layout symbols using a thin, clear diazo film (Fig. 5.29). The artwork for these symbols is drawn on drafting film with a suitable number of symbols per sheet. The symbols closely approximate to the

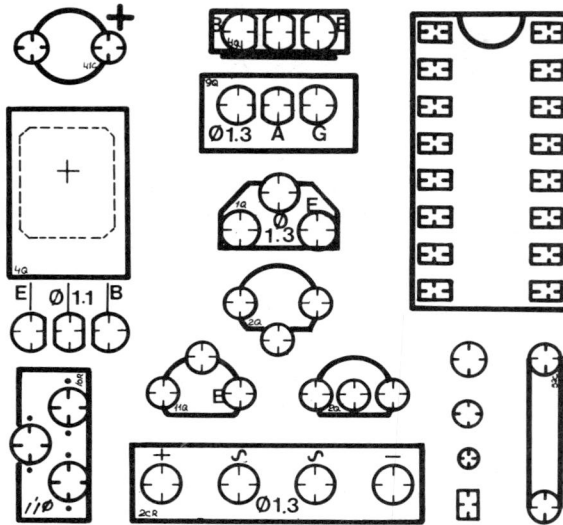

Fig. 5.29 Examples of semi-sticky layout symbols.

outline of the components. The solder pads are drawn to the correct size and geometrically positioned. It should be noted that the solder pads are not provided with an inner circle as is the case with ordinary solder pads. Also, they are not filled and are not provided with a centring cross. If they were, the draughtsman would have trouble when taping the actual solder pads as the intersections of the grid lines would be hidden by the layout symbols.

When the artwork is complete, the necessary number of diazo-contact prints are made. These prints are sprayed on the reverse side with a semi-sticky glue such as 3M's Spray Mount. After the glue has dried, the symbols are cut out using scissors and they are then filed, i.e. they are stuck to a sheet of relatively thick drafting film.

These 'home-made' layout symbols can carry various information, e.g. an indication of the emitter locations, polarities, and diameters of the solder holes. Eventually the PCB designer can establish a complete 'library' of the components normally used.

The type of glue used makes it possible to position and remove the layout symbols at least 25 times. Also, there are no problems with shadows or dark backgrounds on the contact prints of the component layer. The surface of the drafting film on which the layout symbols are to be placed need not be glossy. In contrast to the Puppets' System, the PCB designer can use a drafting film which is matt on the upper side. This makes it possible for the contour of the board as well as the mounting holes etc. to be pencilled on the surface.

If layouts have to be prepared for a large number of standard boards, an alternative method, based on diazo prints contact-printed from a master containing all the common details, can be used, thereby avoiding the need to use a pencil.

The use of the layout symbols is in accordance with the method outlined in Section 5.22.3.

5.22.5 Information Given in Layout Sketch

The layout sketch gives the draughtsman all the necessary information regarding the location, shape and size of the solder pads and the routing, width and spacing of the conductors. In the interests of simplicity, as few pad sizes as possible should be used. Frequently there will be a certain size which covers perhaps 90% of all the solder pads required. It is necessary therefore to indicate the deviations from standard where these occur. To make it possible for the draughtsman to finish the master drawing (also called the engineering drawing) of the board, the layout sketch must indicate the diameters of the holes. There is likely to be a distinct relationship between the sizes of the solder pads and the sizes of the holes themselves. For example, a 1.39-mm (0.055-in) solder pad will nearly always have a solder hole of 0.8 mm (0.031 in) whereas a 1.98-mm (0.078-in) solder pad may have a solder hole of 1.1 mm (0.043 in) or 1.3 mm (0.051 in). Again, it is necessary to indicate only the deviations from the standard.

In addition to the above information which relates directly to the board, the draughtsman needs other information which must be indicated on the sketch. First, the mechanical dimensions necessary for correct taping of the board's contour and a precise location of the various mounting holes are necessary. In this connection, there is a fundamental difference between printed circuit boards and normal mechanical parts. The latter are manufactured against measurements given on the workshop drawing and, therefore, the drawing need not be precisely to scale. In the case of printed circuit boards the artwork, i.e. the finished taping of the pattern, is the basis of manufacture of the board. The location of a mounting hole is normally indicated by means of a pad which has to be located very precisely. When the board is digitized (see Section 3.8.2d), the coordinates of the mounting holes are measured in the same way as the solder holes, so that the dimensioning of such holes is superfluous. Similarly, the contour of the finished board is determined by the corner marks on the artwork which indicate the sawing or milling lines. For this reason, the draughtsman must know the mechanical dimensions so that the contour can be accurately drawn with respect to the mechanical mounting holes.

In some cases, however, a reference system is introduced in order to tie the contour to the finished board more precisely than can be achieved using the corner marks. This is described in detail in Section 8.9.3 and is shown in Fig. 8.46.

This section is concluded by Checklist Number 3 which lists all the matters concerning the taping for which the PCB designer is responsible to the draughtsman.

Checklist Number 3. Drawing Considerations

1. Has any special machining of the board, e.g. cut-outs or slots, been indicated?
2. Are all the necessary mechanical dimensions stated?
3. Have all critical tolerances been assessed?
4. Have constructional factors been taken into consideration when determining the size of the board?
5. Has the introduction of a reference system been considered?
6. Are all the conductors at least 0.5 mm (0.02 in) from the edge of the board?
7. Have solder pads along the board edge been located so that they do not interfere with the board guides, in case the board is to be plugged into a rack?
8. Have all 'forbidden' areas been indicated?
9. Do the conductors have standardized widths?
10. Are special conductor widths or shapes clearly indicated?
11. Have special minimum spacings been stated?
12. Do the solder pads have standardized sizes?
13. Does the conductor routing allow for conductor width and minimum spacing, where a number of conductors are run through a narrow gap.
14. Do the solder holes have the preferred diameters?
15. Do the sizes of the solder holes match the diameters of the terminal leads?
16. Are the solder pads located at the intersections of the grid lines?
17. Have the locations of the solder pads, which do not coincide with the intersections of the grid lines, been indicated?
18. Have any via holes been located under components?

6

Generation of Precision Artwork. Theory

6.1 Introduction

It has been pointed out earlier that the documentation package which forms the basis of the finished boards, must be compiled in such a manner that technical difficulties in production are minimized. We shall in this chapter discuss the most important conditions which must be met in the work of the PCB designer and the draughtsman to ensure that it embraces no unnecessary difficulties in manufacture. Most important is the accuracy of the artwork, especially in the case of double-sided boards, where a small displacement of the pads of the solder side with respect to the pads of the component side may give rise to serious manufacturing problems. On the other hand, it is of minor importance if a group of solder pads (a so-called 'cluster') is located very precisely with respect to the contour of the board as long as the component in question is not tied mechanically to any part of the equipment external to the board. This applies to most components such as resistors, capacitors, and semiconductors. The exceptions are switches, potentiometers, and connectors; these have to be accurately placed in relation to the profile of the board as do any holes to be used for mounting the board in assembly of equipment. Within the individual group of pads it is required that the pads be mutually located with a reasonable tolerance.

Also important is the taping of the artwork. There are special requirements for the minimum width of the conductors and the annular rings of the solder pads. Similarly, there are requirements for the minimum spacing between conductors, both over a short distance and also over relatively long runs.

In this chapter we analyse the requirements for precision artwork taking into account the practical tolerances for preparation of the artwork itself and for the manufacturing processes which jointly enable the board to be correctly produced. From this analysis, Design Rules are formulated for guidance in artwork preparation. To avoid confusion, the discussion is limited to plated through boards, but the Design Rules will be applicable to conventional boards, i.e. non-plated through boards. The practical preparation of the artwork is described in Chapter 8.

6.2 The PCB Department's Place in the Organization

Usually the PCB drawing office is started as a little corner of the mechanical design department, and when the little corner grows larger it frequently remains part of the mechanical design department, without specialist training of staff. Furthermore, the requirements of work flow in a larger office may lead to frequent changes of the individuals engaged in the work regardless of their understanding of the special problems involved.

There are good grounds, therefore, for avoiding this type of organization. A quite different approach will be found in a PCB department separate from the mechanical design department. The need for this separation resides in the way in which the drawings are used in the factory. Workshop drawings, even if they are not strictly accurate, are nonetheless useful so long as the dimensions, tolerances, and finishes are correctly indicated. Such drawings do not need to be dimensionally correct nor stable. Printed circuit board artworks, on the other hand, demand the highest standards of precision and stability since, as has been described earlier, the pattern is transferred to the board by a photographic process and any inaccuracies are directly transferred to the product. The taping of the PCB artwork therefore requires quite a different procedure from that normally followed by the mechanical design department.

It is not always realized how important are meticulous accuracy and a good understanding of the problems of manufacture to the preparation of satisfactory artwork. Knowledge of manufacturing problems and the establishment of a well thought-out procedure are both essential. Many difficulties experienced in board production stem from failure to meet these requirements.

In addition, it is necessary for the designer and the draughtsman to be able not only to assimilate the advances in the technology of this fast-moving field but also to make their own contribution to these advances.

In achieving these results, the identification of the printed circuit activity as a separate and specialist department within an electronics company is of prime importance.

Printed circuit boards are a custom-designed product with widely varying requirements. For this reason, an independent PCB department has as one of its prime duties to be fully familiar with manufacturing processes and problems, and technological developments. Only in this way can advances in technology be utilized, whether these be improved manufacturing methods or newer types such as multilayers, flexible circuits, or boards bent in two or more planes.

Today, printed circuit boards constitute a very important element in nearly every piece of electronic equipment, and the price of the boards will frequently amount to a high percentage of the total cost of the components. An optimization of the PCB documentation package should lead to a significant cost reduction due to easier manufacture and consequently a higher yield. Added to this are the savings obtained in the customer's own production department when the board is manufactured in close agreement with the specification, difficulties in assembly of components being avoided.

In short, professional preparation of drawings, artworks, and specifications can yield a great cost advantage. Consequently, every electronics company should aim to enhance the prestige and authority of its PCB department to the highest possible level.

6.3 General Methods for the Preparation of Artwork

Since the quality of the artwork has so much influence on manufacture and on the finished boards, the artwork must be prepared with an accuracy which matches the complexity of the board and the scale ratio. In this section various methods of preparing the artwork are described and the advantages and disadvantages of each method indicated.

The artwork displays only those items which have to be reproduced as a copper pattern in the manufacture of the board. It includes all solder pads and conductors true-to-scale in respect of both their dimensions and clearances and their location on the board. Also included are lines locating the outline of the complete board in relation to the conductor pattern, any identifying symbols and test patterns which may be required, and a 'plating bar', if necessary, for selective plating of contacts. The 'master film' is an actual-size photographic reproduction of the artwork on a highly stable film.

In certain cases, a dimensionally stable copy of the original artwork is made by contact printing. This copy is called 'permanent artwork' to distinguish it from the original artwork. Permanent artwork is not subject to accidental displacement of any part of the pattern. The use of such copies will be shown later to offer significant advantages.

A detailed description of the most suitable procedures for preparation of the artwork is given in Chapter 8. Here the following basic methods are discussed:

A. Drawing with ink
B. Taping with special pre-cut symbols and tapes
C. Cut-and-strip method

By combining the methods A and B and using different drawing materials in conjunction with contact-printing methods, it is possible to develop procedures which will be discussed in detail in what follows. Finally, the cut-and-strip method shall be discussed.

6.3.1 Drawing with Ink on Bristol Board

This method was originally based on the use of Bristol board as the drawing material, usually printed with a blue grid which is not reproduced photographically and therefore does not appear on the master film. The method is very troublesome as the circumference of the solder pads and possibly also the centre holes are drawn by means of bow compasses, whereafter the solder pads are filled with ink. The conductors are drawn by means of an ink pen either giving the desired width directly or as double lines which are then filled with ink. Artwork for double-sided boards can be made simply by pricking the centres of the solder pads through the Bristol board, whereafter the solder pads and the conductors are drawn on the other side.

The method was commonly used in the early development of PCB technique, but as appears from the above description, the method is especially time consuming. In addition, it is subject to inaccuracy arising from the instability of the Bristol board under varying conditions of temperature and humidity. Typically, the length can change by 0.5 mm (0.02 in) over a length of 500 mm (20 in), i.e. 0.1%, due to a simultaneous change in temperature and humidity of 10°C and 20%, respectively. It is not entirely possible to compensate for the variations in length when the artwork is photographically scaled down since their extent can vary considerably in different directions on the board. Another disadvantage is that the solder pads and the conductors do not always have clearly marked edges. At some places, the Bristol board causes the ink to flow a little, and if rectifications are to be introduced, either caused by changes in the circuitry or by sheer bad luck at the drawing table, the erasing operation will often cause so rough a surface that the lines become smeared or ragged. It is also difficult to keep the conductor width constant, and an estimated variation of 0.1–0.2 mm (0.004–0.008 in) is to be expected. Concerning the tolerance on the location of the individual constituents of the pattern such as solder pads and conductors, displacements up to 0.3–0.5 mm (0.012–0.020 in) can occur.

The conclusion is that the method is unsatisfactory and must be discarded as being inaccurate and troublesome.

A variation of the method is to replace the Bristol board by tracing vellum which is placed on top of a precision grid but the result is equally unsatisfactory. It is a little better if a stable polyester drawing film is used, but the pattern suffers from the same inaccuracies in location of the individual constituents.

6.3.2 Taping on Bristol Board

The appearance of special symbols and tapes, either self-adhesive or of the transfer type, has made it possible to prepare the artwork in an improved manner. Figure 6.1 shows various pads as well as groups of pads. Bristol board printed with a blue grid has been used with these symbols but is not recommended because of its instability, as mentioned in Section 6.3.1. Another important disadvantage when using symbols and tapes with Bristol board lies in the photographic process. This is shown schematically in Fig. 6.2 where the artwork is front illuminated at an angle of 45° from each side. The tapes and symbols, however, have a certain thickness, which owing to the oblique light, throws a half-shadow and this in turn makes the negative a little blurred. The conclusion, therefore, is that the use of Bristol board should be discarded in favour of other materials.

6.3.3 Taping on Polyester Film

The drawing material used most today is a polyester film whose dimensional stability is ensured by special precautions, including heat treatment, in its

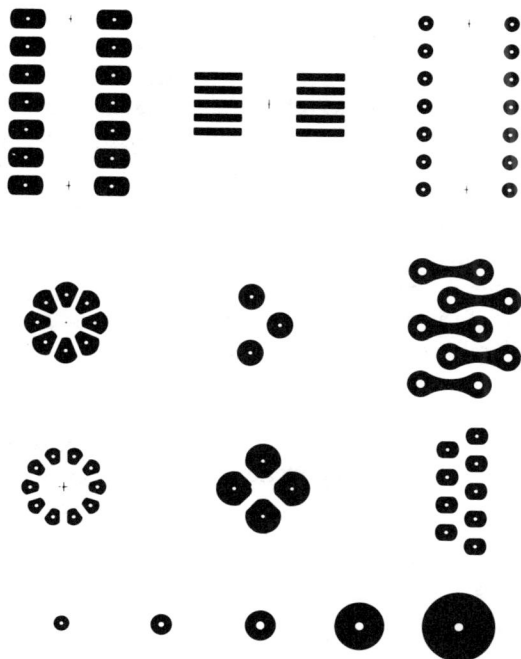

Fig. 6.1 Solder pads and groups of pads.

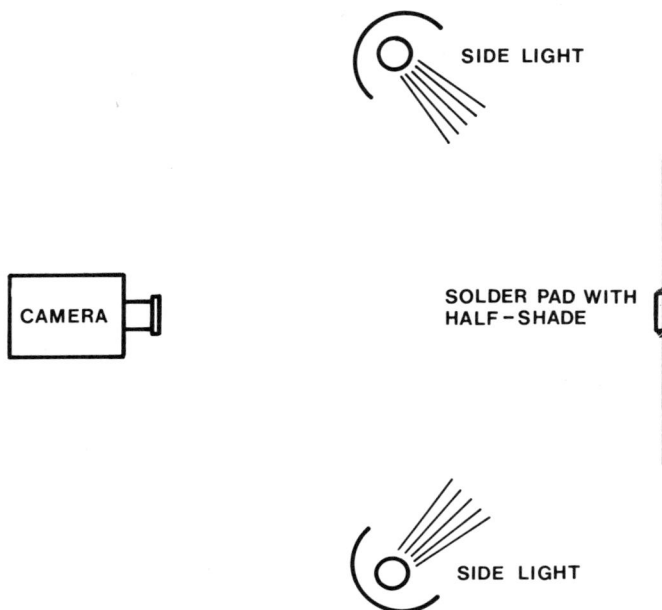

SIDE LIGHT

CAMERA

SOLDER PAD WITH
HALF-SHADE

SIDE LIGHT

Fig. 6.2 Lighting an artwork.

manufacture. Typically, such a film displays a stability in length of 0.0011% per 1% change in relative humidity, and 0.0017% per 1°C change in temperature. Simultaneous changes in the relative humidity and the temperature of 20% and 10°C, respectively, give a change of 0.2 mm (0.008 in) over a total length of 500 mm (20 in), which is roughly $2\frac{1}{2}$ times less than with Bristol board. After ageing over 5 years, the residual change in length amounts to approximately 0.02%, i.e. 0.1 mm (0.004 in) over a length of 500 mm (20 in). This change is quite tolerable and changes due to any further ageing can be ignored. Polyester films can be obtained in many thicknesses and with a glossy or matt surface. The latter is an advantage when drawing with ink, as the ink adheres better to the matt surface. The films can, furthermore, be obtained with a pre-printed blue grid which facilitates the work a little. The draughtsman should, however, exercise caution in their use since their accuracy may be suspect unless specified. It is, in fact, doubtful if the extra cost of a high-precision pre-printed film is justified. The following discussion assumes that the drafting film is not so printed. Instead, a very accurate grid is used as a permanent underlay on the light table. The polyester film is placed on top and the permanent grid visible through the film is used to aid the location of the pads and conductors. Films, not under 0.1 mm (0.004 in), preferably 0.18 mm (0.007 in), are recommended since they could otherwise become puckered as a result of shrinkage of the conductor tapes, caused by unintentional stretching.

6.3.4 Taping of Two Separate Sheets of Artwork

The artwork for the other side of a double-sided board can be obtained by placing pads on another sheet of polyester film in strict accordance with the solder pads of the first film, followed by taping the conductors of the second side.

In Section 6.4.1, however, it is shown that to place pads on two sheets of film in such a way that all pairs of pads are in accurate register all over the board (i.e. the pads are located exactly concentrically) is a difficult task even for a trained draughtsman. It should always be remembered that if a board embodies say 1000 pads, their location demands great concentration. Nobody can expect the draughtsman to regard it as his exalted mission in life to focus sharply on each separate pad of the 1000 pads to check whether it has been precisely placed. Nevertheless, the requirements for accuracy in the artwork become more demanding year by year, owing chiefly to the growing complexity of the circuits to be placed on the boards. It becomes necessary to achieve a progressively higher component density and as a consequence, the use of very small solder pads. For this reason, pads with annular rings having widths down to 0.3 mm (0.012 in) in a 1:1 scale are commonly used today. It is easy to understand that if a manufacturer is to be given a fair chance to avoid breaking the annular rings, the artwork must not take up more than a small fraction of the tolerance range (see Sections 6.4.4 and 6.5.4b). Considerable ingenuity, therefore, has been displayed in finding methods of eliminating the difficulties in securing accurate registration.

6.3.5 Taping of a Common Pad Master and Two Conductor Sheets

In this method, the solder pads are applied on one sheet of film, which is called a pad master. A piece of film is then placed on top of the pad master and taped with the conductors of the first side. The pad master is turned upside down, a new sheet of film placed over it and the conductors on the second side taped (see Fig. 6.3). It

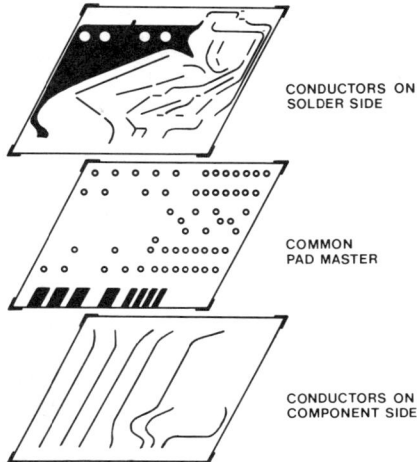

CONDUCTORS ON
SOLDER SIDE

COMMON
PAD MASTER

CONDUCTORS ON
COMPONENT SIDE

Fig. 6.3 Common pad master and separate conductor sheets.

stands to reason that when the artwork is scaled down by the photographic reduction, it is necessary to register very carefully each conductor sheet, in turn, with respect to the pad master. This is one of the weak points of the method. Another difficulty arises in any rectification of the artwork which may be required since two sheets have to be worked on simultaneously in correcting either side of the board. In order to do this the pad master and one conductor sheet are placed on the light table and carefully registered, i.e. they are carefully aligned and affixed with respect to each other. It is necessary to work in turns on the two sheets, if any of the solder pads are to be moved. Therefore, the upper film is either removed completely or it can be bent backwards if it is fixed only along one edge. In the case of a large artwork, either method is troublesome but especially the latter since some of the conductors can be displaced on rolling the film back. This can happen without the draughtsman being aware of it and the resulting fault may not come to light until the boards are manufactured.

6.3.6 Red/Blue Taping on Polyester Film

Another method is based on two-coloured artwork prepared on a single sheet. The colours are separated by a special filtering technique during the photographic process in such a way that a 'blue' side and a 'red' side are obtained. All parts of the pattern common to both sides, i.e. the pads, edge connector, corner marks, registration marks, etc., are taped with black symbols. The conductors of the solder and component sides of the board are taped with red and blue tapes, respectively. The most convenient way is to apply the red and the blue tapes on the same side of the pad master, but on the other hand it becomes very complicated if alterations have to be made. It seems more practical, therefore, to apply the red and blue conductors separately on the two sides, although the tapes on the side turned downwards are liable to be damaged during handling. The two colours are separated by the photographic process. The red colour is dropped by suitable filtering and an image of the blue pattern and the black symbols is printed on panchromatic film. In similar fashion, the blue colour is dropped, and an image of the red pattern plus the black symbols is printed on orthochromatic film.

The method is very attractive as there is no chance of misregistration between the two sides of the board. There are, however, certain disadvantages militating against its general use. The red and blue tapes are usually polyester or polyvinyl material and therefore do not stretch to follow curves and bends as do the usual black crêpe tapes. All curves and bends must, for this reason, be made by means of special pre-cut elbows and corners which considerably complicates the taping process. In addition, a small narrowing of the conductors occurs in the photographic process at the crossings of the red and blue tapes. The largest problem, however, arises from the use of films of two different types, namely, panchromatic and orthochromatic. The base material of such films will normally be manufactured in different production batches which may not be of identical stability, and furthermore, emulsion layers of the two types of film are of different thicknesses. It is, therefore, not unlikely that the 'red' and the 'blue' copies will differ after being developed, rinsed and dried. The photographic process in itself sets severe requirements on the colour correction of the optical system, and also on the experience and careful attention of the photographer.

6.3.7 Taping of Pad Master and Contact Printing on Photographic Films

A better method is to tape the pad master as described in Section 6.3.6. Two identical contact prints of the pad master are then made and used for the taping of the conductors of the component and solder sides. In this way, perfect registration of the two sides is obtained which implies a higher degree of manufacturability (see Fig. 6.4).

The contact prints can be made on photographic films such as Kodalith Ortho (trademark of Eastman Kodak Company), or Cronaflex (trademark of E. I. du Pont de Nemours & Co., Inc.), both on a polyester base.

The finished film has a stability in length of 0.0016% per 1% change of the relative humidity and 0.0027% per 1°C change in temperature. Simultaneous changes in relative humidity and temperature of 20% and 10°C, respectively, give a change of 0.3 mm (0.012 in) over a length of 500 mm (20 in). The variations are not quite linear but the figures indicate the magnitudes involved. Cellulose triacetate should not be used as a film base since it is less stable than polyester base. Polyester films are usually obtained in thicknesses of 0.003 in (0.076 mm)

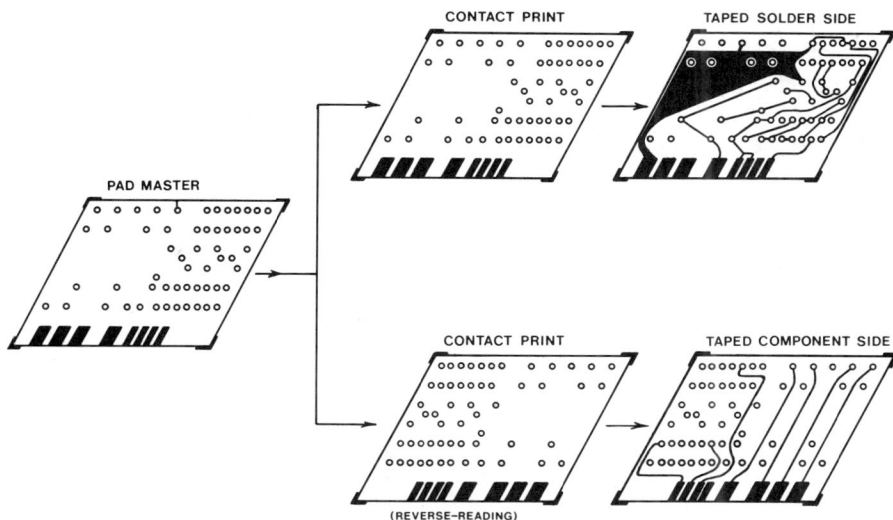

Fig. 6.4 Contact printing of pad master and taping of contact prints.

and 0.007 in (0.18 mm). For artworks to be taped, the thick film should be used as it prevents the finished artwork from puckering due to shrinkage of the taped conductors. The change in length due to changes in the relative humidity as given above applies to a 0.007 in (0.18 mm) thick film. For a 0.003 in (0.076 mm) thick film the value is 20–30% higher, depending on type and make.

Ageing occurs to a similar extent as in plain polyester films, i.e. 0.02% after 5 years. A small permanent dimensional change occurs in the photographic processing and is dependent on how well the photographer observes the process conditions. In the worst case, this should not exceed 0.02%.

6.3.8 Taping of Pad Master and Contact Printing on Diazo Films

Most electronics companies do not have in-house facilities for working with photographic materials, at least not for materials in the size necessary, for example 600 × 900 mm (24 × 36 in). It will therefore be necessary to send the pad master to a professional photographer with attendant delays which may amount to some days. This slows the work so much that if any serious volume of work is to be handled, a method must be found which avoids such delays.

The solution is to invest in contact-printing equipment and to use a diazo-coated polyester film, i.e. a contact film which is sensitive to a light in the blue–violet/ultraviolet range. The diazo film has the same properties with respect to the relative humidity and temperature as stated in Section 6.3.3 which means a somewhat better stability than that of photographic contact films. Moderate daylight has no influence on the diazo film for which reason it is not necessary to use a dark-room. The development is performed by means of ammonia vapour, which means that it is possible to make use of the developing section of the normal diazo-printing machine. As no water is involved in the developing process no

148

change in length occurs, and the copy is ready for immediate use after the development. No over-development can take place which results in great latitude in the copying process. The light-sensitive diazo layer gives a grainless image with a clear edge definition and a high resolution, theoretically above 500 lines per mm (1250 lines per 0.1 in). The quality of multi-generation copies is so high that no practical difference can be detected between first- and fourth-generation copies (see Fig. 6.5).

A light source in the blue–violet/ultraviolet spectrum is used, i.e. light with a wavelength in the range 350–450 nm, commonly derived from metal halide, mercury vapour or xenon lamps (see Section 7.2). Since sharp contour definition requires good contact between the original and the diazo film, a vaccuum printing frame is used.

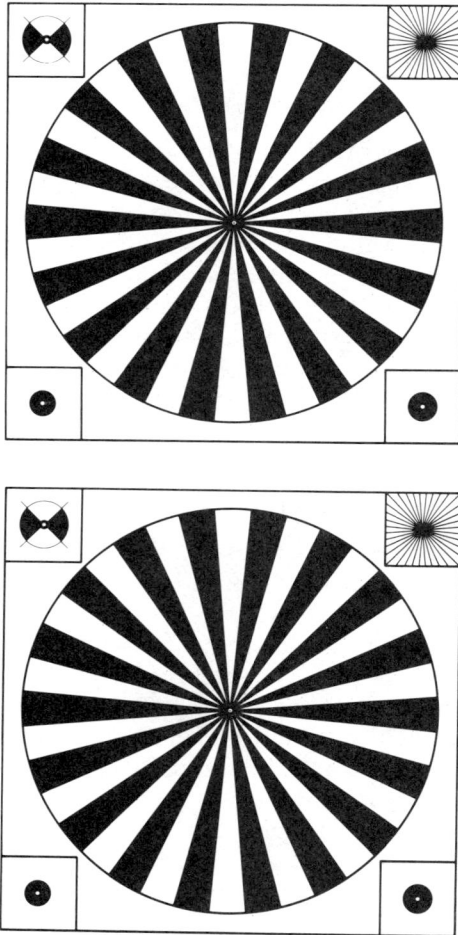

Fig. 6.5 Multi-generation diazo prints. Top: first print; bottom: fourth print.

When the PCB department has a contact printing equipment at its disposal, the work flow will be as shown in Fig. 6.6. As the conditions depend to a very high degree on the equipment used, it is possible to give only a rough indication of the time necessary for the printing operation. The pad master is placed together with the diazo film in the vacuum frame and exposed for approximately 3 min. The

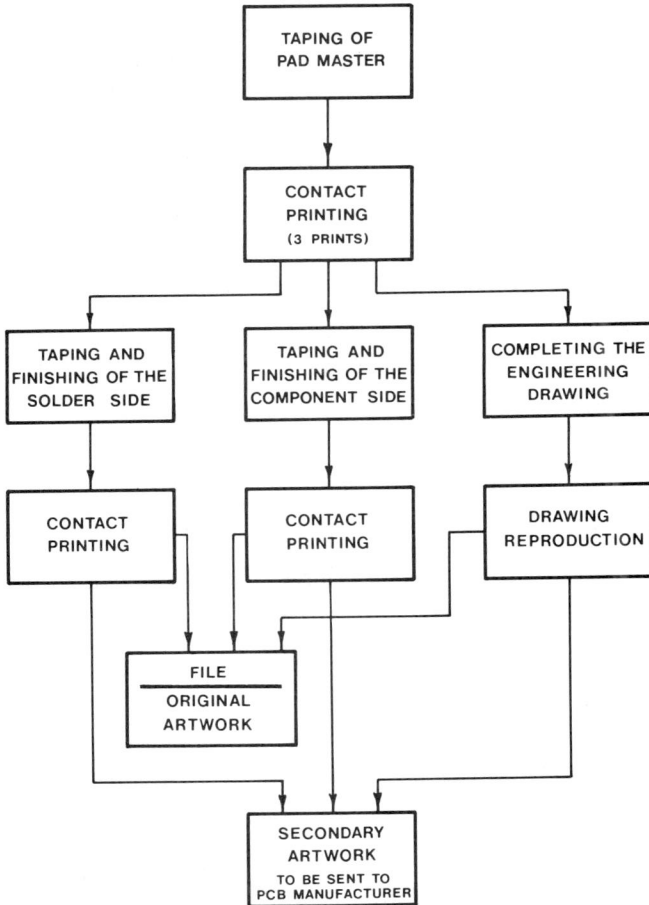

Fig. 6.6 Work flow of the PCB department.

process is repeated with another diazo film, so after approximately 8 min it is possible to bring the exposed copies to the developer. If the diazo printing machine is located in another department, this part of the process is estimated to take about 5 min including the walk to and fro. Including the time wasted the draughtsman has two copies ready for taping in less than 15 min. The great advantage of being able to carry on the work is quite obvious. The two contact prints are completely identical and will remain so provided they are stored and handled under the same conditions of temperature and humidity. Typically, any misregistration which can be detected over a length of 750 mm (30 in) is within the uncertainty of the measurement—say 0.03 mm (0.0012 in). A detailed description of the contact printing technique is given in Chapter 7.

The application of the diazo film is not limited to printing the pad masters. Since a taped artwork is always liable to displacement, or loss of a symbol or a conductor, a secondary artwork should always be printed and filed as a permanent artwork. It is possible to perform rectifications of the secondary artwork, but if in the design stage larger changes have to be made, these should be done on the taped artwork. Later, as the design approaches finality and only minor changes can be expected, these can be carried out on a secondary print and the taped master is then no longer required and can be destroyed.

Finally, a secondary artwork can be sent to the PCB manufacturer in a much cheaper and faster manner than a taped original, especially if it is sent abroad. The despatch of such copies is mentioned in detail in Section 7.3.8, but the author wishes at this stage to stress the great advantage of not having to send original artwork away from the design department. It could be lost or damaged, and it must be remembered that the artwork often represents an investment which is measured in hundreds of hours, i.e. a value of hundreds of pounds. In addition, demands for changes often arise just as the artwork is finished and despatched to the PCB manufacturer. With secondary artworks, it is easy to despatch an up-issued copy but in the case of an original artwork, this has first to be returned, and if it has been sent abroad one or two weeks are easily lost.

The conclusion is that by the use of the diazo technique, the PCB department reaches a higher technical level and is able to prepare a precision artwork of great reliability and convenience. The author has had extremely good results by using this technique, and strongly recommends it.

6.3.9 MCS (Master Circuit System) Artwork

This is an alternative to the more usual practices of taping the artwork, and as such, warrants a brief description. The artwork is prepared by a special and patented method based on the projection of the layout sketch on a large matt glass plate on which the actual taping is carried out. When the taping is completed, the pattern is photographed. The preparation of MCS artwork is offered as a service based on the clients' layout sketches which, for this purpose, need not be particularly accurate.

Briefly, the work procedure is as follows: The layout sketch is located in the optical system where it is placed so that the solder pads coincide fairly well with a precision grid which is an essential part of the system. The sketch and the precision grid are projected upon a matt glass plate as tall as a man, and the taping of the artwork is performed on the rear side of the glass plate. Here it is possible to align all the solder pads exactly in accordance with the intersections of the grid lines, and as the scale is rather large, it is possible to tape the conductors accurately. When the taping is finished, the layout sketch and the precision grid are replaced by a photographic film. A light arrangement behind the glass plate is switched on, the film is exposed and, on development, yields a master artwork in a 1:1 scale.

One advantage of the method is that optical distortion in projection is cancelled in the photography since the same optical system is used for both processes. The result is an extremely accurate artwork in the form of a 1:1 master film.

Rectifications are introduced by simultaneous projection of the layout sketch of the rectified area, the precision grid and a positive film, where the area to be rectified is cut away. The area to be rectified is taped on the glass plate as described above. The subsequent photography is performed as a combined contact printing of the positive film and a photography of the taped artwork on the matt glass plate.

The described procedure can also be followed if a CAD board accomplished by a service bureau has to be rectified. To perform even small rectifications on a CAD system is a very costly matter as the complete artwork—and not only the rectified area—must be photoplotted, and this involves expensive computer time. The cost can easily amount to £100–200 or more, to which must be added the loss due to the queueing time. The MCS system is an excellent solution, although the amended master pattern is no longer available in the filed CAD documentation. Subsequent CAD rectifications require that the CAD file be up-dated (see Section 9.3.6f).

In the event that someone wishes to install an MCS system, it should be noted that the method is patented and that expensive special equipment is required and that a special know-how has to be built up. Furthermore, it is necessary to install a dark-room and to employ a photographer and this would be against the policy of many electronics companies.

A company without a PCB department or a PCB drawing office can take advantage of the MCS system, and for the established PCB department it means a welcome relief in the case of peak loading. It should be noted, however, that the artwork in question is tied to the MCS system as it is difficult to introduce other than fairly small and uncomplicated changes without using the MCS equipment.

6.3.10 Cut-and-strip Method

We shall now complete the description of the general methods for preparation of artwork by a brief account of the cut-and-strip method. This depends on the use of a special foil consisting of a clear, stabilized polyester-base film with a red coating, opaque to ultraviolet light, but transparent to the eye. The red coating has a rather low adhesion to the base, so by means of a knife the pattern can be cut through the coating which then is peeled off, leaving a negative pattern.

The artwork is normally made in a 1:1 scale, if the pattern is not too complicated, and as the resulting film is a negative, the method is very suited for making prototype boards on a laboratory scale. If, however, the board is even slightly complicated, the method is not particularly expedient as it is too slow and inaccurate, and also fairly fatiguing to the operator.

The pad outlines are cut by means of modified pad cutting bow compasses using a small knife. When all pads are cut, they are lifted and peeled off by means of a knife or tweezers. The conductors are cut by means of an adjustable double knife and peeled off in the same way as the pads.

6.4 Determination of Optimum Artwork Scale

One of the most important decisions of the PCB department is the determination of the scale in which the artwork is to be prepared. In Section 5.12 a scale of 4:1 was regarded as particularly suitable, and in this section we shall throw light on the various factors which lead exactly to this scale. The object is to work in so large a scale that all the unavoidable inaccuracies resulting from the taping of the artwork become so small on the scaled-down master film, that a satisfactory quality is obtained in the finished boards. The inaccuracies must be seen in relation partly to reasonable working conditions for the draughtsman, and partly to the inherent form tolerance of the symbols and possible creepage of the taped conductors. The treatment is especially directed to methods previously described, viz. taping of two separate artworks as described in Section 6.3.4, and taping of a pad master and the subsequent contact printing on diazo films which were described in Section 6.3.8. It is very important to know how accurately a solder pad can be located or a conductor can be applied, and also how much a conductor is liable to move, in the course of time, due to internal tension in the tape. In assessing these matters we assume that the boards have a rather high packaging density based on modules of 0.1 in, and also, that it has to be possible to lay down a conductor between any pair of pads having a centre-to-centre distance of 0.1 in. Based on these assumptions as well as the accuracy required on the finished master film, it is now possible to determine the scale which should be used.

6.4.1 Positional Accuracy of a Pad

We consider first a solder pad having a centre hole. A very usual diameter of the centre hole is 1.5 mm (0.06 in) and the pad has to be located with its centre directly over an intersection of two grid lines as shown in Fig. 6.7. If the width of the grid lines is about 0.1 mm (0.004 in) the naked eye will normally be able to discern an asymmetry of 0.05–0.1 mm (0.002–0.004 in). It should therefore be possible for a fairly skilled draughtsman to apply a pad with an accuracy of ±0.1 mm (±0.004 in). For reasons of economy, a precision grid under the drafting film is often used in preference to a drafting film with a printed grid, resulting in a further possible displacement of the pad due to parallax. This is illustrated in Fig. 6.8 which shows a cross-section through the grid and the polyester film with the pad applied. Under ideal conditions, the eye should be kept in a position exactly over the pad, but in practice this will not always be the case. It is assumed that the eye is displaced 50 mm (2 in) with respect to the centre line, and at a distance of

LINE THROUGH THE IMAGINARY
CENTRE OF SOLDER PAD

SOLDER PAD

LINE OF GRID PATTERN

GRID

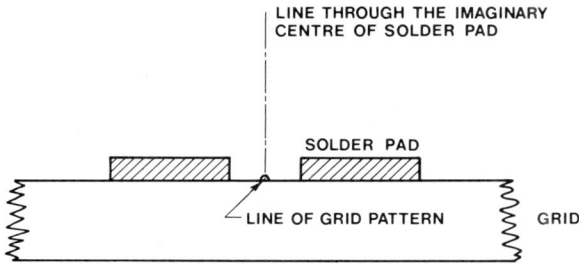

Fig. 6.7 Application of solder pad.

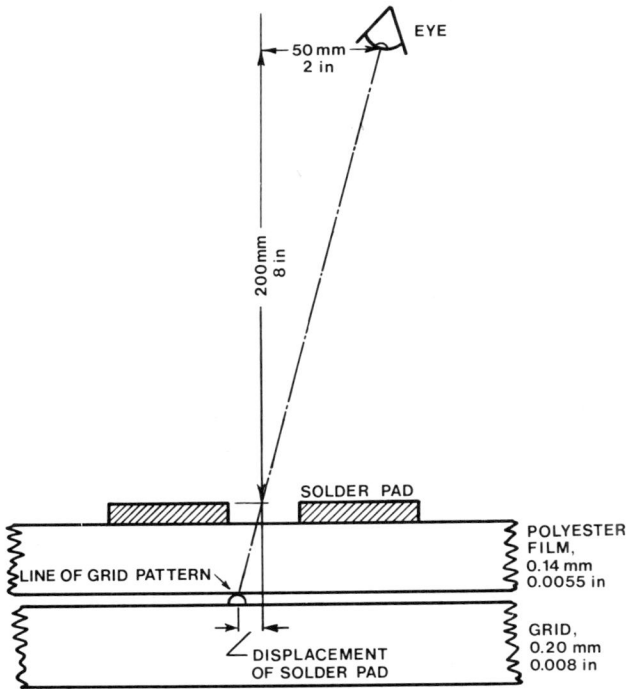

EYE

50 mm
2 in

200 mm
8 in

SOLDER PAD

POLYESTER
FILM,
0.14 mm
0.0055 in

LINE OF GRID PATTERN

GRID,
0.20 mm
0.008 in

DISPLACEMENT
OF SOLDER PAD

Fig. 6.8 The parallax error.

200 mm (8 in). If the thickness of the polyester film and the solder pad is 0.14 mm (0.0055 in) and 0.04 mm (0.0016 in) the possible displacement of the pad is:

$$(0.14 + 0.04) \times \frac{50}{200} = 0.045 \sim 0.05 \text{ mm } (0.002 \text{ in})$$

If the precision grid—in order to reduce wear and tear on the printed surface—is placed with the printed face towards the light box, the parallax error is increased. As the precision grid, for reasons of stability, is usually printed on fairly thick polyester films, often 0.20 mm (0.008 in) thick, the possible displacement of the pad under the same conditions as above becomes:

$$(0.14 + 0.004 + 0.02) \times \frac{50}{200} = 0.095 \simeq 0.10 \text{ mm } (0.004 \text{ in})$$

In assessing the minimum distance between a solder pad and a conductor, the unavoidable eccentricity of the pad which may be as much as 0.05 mm (0.002 in) must be taken into account. Pads which have been wrongly applied and removed from the board are inevitably distorted by this handling and must never be reused since, otherwise, an additional eccentricity would have to be allowed for.

The total positional tolerance of the circumference of a pad is the sum of the inaccuracy of application (the discriminating ability of the eye), the parallax error and the eccentricity. In the practical cases discussed these add up to:

A. A grid printed on the drafting film: approximately ±0.15 mm (0.006 in).
B. A grid located under the drafting film: approximately ±0.20 mm (0.008 in).
C. As B, but the grid is facing downwards, approximately ±0.25 mm (0.010 in).

6.4.2 Positional Accuracy of a Conductor

If a conductor is to be routed between a pair of adjacent pads, it is necessary to apply it very symmetrically in order to maintain the minimum spacing. As before, we shall presume that within the distances occurring in practice, it is possible for the naked eye to discern an asymmetry of 0.05–0.1 mm (0.002–0.004 in). The skilled draughtsman should therefore be able to apply a conductor with an accuracy of ±0.1 mm (0.004 in) inclusive of any parallax error which in this case can arise only from the thickness of the pads and the conductor tape.

The tolerance of the conductor width which normally amounts to ±0.05 mm (±0.002 in) has also to be taken into account. This, however, is not added directly to the tolerance of application, as is explained later.

Finally, a tape may creep a little in the course of time as a result of the relaxation of internal tensions in the tape arising either in manufacture of the tape or by accidental stretching when it is being applied to the board, especially when it is laid down in a curve. This is the most common case and a conservative estimate of the creepage gives a value of 0.1 mm (0.004 in).

6.4.3 Artwork Scale Determined by the Minimum Spacing Criterion

Below we shall consider two pads having a diameter of 1.4 mm (0.055 in), located with a centre-to-centre distance of 0.1 in (2.54 mm), and with a 0.3-mm (0.012-in) wide conductor running between the two pads. The desired minimum spacing between a solder pad and the conductor is 0.35 mm (0.014 in). As the above conditions are representative of current products in the industry, they can be made the basis of an assessment of the scale necessary for preparing the artwork. In this connection, we shall ignore the growth or shrinkage of the pad diameter and the conductor width in the manufacturing processes.

We shall now examine how well the requirements set above are fulfilled in the scales of 1:1, 2:1, and 4:1. Although the tolerances given in the preceding sections may seem fairly large, they can be regarded as worst-case tolerances. The largest part of the pattern will normally be well within the tolerance, so it will be in only a few places that the minimum requirements are violated. On the other hand, it is at these places the PCB manufacturer meets the problems. In day-to-day preparation of artwork it is impractical to check every clearance with a magnifying glass. It is, therefore, appropriate to take as a starting point, the worst-case situation and choose the artwork scale in such a way that, after photographic reduction, the resulting master film will allow some tolerance for inevitable variations which occur in manufacture.

Scale 1:1

Under the nominal conditions shown in Table 6.1 the width of the gap between two pads is 1.14 mm (0.0449 in) and the minimum clearance from pad to conductor 0.5 $(1.14 - 0.30) = 0.42$ mm (0.0165 in). We shall now consider the worst-case situation, but in order not to be too pessimistic we shall assume that one of

Table 6.1 Minimum Spacing as a function of artwork scale (N). (Dimensions in millimetres.)

			N	1:1	2:1	4:1
Positional tolerance of solder pad						
Location	0.10	} 0.20				
Parallax	0.05					
Eccentricity	0.05					
Width of gap, nominally N × 1.14				1.14	2.28	4.56
Displacement of one of the pads		0.20		0.20	0.20	0.20
Width of gap, effective				0.94	2.08	4.36
N × width of tape + tolerance on width of tape (0.05)				0.35	0.65	1.25
2 × minimum spacing				~0.60	1.43	3.11
Minimum spacing				0.30	~0.72	~1.56
Positional tolerance of tape	0.10	} 0.20		0.20	0.20	0.20
Creepage of tape	0.10					
Minimum spacing, worst-case				0.10	0.52	1.36
Minimum spacing, scale 1:1				0.10	0.26	0.34

the pads is located in the correct position. The width of the gap is now reduced by the positional tolerance valid for the pad circumference, which corresponds to case B, Section 6.4.1, so the width becomes $1.14 - 0.20 = 0.94$ mm (0.037 in). Taking the tolerance 0.05 mm (0.002 in) on the width of the tape into consideration, and applying the tape symmetrically, the minimum clearance becomes 0.5 $(0.94 - 0.35) \simeq 0.30$ mm (0.012 in). If the tape is applied with a displacement corresponding to the worst-case tolerance of 0.1 mm (0.004 in) to one side, the minimum spacing is reduced to 0.20 mm (0.008 in). When we imagine the additional, but not unrealistic, occurrence that the tape creeps 0.1 mm (0.004 in), the minimum spacing is reduced to 0.1 mm (0.004 in). It is clear that the PCB manufacturer will find serious production problems under these conditions, if only because a growth of 0.05 mm (0.002 in) along the edges of the solder pads and the conductors in the plating processes can easily occur.

Scale 2:1

We shall now consider what will happen when the artwork is prepared in a 2:1 scale and reduced photographically to the scale 1:1. The nominal conditions are shown in Table 6.1 where all dimensions are exactly doubled with respect to the preceding case. All the tolerances, however, are entered with the same absolute value as before, and under the same conditions as before it is found that the minimum spacing is 0.52 mm (0.0204 in) when measured on the artwork, and consequently 0.26 mm (0.0102 in) on the master film. In order to simplify the case, it is taken for granted that the photographic processes do not change the line width perceptibly. Working in a 2:1 scale is an improvement but the requirement for a minimum clearance of 0.35 mm (0.014 in) is still not fulfilled.

Scale 4:1

Finally, we shall note how all the small inaccuracies are minimized when the scale is increased to 4:1. Again, Table 6.1 shows the nominal conditions, and a simple calculation shows that in the worst-case situation the minimum clearance becomes 1.36mm (0.0535 in) when measured on the artwork, and consequently 0.34 mm (0.0134 in) when measured on the master film. This is very close to the given requirement.

Conclusion

Although some of the assumptions made in the three examples might be subject to argument, the results show clearly that a scale of 1:1 is inapplicable, that a scale of 2:1 can only just be used, but that a scale of 4:1 is extremely applicable and can, therefore, conveniently be standardized for this purpose.

The designer could be tempted to reduce the diameter of the solder pads, for example, by 0.2 mm (0.008 in) in the 1:1 scale, in order to obtain a correspondingly greater minimum clearance. In Section 6.5.4b, however, we shall find that the diameter of a solder pad with a 0.8 mm (0.031 in) solder hole must not be less than 1.4 mm (0.055 in) as used in our examples.

6.4.4 Artwork Scale Determined by the Concentricity of Solder Pad and Solder Hole

In the preceding section, we discussed the artwork scale on the basis of the details of the pattern as seen on one side of the board. If the two sides of the board are derived from a common pad master by contact printing (see Sections 6.3.7 and 6.3.8) there will be no problems of registration between the two artwork sides. This implies, however, that no rectifications involving the application of additional solder pads are performed. Rectifications are bound to happen sooner or later, for which reason determination of the necessary artwork scale is also valid for contact printed artwork. If the draughtsman tapes both sides of the artwork individually, including all pads, there is a risk of so large a misregister of the pads of any pair that the manufacturer will be unable to meet the requirements for the width of the annular ring on both sides of the board. In the manufacture of plated through hole boards (see Section 3.4.2) the holes are drilled as the first process, whereafter the board is sensitized and copper plated to a thickness of 5μm. Whether the board is going to be produced by a pattern-plating process (Section 3.4.3) or by a panel-plating process (Section 3.4.2) the images of the two sides are to be applied in one of the process stages. If the pads on the two sides of the board do not coincide exactly, this would lead to the result shown in Fig. 6.9.

Fig. 6.9 Broken annular rings.

The tape used in the NC drilling machine is made by digitizing the centres of the solder-side pads and it is, therefore, possible to locate the master film in such a position that all the solder pads are placed concentrically with respect to the drilled holes. Unfortunately, there is no guarantee that the same is the case with the pads of the component side. In addition to the artwork inaccuracies, there is a chain of manufacturing tolerances. The object must be that the annular ring is not broken under the image transfer which is done before the plating process. If the image is transferred by a screen-printing process, there is a risk that a little of the printing ink will run down into any holes which have the annular ring broken or nearly broken. This results in voids in the plated walls of the holes. It is, therefore, expedient to use the more expensive photoprinting method for image transfer. If the manufacturing tolerances are known, it is possible to calculate how much is left to cover the inaccuracies of the artwork, which in turn makes it possible to choose a suitable artwork scale.

As before, we shall base the consideration on a solder pad with a diameter of 1.4 mm (0.055 in) and a solder hole of 0.8 mm (0.031 in). This gives a nominal width of the annular ring of 0.5 $(1.4 - 0.8) = 0.3$ mm (0.012 in). By an eccentricity of 0.3 mm (0.012 in) the annular ring is just broken, and below we shall discuss how the various manufacturing tolerances can be accommodated without exceeding this degree of eccentricity.

6.4.4a Oversize of Drilled Hole

The PCB manufacturer has to drill the solder holes slightly larger than 0.8 mm (0.031 in) in order to compensate for the plating. If the thickness of the plating is specified as minimum $25\,\mu$m (0.001 in) copper and minimum $10\,\mu$m (0.0004 in) tin/lead, in total $35\,\mu$m (0.0014 in), the PCB manufacturer is forced to drill the holes of a diameter not less than 0.8 mm + (2 × 35μm) = 0.87 mm ≃ 0.9 mm (0.0354 in). The final determination of the drill size must, however, take the following facts into consideration:

1. The plating thicknesses stated have a tolerance of +100%.
2. Isolated solder holes are subjected to a plating thickness exceeding the +100% tolerance (see Section 4.5.3).
3. The final reflowing process may cause some holes to be more or less blocked, their cross-section having a slight resemblance to an hour-glass.

For these reasons it is tempting for the manufacturer to drill the holes 0.15 mm (0.006 in) larger than the final diameter. In this way, it becomes easier to keep the specification of the finished plated through holes. However, the width of the annular ring as it appears on the image of the board is now reduced by $0.5 \times 0.15 = 0.075$ mm (0.003 in).

6.4.4b Drilling Accuracy

It is evident that the accuracy of the drilling machine plays an important part. In our example, we shall assume that the board is so complicated that it must be drilled by means of an NC drilling machine. In Section 3.8.2d the drilling accuracy was seen to lie within the range 0.025–0.05 mm (0.001–0.002 in). We shall take the worst-case situation and the value of 0.05 mm (0.002 in) will be used.

6.4.4c Accuracy of Master Film

The accuracy of the master film is normally about ±0.05 mm (±0.002 in). This value covers partly the accuracy of the photography including possible changes owing to the development, and the film's instability under varying temperature and humidity. It is obvious that in order to keep an accuracy of ±0.05 mm (±0.002 in) the film must be stored and used under controlled environmental conditions. Since the master film of the solder side was the basis of the digitizing, the solder holes conform accurately with the pads of the solder side. On the component side, however, there is a risk of a gradually increasing eccentricity between the solder holes and the pads, the worst case being 0.05 mm (0.002 in).

6.4.4d Photographic Changes of Pattern

Depending on the development process, the width of the conductors and the diameter of the pads may be subject to slight variations, unlikely to exceed ±0.05 mm. We have, therefore, to calculate for a possible reduction of the width of the annular ring, amounting to 0.025 mm (0.001 in).

6.4.4e Misregistration

Finally, we shall consider the misregistration of the master film of the component side with respect to the board. Here we must calculate with a possible displacement of ±0.05 mm (±0.002 in), most frequently originating from an erroneous location of guide holes and pins of the master film and the board, and from play between the guide holes and pins.

6.4.4f Total Reduction of the Width of the Annular Ring

We are now able to sum up the various contributions. Because the number of sources of faults involved is fairly limited, and the contributions a, d and e have the same weight for all holes of the board, there are very good reasons for performing the summing up with coinciding signs:

a.	Oversize of drilled holes	0.075 mm	(0.003 in)
b.	Drilling accuracy	0.05 mm	(0.002 in)
c.	Accuracy of master film	0.05 mm	(0.002 in)
d.	Change of pattern	0.025 mm	(0.001 in)
e.	Misregistration	0.05 mm	(0.002)
	Total reduction of width of the annular ring	0.25 mm	(0.01 in)

6.4.4g Conclusion

Since a nominal design width of 0.3 mm (0.012 in) has been chosen for the annular ring and the manufacturing processes in the worst case take up a further 0.25 mm (0.01 in), there remains only 0.05 mm (0.002 in) to cover any misregistration between the pads of the two sides of the artwork. If we assume that the pads are applied to the component side, concentrically with the pads of the solder side this would correspond fairly well to case C in Section 6.4.1. The positional tolerance of the pads on the component side with respect to those of the solder side becomes ±0.25 mm (±0.01 in), and the necessary artwork scale is therefore determined as:

$$\frac{0.25}{0.30 - 0.20} = \frac{0.25}{0.05} \sim 5:1$$

It is possible to manipulate a little the size of the error sources, especially the summing up, so it is also possible to reach a total error of 0.20 mm (0.008 in) instead of 0.25 mm (0.01 in) (see the preceding section). Under this condition the necessary artwork scale could be determined as:

$$\frac{0.25}{0.30 - 0.20} = \frac{0.25}{0.10} \sim 2.5:1$$

The author's conclusion is, that an artwork scale of 2:1 can only just be accepted, whereas it is far better to use a scale of 4:1, unless it is impossible in practice, cf. Section 5.2.3.

6.5 Selecting Artwork Symbols and Tapes

Now we shall deal with the design of the symbols used in the preparation of artwork, i.e. the pads and tapes. The suppliers offer a very great selection of symbols in many different shapes and sizes, and tapes in many different widths. If some degree of standardization of the symbols and tapes is not adopted and directions for their use not worked out, the artworks will inevitably lack uniformity in appearance, and the number of types kept in stock will be too large and make rational and economical purchasing impossible.

6.5.1 Minimum Spacing Determined by Artwork Considerations

It is necessary to determine the design value of the minimum spacing between conductors since many of the details of the pattern depend on the value chosen. The choice of plating method determines the change in pattern, namely, an increase in width by pattern plating and a reduction in width by panel plating. In turn, these changes depend on the method of image transfer (screen printing or photo printing), the thickness of the copper foil, and the etchant used. Within the combinations of board types and manufacturing methods occurring in practice, it seems reasonable to estimate that the changes in width of the constituents of the finished pattern will be less than approximately ± 0.13 mm (± 0.005 in). As these changes imply corresponding changes in the spacings between conductors and solder pads, the minimum spacing is also subject to variation within this limit. In order to supply the reader with more definite information, Table 6.2 lists the various width tolerances applicable to the different manufacturing processes.

If the circuit has to accommodate the maximum number of conductors in a given board area it is necessary to give careful attention to the above manufacturing tolerances, so that the draughtsman must have fore-knowledge of the manufacturing processes in order to be able to make the appropriate allowances in the preparation of the artwork. The manufacture of the board is, therefore, tied to these processes which, in turn, limit the choice of available PCB manufacturers. Nonetheless, the artwork must, wherever possible, be designed so that any manufacturer, regardless of the processes he is using, can meet the requirements regarding conductor width and spacing.

The determination of the minimum spacing requires that the tolerances, both of the artwork and of the manufacturing processes, are taken into account. In Section 6.4.3, the possibilities of realizing a minimum design spacing of 0.35 mm (0.014 in) were examined and an artwork scale of 4:1 decided upon. By a 0.13-mm (0.005-in) reduction of the finished board's minimum spacing due to the plating growth, the effective minimum spacing, as measured on the finished board, becomes 0.22 mm (0.0087 in) which is fully acceptable by the PCB manufacturer.

Table 6.2 Typical adjustments (in micrometres; $1\ \mu\text{m} = 0.001$ mm) to conductor widths in order to achieve desired conductor width on finished boards.

Plating Method*		Etchant** and board type								
		Conventional board			Plated through board (Sn/Pb)			Plated through board (Au/Ni)		
		FC	AP	CS	FC	AP	CS	FC	AP	CS
17.5 μm ($\frac{1}{2}$ oz) Copper										
Panel plating	Sc Pr	–	–	–	–	0	+25	−25	0	+25
	Ph Pr	–	–	–	–	+50	+75	+25	+50	+75
Pattern plating	Sc Pr	−75	−50	−50	–	−100	−100	−100	−75	−75
	Ph Pr	0	0	0	–	−50	−50	−50	−25	−25
35 μm (1 oz) Copper										
Panel plating	Sc Pr	–	–	–	–	+25	+50	0	+25	+50
	Ph Pr	–	–	–	–	+75	+100	+50	+75	+100
Pattern plating	Sc Pr	−50	−50	−25	–	−100	−75	−75	−75	−50
	Ph Pr	0	0	+25	–	−50	−25	−25	−25	0
70 μm (2 oz) Copper										
Panel plating	Sc Pr	–	–	–	–	+50	+75	+25	+25	+75
	Ph Pr	–	–	–	–	+100	+125	+75	+100	+125
Pattern plating	Sc Pr	−25	0	+25	–	−50	−25	−50	−25	0
	Ph Pr	+25	+50	+75	–	0	+25	0	+25	+50

* Sc Pr, Screen printing; Ph Pr, Photo printing.

** FC, Ferric chloride; AP, Ammonium persulphate; CS, Chromic–sulphuric Acid.

Prepared from IPC-D-310 by permission of the Institute of Printed Circuits.

If the example in Section 6.4.3 is re-calculated without regard to the worst-case tolerance, the design value of the minimum spacing is found to be approximately 0.4 mm. (0.016 in). We can now summarize the values found and state the following design rule:

Design Rule Number 1

Minimum spacing between a conductor and a pad (or a row of pads):

	4:1	1:1
Design value	1.60 mm (0.063 in)	0.40 mm (0.016 in)
Accepted value	1.40 mm (0.055 in)	0.35 mm (0.014 in)

By the accepted value given above and in the design rules to follow is understood the actual value measured on the artwork (4:1) and the master film (1:1), respectively, by means of a magnifying glass (minimum × 7 magnification).

If two or more conductors are running in parallel over longer distances it is desirable to increase the minimum spacing a little as it otherwise might be difficult for the PCB manufacturer to achieve a 100% removal of the plating resist. As has been mentioned in Section 3.6, a residue of plating resist, even if it is very thin, acts as an etch resist and causes a short-circuit between adjacent conductors. It is, therefore, recommended that the spacing between parallel conductors be increased, leading to the design rule given below:

Design Rule Number 2

Minimum spacing between parallel conductors:

	4:1	1:1
Design value	2.00 mm (0.079 in)	0.50 mm (0.020 in)
Accepted value	1.60 mm (0.063 in)	0.40 mm (0.016 in)

6.5.2 Minimum Spacing Determined by Voltage Considerations

Up to this point only the needs of the artwork preparation and of the production processes have been taken into consideration. It is quite clear that the minimum spacing must be so designed that no voltage flash-over takes place between conductors due to insulation failure. Firstly, we shall examine the secondary circuits, i.e. circuits with no direct connection to the supply mains. MIL-STD-275C and the IEC recommendation 326, indicate the insulating distances, as a function of voltage, necessary for non-insulated and insulated boards used at heights up to 3000 m and over 3000 m above sea-level, respectively (see Fig. 6.10). By insulated boards is understood boards covered on both sides by an insulating layer, for example, an epoxy lacquer. These recommendations seem very conservative in comparison with widely used current practice in the manufacture of densely packed printed circuit boards. The smallest insulating distance indicated in these specifications for non-insulated boards is 0.625 mm (0.0246 in) which covers d.c. voltages up to 150 V. A smaller insulating distance 0.25 mm (0.01 in) is permitted if an insulating layer is applied to the board.

The application of an insulating cover layer does, however, involve extra cost, and can cause other problems, such as wrinkling of the insulating layer when the underlying tin/lead melts in the automatic soldering process. It is, of course, possible to plate selectively so that only the pads are tin/lead plated, the conductors being nickel plated, so that wrinkling does not occur. Unfortunately, this adds to the cost of the finished board due to the extra masking operation necessary. In practice, innumerable boards with considerably smaller insulating distances and with no insulating cover layer have been used. In two drafts of IEC recommendations concerning insulation in low-voltage equipment, 28A (Secretariat) 5 and 28 (Secretariat)

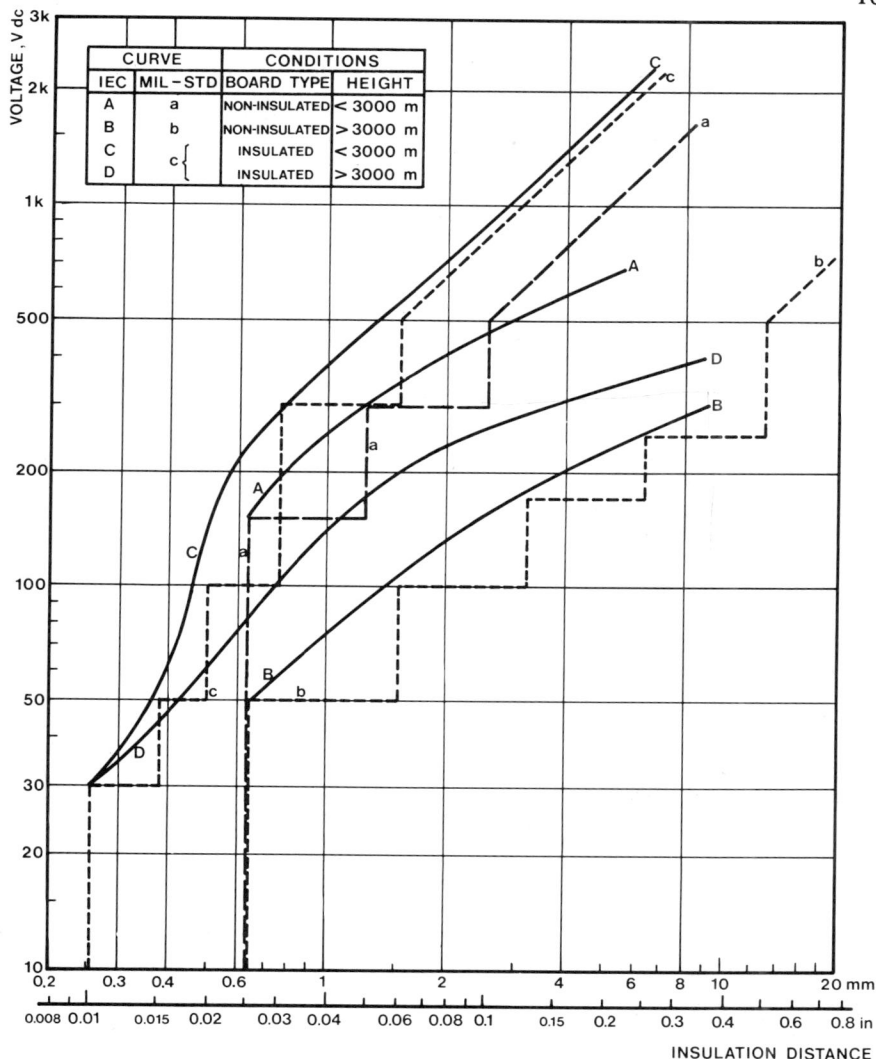

Fig. 6.10 Insulation distance as a function of the voltage for various applications. (IEC 326 and MIL-STD-275C.)

54, considerably smaller insulating distances are stated. These depend, of course, on the category of insulation, and the following description is based on a category relevant for ordinary civil electronic equipment used under clean and protected conditions, i.e. electronic measuring instruments, components and equipment and installations for telecommunication. The curve in Fig. 6.11 has been drawn in accordance with the chosen category and indicates the insulating distance as a function of the voltage. The insulating distance, i.e. the minimum spacing is understood as the distance which is measured on the finished board. As has been previously shown, a

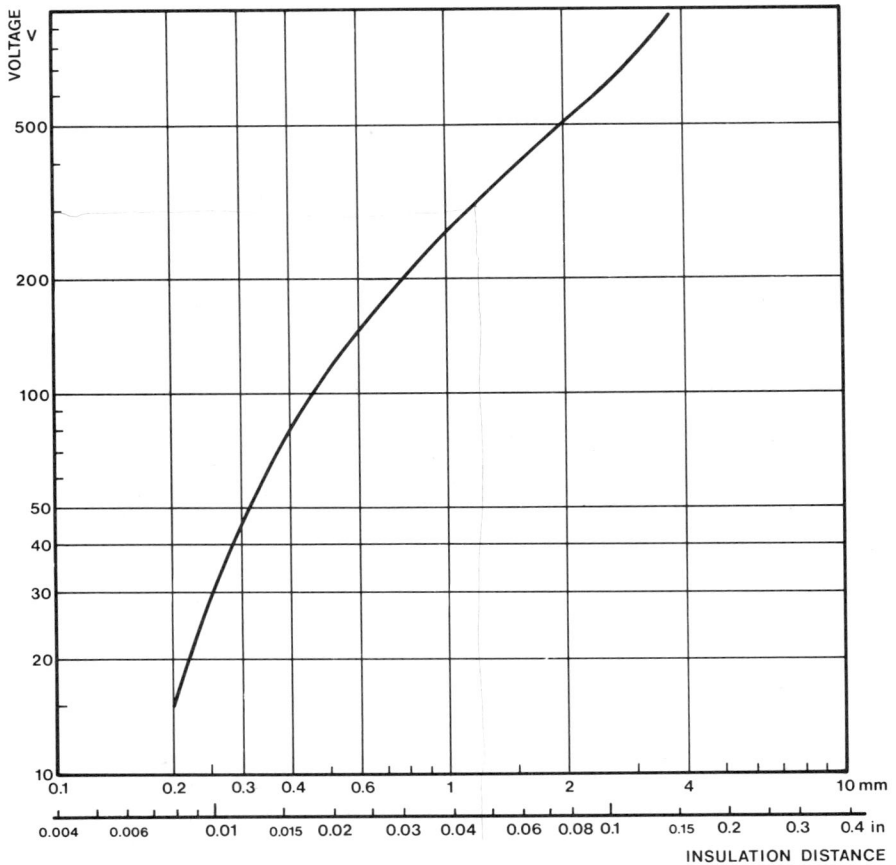

Fig. 6.11 Insulation distance as a function of the voltage. (IEC draft.)

growth in width of 0.13 mm (0.005 in) due to the plating process, and a corresponding reduction of the spacing, is to be expected under worst-case conditions. For this reason, the PCB designer and/or the draughtsman must compensate for this reduction by adding 0.13 mm (0.005 in) to the insulating distance found by means of the curve. Vice versa, it is possible to determine the maximum voltage corresponding to the minimum distance determined as per Design Rule Number 1. The curve is simply entered with the minimum spacing minus 0.13 mm (0.005 in). This leads to the following design rule:

Design Rule Number 3 (A)

The accepted value of the minimum spacing of 1.40 mm (0.055 in) in the scale 4:1, and 0.35 mm (0.014 in) in the scale 1:1 applies for d.c. voltages up to 20 V.

In the case of d.c. voltages higher than 20 V, the insulating distance is determined from the curve in Fig. 6.11, 0.5 mm (0.02 in) being added to the 4:1 value, and 0.13 mm (0.005 in) to the 1:1 value.

It might seem fairly academic to introduce the said correction. For an uncorrected minimum spacing of 0.35 mm (0.014 in) in the scale 1:1, the curve shows a voltage of 60 V, which is three times that taken in formulating Design Rule Number 3A. One of the PCB designer's most important tasks, however, is to ensure that the boards will function reliably even when the finished board lies at the unfavourable end of the tolerance range.

In many types of equipment, the mains are directly connected to conductors on the board and, for safety reasons, such conductors must be designed with insulating distances of a higher order of magnitude. Referring to the IEC recommendation 348 *Safety Requirements for Electronic Measuring Apparatus*, the effective safety distance (creepage distance) for class I is 3 mm (0.12 in) for a.c. voltages from 60 V up to 250 V. In the case of printed circuit boards, since the position of the conductor cannot be changed, the safety distance may be reduced to 2 mm (0.079 in) for all pads and conductors carrying the mains voltage. National safety regulations can, however, set more severe requirements than IEC. This is the case with the UL safety regulations (UL, Underwriters Laboratories Inc., USA) where a safety distance of 1/16 in (1.6 mm) is required for a.c. voltages up to 125 V and 3/32 in (2.4 mm) for a.c. voltages up to 250 V. If export to countries which recognize UL specifications is to be safeguarded, it will be necessary to use a 3/32-in (2.4-mm) safety distance, even when the transformer is driven from 115 V a.c. The transformer will often have a 220/240 V terminal which will be connected to the printed circuit board. The minimum distance must be maintained with all tolerances in the most unfavourable conditions for which reason it is necessary to include an allowance for possible plating growth of 0.13 mm (0.005 in) and to take into consideration unwanted isolated projections at the circumference of a pad or along a conductor edge (see Figs. 4.4 and 4.11) which may extend to about 0.20 mm (0.008 in). If it is assumed that two projections do not occur exactly opposite to each other, the safety distance must be 2.4 + 0.13 + 0.20 = 2.73 mm (0.107 in), but in order to have an extra margin for a possible growth in width owing to the automatic soldering and also to allow for the very rare case that two projections occur exactly opposite each other, it is reasonable to chose a minimum spacing of 3 mm (0.118 in). This leads to the following design rule:

Design Rule Number 3(B)

In the case of a.c. mains voltages up to 250 V, the safety distance is 12 mm (0.47 in) in the 4:1 scale, and 3 mm (0.118 in) in the 1:1 scale.

6.5.3 Minimum Spacing Determined by Soldering Considerations

Automatic soldering is so generally used that spacings should be determined on the assumption that this method will be used. The most serious problem connected with automatic soldering is solder bridging between adjacent conductors or between a conductor and a pad. Such solder bridges might damage the assembled board seriously if they have not been removed before voltage is applied to the board. Inspection and removal are, however, expensive and the designer must, therefore, be aware of this problem from the beginning of the layout and run the conductors of the solder side so that they do not invite solder bridging. The most important thing is to avoid large deviations between the direction of soldering and the direction of the conductors which is expressed in the design rule stated below:

Design Rule Number 4

The accepted minimum spacing of 1.60 mm (0.063 in) in the scale 4:1, and 0.40 mm (0.016 in) in the scale 1:1 applies to parallel conductors, and to a conductor and adjacent pads, if the angle between the direction of the conductors and the direction of travel on the soldering machine is less than 15°. If the angle is larger, the minimum spacing must be increased to 2.40 mm (0.094 in) in the 4:1 scale, and to 0.60 mm (0.024 in) in the 1:1 scale.

This design rule creates additional difficulty for the designer and the draughtsman since an extra parameter is introduced. Even so, full security against solder bridging is not obtained. A more radical solution is to apply a solder mask on the solder side and forget about Design Rule Number 4. The solder masks are dealt with in Section 6.9.

6.5.4 Pad Size

A pad must fulfil two essential requirements: It shall be possible to solder a component lead to the pad, and the pad shall remain attached to the board.

6.5.4a Diameter of Solder Hole

To permit assembly and soldering, the solder hole has to be somewhat larger than the component lead, and experience has shown that an oversize of 0.2–0.5 mm (0.008–0.02 in) is satisfactory. Too close a fitting makes the mounting of the component unnecessarily difficult and expensive, especially in the case of multi-lead components. Furthermore, the quality of automatic soldering suffers if the leads fit too tightly, especially when the hole is plated through. This is due to the fact that the gases formed during the soldering will find it difficult to force their way out and may contribute to the formation of blow holes and porous joints.

In order to keep the price down it is advisable to keep the number of different hole sizes as low as possible, as every change of drills adds to the cost. The above range of oversize of the solder hole should, therefore, be fully utilized to accommodate as many sizes of component leads as possible with the same drill size.

The task is now to choose as few hole sizes as possible and yet to cover the lead diameters occurring in practice. This is shown graphically in Fig. 6.12. The upper half of the diagram shows how the range of utilization of five different hole sizes is located 0.2–0.5 mm (0.008–0.02 in) below the nominal diameter D_N. In addition to this is indicated the minimum diameter D_M which is determined on the basis of the tolerances of non-reflowed plated through holes, i.e. ±0.10 mm (±0.004 in) for $D_N \leqslant 0.8$ mm (0.031 in), and ±0.13 mm (±0.005 in) for $D_N > 0.8$ mm (0.031 in). The lower part of

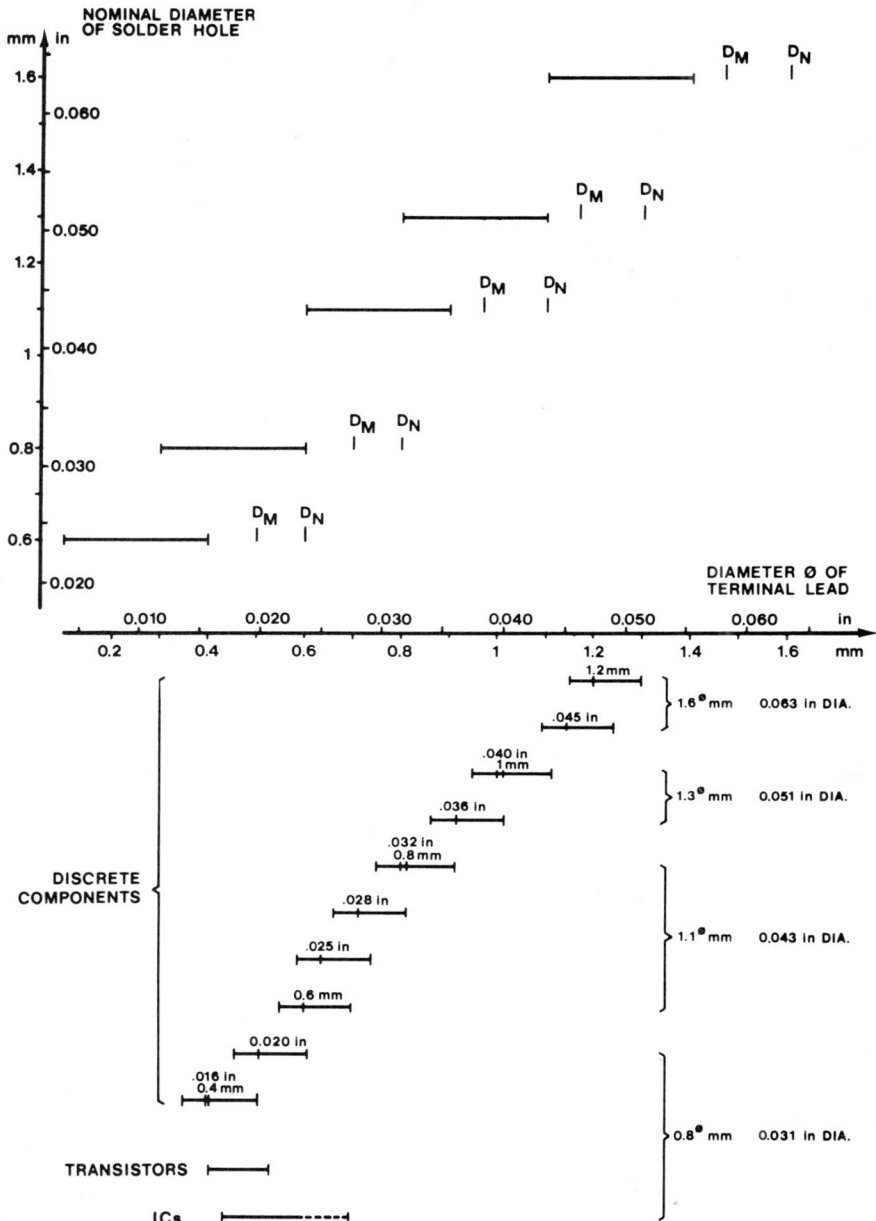

Fig. 6.12 Determination of hole diameter as a function of lead diameter.

the diagram in Fig. 6.12 shows the nominal diameter and the tolerance range for a large number of discrete components of American and European origin. The diameter tolerance is +0.10 and −0.05 mm (+0.004 in and −0.002 in) which covers all relevant cases. For certain American components the tolerance range is as much as ± 0.005 in (± 0.13 mm), but even this fairly large tolerance range can be fitted into the system. It appears that discrete components are covered by four hole sizes only, namely, 0.8, 1.1, 1.3, and 1.6 mm (0.031, 0.043, 0.051, and 0.063 in, respectively).

The vast majority of transistor types and all dual-in-line ICs are covered by 0.8 mm (0.031 in) holes. The thickness and width ranges of IC terminals are given in the JEDEC specifications as 0.204–0.381 mm (0.008–0.015 in) and 0.381–0.584 mm (0.015–0.023 in), respectively. The maximum diagonal is, therefore, 0.697 mm (0.027 in) which just fits into a 0.7 mm (0.028 in) hole corresponding to the minimum size of an 0.8 mm (0.031 in) hole. Fortunately, most manufacturers of ICs specify considerably smaller terminals, for example, 0.20–0.30 mm (0.008–0.012 in) and 0.38–0.51 mm (0.015–0.020 in) for the thickness and width, respectively. This gives a maximum diagonal of 0.59 mm (0.023 in) which implies a rather easy fitting into the nominal 0.8 mm (0.031 in) hole.

The above discussion is summarized in the following design rule:

Design Rule Number 5

Use as few hole sizes as possible, and let the nominal diameter of the solder hole be 0.2–0.5 mm (0.008–0.020 in) larger than the nominal diameter or diagonal of the terminal lead in a 1:1 scale.

The PCB designer must bear in mind that if a snap-in lead form is used it may be necessary to reduce the range of clearance in the solder holes depending on the lead-forming machine. Otherwise, the leads may fall out of the holes when the boards are transferred to the soldering machine.

The diameter of plated through holes should not be much less than 50% of the thickness of the board, as smaller holes are difficult to plate satisfactorily. The most usual thickness of board is 1.6 mm (0.063 in) and the minimum hole size recommended is 0.6 mm (0.024 in). Holes of this size are justified only when used as via holes in boards with a very high packaging density, as the diameter of the pads can be reduced correspondingly. In the case of boards with a lower packaging density it will be an advantage to specify the same diameter for the via holes as for the IC holes, i.e. 0.8 mm (0.031 in).

6.5.4b Diameter of Solder Pad

In Sections 6.4.4f and 6.4.4g we found that the nominal width of the annular ring cannot be less than 0.3 mm (0.012 in) if all the normally occurring tolerances of the manufacturing process as well as the tolerances of the artwork are to be covered, while still ensuring that the annular ring is not broken at the image transfer. It is also desirable, though not absolutely necessary, that it is not broken on the finished board. We shall now see what happens to the annular ring in the subsequent processes. During plating the annular ring grows inwards, by the amount of the plating thickness, viz. 25 μm (0.001 in) copper and 10 μm (0.0004 in) tin/lead, to which is added the plating tolerance which, in this case, is set to $+75\%$, i.e. in total, 0.062 mm (0.0024 in). The outer diameter of the pad might either be increased or reduced depending on the manufacturing processes as described in Section 6.5.1. Taking the worst case, we have a shrinkage of 0.13 mm (0.005 in) which means that the outer radius of the pad is reduced by 0.065 mm (0.0025 in) resulting in practically no change in the width of the annular ring. Usually the width will increase a little. It follows that the finished board does not set requirements to the annular ring which cannot be fulfilled with a minimum width of the annular ring of 0.3 mm (0.012 in).

It must be realized that an annular ring having a width of 0.3 mm (0.012 in) is just on the limit of what can be achieved by ordinary manufacturing practice. Annular rings of smaller width require special precautions which will be reflected in a higher cost. In particular, the efficiency of the very expensive NC drilling machine drops drastically. Owing to the nature of the base material, small drills are liable to be deflected to one side to an extent dependent on the number of boards in the stack. The drilling accuracy of 0.05 mm (0.002 in) previously given applies to a stack height of three boards, but if the annular ring is very small, it may be necessary to reduce the stack height to two boards or perhaps one.

We have found that the width of the annular ring ought to be as large as is consistent with the complexity of the board. This leads to the use of two categories of solder pads, the first comprising the minimum pad sizes having annular rings of 0.3–0.6 mm (0.012–0.024 in) and the second category comprising the preferred pad sizes having annular rings of 0.6 mm (0.024 in) and above.

It is an advantage to use as few pad sizes as possible: prices of supplies, a reasonably small number of types to be carried in stock, and reasonable convenience in taping are some of the reasons. As shown in Fig. 6.12 it is normally possible to use only three sizes of the solder holes, namely 0.8 mm (0.031 in), 1.1 mm (0.043 in) and 1.3 mm (0.051 in), whereas 1.6 mm (0.063 in) is very rare and therefore left out of the diagram shown in Fig. 6.13. The diagram is entered with the nominal diameter of the

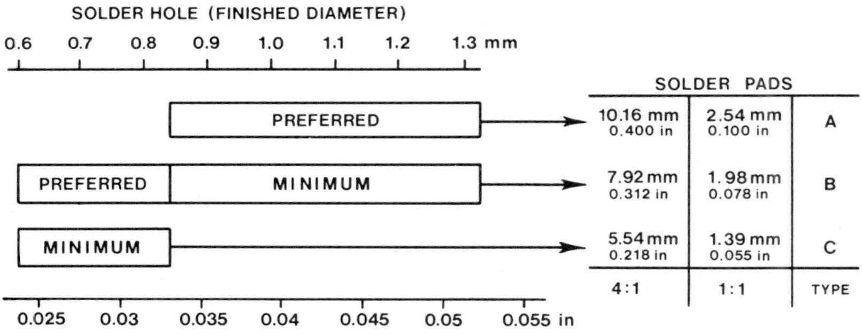

Fig. 6.13 Determination of solder pad size.

solder hole, for example 1.1 mm (0.043 in). This gives a preferred solder pad size of 2.54 mm (0.1 in) and a minimum solder pad size of 1.98 mm (0.078 in), both in the 1:1 scale. The upper end of the ranges is determined by the criterion of an annular ring of 0.3 mm (0.012 in) and 0.6 mm (0.024 in), respectively. Note, that only three different pad sizes are used. The intermediate pad can appear either as a minimum pad or a preferred pad according to the size of the solder hole.

A convenient designation for such selected pads is to call them A-pads, B-pads and C-pads. For the sake of correct taping of the pads, it is necessary to indicate the size on the layout sketch. When this is worked out, it is necessary under all circumstances to assess the size of the pads as they have a very high influence on the conductor routing (see Section 6.6.1). It seems most practical to draw the pads by means of a template in the right scale, i.e. they are drawn with a diameter of 10 mm (0.4 in), 8 mm (0.3 in), or 6 mm (0.24 in). If puppets or layout symbols are used, as described in Sections 5.22.3 and 5.22.4, respectively, the pads come automatically and in the right sizes.

The above considerations lead to the following design rule:

Design Rule Number 6 (Valid for plated through boards)

> The minimum size of a pad must be chosen so that the nominal width of the annular ring is larger than 0.3 mm (0.012 in) in the 1:1 scale. The preferred size must be chosen so that the nominal width of the annular ring is larger than 0.6 mm (0.024 in) in the 1:1 scale.

The above discussion applies to plated through holes. Such holes normally have fairly small solder pads as the adhesion between the pads and the boards depends mainly on the metal plated in the hole which, for this purpose, can be regarded as a rivet.

The matter is quite different in the case of conventional boards, sometimes called 'print-and-etch' boards, i.e. single-sided and double-sided boards with no plated through holes. Here it is the adhesion between the pads and the flat surface of the board which is decisive, and it is quite evident that a solder pad having an annular ring of 0.3 mm (0.012 in), or even 0.6 mm (0.024 in) must be marginal in this repect. It should be borne in mind that it is not sufficient for the pads to survive the assembly and soldering processes. The board must also be able to stand repair whether this is carried out in the test phase, or later as an ordinary service repair. It is usually during repair work that the solder pads lift from the board as a result of a combination of too small an area and excessive desoldering time.

As a rule of thumb, the area of a solder pad ought not to be less than 5 mm^2 (0.0078 sq. in). This implies that the diameter of solder pads having solder holes of 0.8 mm (0.031 in) shall not be less than 2.64 mm (0.104 in), or for convenience 2.54 mm (0.1 in) in the 1:1 scale. It is easily seen that non-plated through boards are not particularly suited to carry ICs, although the use of pads other than circular enables them to be embodied in some instances. This usually occurs when non-plated through boards are adopted for reasons of cost. In these cases, the pad area is increased by using a rectangular form or by providing 'ears' on the pads which help to secure them to the base material. This technique is also commonly used in the layout of flexible circuits. Figure 6.14 shows some examples of such pads. A further improvement is obtained by applying a solder mask, which in addition to preventing solder bridging, also locks the pads in place.

When laying out non-plated through boards the design rule stated below should be followed as far as possible:

Fig. 6.14 Solder pads with 'ears'.

Design Rule Number 7 (Non-plated through boards)

The area of non-plated through solder pads shall not be less than 5 mm² (0.0078 sq. in). A smaller pad can be used if it is provided with ears or locked by means of a solder mask.

6.5.5 Rectangular Solder Pads

Rectangular solder pads are used with advantage when dual-in-line ICs have to be accommodated. It will frequently be necessary to run a conductor between a pair of such pads, and, as previously described in Section 6.4.3, a reasonable choice is solder pads with a diameter of 1.4 mm (0.055 in) and a conductor with a width of 0.3 mm (0.012 in). This gives a nominal minimum spacing of 0.42 mm (0.0165 in) with an accepted value of 0.34–0.35 mm (0.013–0.014 in), all values given for the 1:1 scale. The width of the annular ring is 0.3 mm (0.012 in) as the solder hole is 0.8 mm (0.031 in). We have, however, also shown that this width is critical for the PCB manufacturer, and in order to improve the conditions a little, the use of rectangular solder pads is recommended. Later, we shall compare different sizes of rectangular solder pads, but, at this point, it can already be said that the most practical size is 1.4 × 2.0 mm (0.055 × 0.079 in). Such rectangular pads located in a cluster for ICs are shown in Fig. 6.15. Although a rectangular pad does not have a real annular ring,

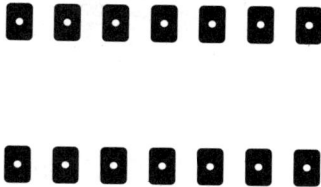

Fig. 6.15 Pads located in a cluster for a 14-lead IC.

we shall, for the sake of simplicity, use this designation. The width of the 'annular ring' shall therefore be understood as the distance from the rim of the solder hole to the edge of the pad in a specified direction. For the dimension 1.4 × 2.0 mm (0.055 × 0.079 in) we find that the width of the 'annular ring' measured transversely and longitudinally is 0.3 mm (0.012 in) and 0.6 mm (0.024 in), respectively, which is consistent with Design Rule Number 6.

These solder pads have advantages when the pattern is screen printed. The screen stretches slightly in the direction of travel of the squeegee, cf. Section 3.2.1. If the board can be located so that the printing takes place in the longitudinal direction of the pads, any displacement will be unimportant due to the larger width of the 'annular ring' in this direction.

As the width of the pads is 1.4 mm (0.055 in) it is possible to run a 0.3-mm (0.012-in) wide conductor between a pair of pads having a centre-to-centre distance of 0.1 in (2.5 mm) and achieve an accepted (measured) minimum spacing of 0.35 mm (0.014 in). Frequently, it is necessary to route a larger number of conductors longitudinally between the two rows of pads which make up the IC cluster. The dimensions chosen allow for six conductors, each having a width of 0.30 mm (0.012 in) and a nominal spacing of 0.55 mm (0.0216 in) in the 1:1 scale. This complies very well with Design Rule Number 2. The conditions appear in Fig. 6.16, which

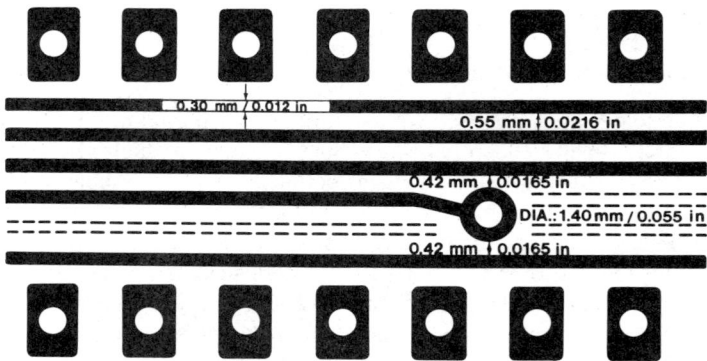

Fig. 6.16 Longitudinal routing of conductors between IC pads.

also shows how, by cancelling one of the conductors, it is possible to provide sufficient space for a via hole, the pad having a diameter of 1.4 mm (0.055 in) and a minimum spacing of 0.42 mm (0.0165 in). This is in accordance with Design Rule Number 1.

The rectangular pad described above is designed by the author and used for dual-in-line IC clusters. These are made to customers' specifications and do not appear in the catalogues of drafting aids. The standard line of other IC clusters do not fulfil the requirements which, in the opinion of the author, must be met in order to gain the greatest ease of manufacture and flexibility in the design stage. For this reason it is interesting to examine the standard types and indicate their weaknesses. A selection of available types (described below) is set out in Table 6.3 showing the essential dimensions on the 1:1 scale.

Type 1: This has already been dealt with and needs no further comments.
Type 2: The pad is so narrow that it is nearly impossible to find space for a solder hole of 0.8 mm (0.031 in).
Type 3: The pad is a little wider but a solder hole of 0.8 mm (0.031 in) allows for a width of the 'annular ring' of only 0.24 mm (0.0094 in). Based on the

previous considerations, as well as practical experience, we must characterize this type as being a little too narrow to ensure a reliable and economical manufacturing process.

Type 4: Here the pad is so wide that the spacing between the pads becomes so little that a 0.30-mm (0.012-in) wide conductor cannot be run between two pads. This is a sufficient reason for characterizing the type as inapplicable.

Type 5: The pad is far wider than the pad type 4, and is, therefore, inapplicable. It has, however, an area of approximately 5 mm^2 (0.0078 sq. in) which permits its use in non-plated through boards, if the designer accepts that a conductor cannot be run between two pads.

Type 6: The pad is circular and gives a spacing between the pads which makes it possible to run a conductor between two pads, if the requirements to the width of the conductor and the minimum spacing can be waived. It should be noted that if a solder mask is to be applied to the board, this type of pad will make it impossible for the solder mask to cover a conductor running between two pads, which, in turn, makes the solder mask rather pointless. This will be dealt with in Section 6.9.

Type 7:
Type 8: } The same comments as stated for type 4 are valid.

Table 6.3 Dimensions of solder pads for ICs.

Dimension (mm)	Type*							
	1	2	3	4	5	6	7	8
Width of solder pad or diameter	1.40	1.09	1.27	1.78	2.03	1.52†	1.90†	2.16†
Length of solder pad	2.00	3.17	3.17	3.17	2.54			
Spacing between two pads	1.14	1.45	1.27	0.76	0.50	1.02	0.63	0.38
Spacing between two rows	5.62	3.44	4.44	4.44	5.08	6.10	5.71	5.46

* A full description is given in the text.
† These values are diameters.

Finally, attention is drawn to some types of IC clusters which have short pieces of conductor placed in all the spacings between the pads. From the manufacturer's point of view these critical spots contribute nothing to the performance of the board which unnecessarily loads the manufacturer's inspection and touch-up departments. It has been reported that such an IC-cluster design is responsible for many faults which in turn are reflected in higher prices. The application of such symbols is therefore rejected although they might appear to offer some convenience for the draughtsman.

6.5.6 Other Pad Shapes

A large number of solder pads in various shapes and sizes is in existence and some examples are shown in Fig. 6.17. It pays, however, to rationalize and use as limited a selection as possible.

174

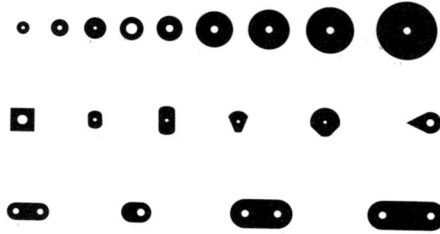

Fig. 6.17 Examples of solder pads.

Teardrop pads have the advantage over circular pads of a smoother transition between the conductor and the pad. This should lead to an easier plating and etching but if modern manufacturing equipment is used, this should be of no real importance. The use of teardrop pads unfortunately complicates the taping of the artwork since the direction of the conductors has to be fairly accurately determined before the pads can be placed. The most practical procedure is to place all the pads and then tape all the conductors, making it difficult to ensure that all the teardrop pads point in the correct direction. An even greater problem arises when the solder side and the component side are derived from a common pad master by contact printing. In this case, the application of teardrop pads will be in conflict with the ordinary XY-taping practice whereby conductors are taped in the X-axis direction on one side of the board, and in the Y-axis direction on the other side. The author sees a desirable increase of the area of non-supported hole pads (non-plated through holes) as the only justification of teardrop pads. It is possible to get 'half teardrops' which can be used with all sizes of solder pads and conductors should this be felt necessary. The application is shown in Fig. 6.18.

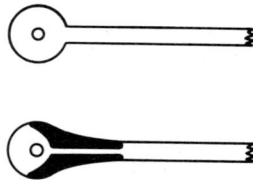

Fig. 6.18 Application of 'half teardrops'.

Certain standards for assembly require that the component leads are clinched as illustrated in Fig. 6.19. In this case oval pads with either a central or an offset hole can be used. As in the case of the teardrop pads, the draughtsman has a problem regarding artwork of the two sides as produced by contact printing from a common pad master. A solution could, in both cases, be to apply circular pads on the common pad master and then to superpose teardrop pads or oval pads on the contact-printed artwork for solder side of the board. For this purpose, transparent, red symbols, preferably with a centre hole a little larger than that of the circular pads, should be used. The larger centre hole provides for the use of the original centre hole when digitizing the board, so

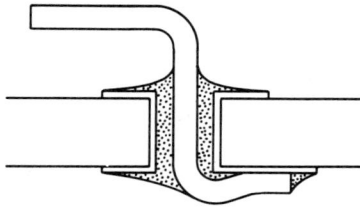

Fig. 6.19 Clinched component lead.

that the registration between the drilled holes and the pads of the component side is not compromised. The transparency makes it far easier to apply the pads and to check the concentricity of the centre holes.

6.5.7 Influence of Pad Size on Solderability

Printed circuit boards which have been soldered automatically by means of a soldering machine often show a great difference in the extent to which the solder holes are filled, although they may be the same type as in the case of dual-in-line ICs. These variations in filling are shown in Fig. 6.20 and are chiefly due to differences in the sizes

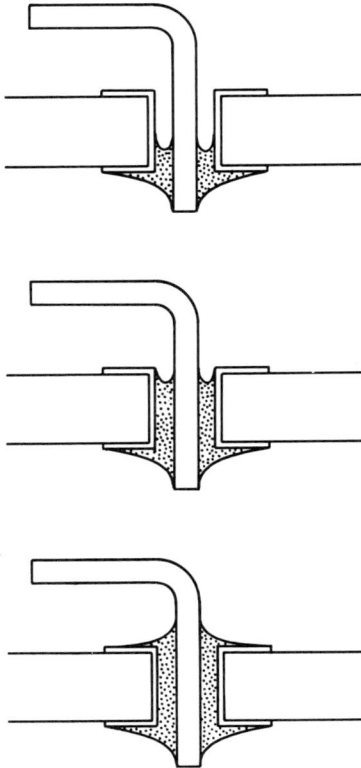

Fig. 6.20 Filling the solder hole.

of the pads on the two sides of the board resulting in varying degrees of heat transfer away from the pads on the component side.

During the automatic soldering, the heat from the solder has to travel up through the solder hole, due partly to the fluid solder which is sucked up through the hole, and partly to conduction through the copper plating of the hole and through the component lead. If the pad on the component side is large, or wide conductors are connected to the pad, these act as heat sinks and cause the temperature up through the hole to drop. When this happens the solder can freeze before the hole is completely filled.

In the purchase specification (Section 4.5.3) a maximum thickness of the tin/lead plating was specified. The reason was to avoid incomplete filling of the hole. The explanation was given under the comments.

A complete filling is desirable for various reasons: Firstly, it ensures a higher reliability of the finished board. As the plated copper wall of a plated through hole has a lower thermal coefficient of expansion than the base material and is somewhat brittle, there is some risk that cracking of the hole wall may take place during soldering. If the solder does not fill the solder hole completely as a plug, the cracking may manifest itself as an intermittent interruption of the connection between the solder and component sides, a failure which is rather difficult to locate. Secondly, the inspection of solder joints on the component side is facilitated when the solder is sucked right up through the hole.

It is worthy of note that some PCB manufacturers test the solderability of plated through holes with any conductors leading from the pads removed beforehand.

More complete and consistent filling of solder holes can be achieved by preparing the artwork in accordance with the following design rule.

Design Rule Number 8

Use at least the same size of pads on the solder side as on the component side, and take care to reduce the heat transfer away from the pads on the component side.

Figure 6.21 shows various typical examples of the utilization of Design Rule Number 8. The component side is shown at the top of the sketches.

(a) This shows how the heat sink action is reduced on the component side.
(b) The component side is unchanged, whereas the area of the solder pad on the solder side is enlarged, either by applying a large pad (cf. the remarks at the end of Section 6.5.6 concerning the importance of a larger centre hole than in the original pad), or by applying dummy 'wings'.
(c) The heat sink action can also be reduced by moving conductors away from the component side.
(d) A typical zero-point for the circuitry of the component side is shown in this sketch. Improved soldering conditions are achieved by applying a ring around the solder pad and connecting the pad by means of one or two straps to the ring, which forms the real zero-point. The ring is simply made by applying a piece of self-adhesive, transparent, red litho-tape, cutting away the excess parts of the litho-tape by means of bow compasses fitted with a knife.

(e) Here is shown a via hole which connects a ground plane to a ground conductor on the solder side. The ground plane is normally taped on the artwork by means of a self-adhesive, transparent, red film, in which a cut-out for the pad is made by means of bow compasses fitted with a cutting knife.

Note: In the cases where the ground plane is situated on the solder side of the board, it is also advisable to make a similar cut-out around the pad. This will facilitate repair work tremendously as the solder joint is far easier to heat.

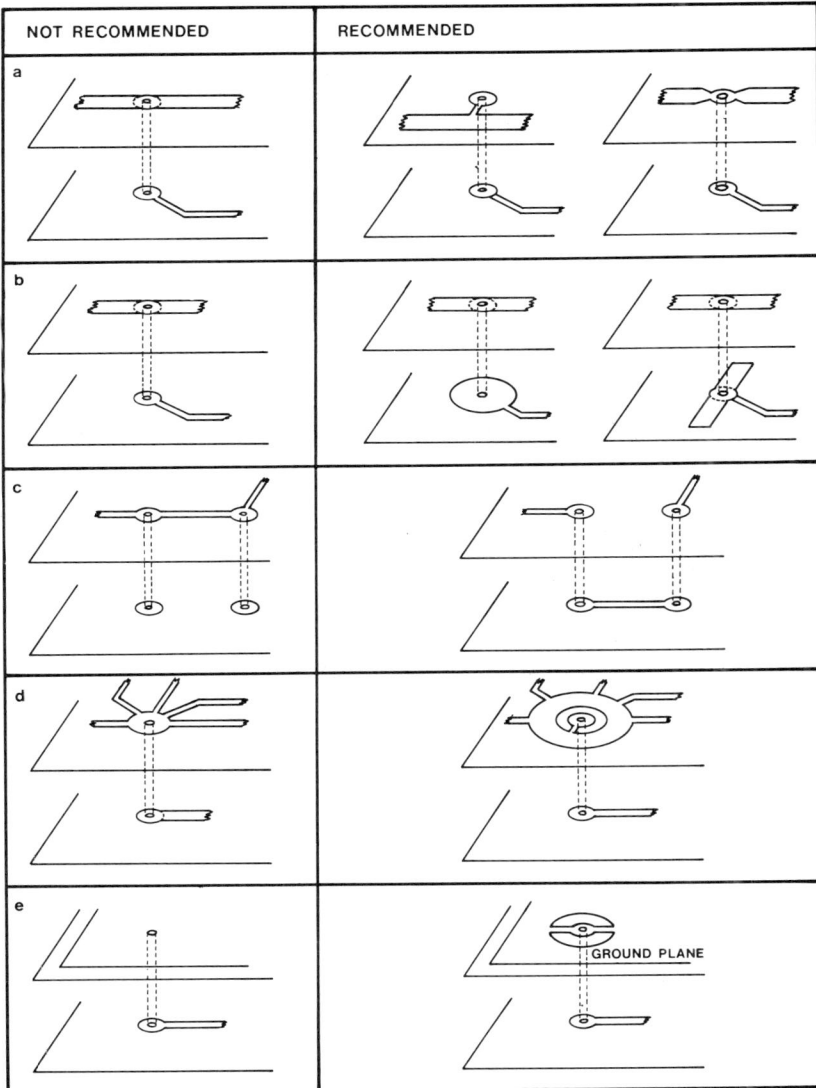

Fig. 6.21 Influence of solder pads on solderability. See text.

Reasonable attention to these guidelines will, in most cases, ensure satisfactory soldering and only in extreme cases will a change of design procedure be necessary. The author has found that an adequately soldered board results if the soldering machine operates correctly, the solderability of the board is not deficient and, in the more difficult cases, departures from the guidelines are not permitted. These cases include a wide voltage supply conductor or bus (sketches a and b, Fig. 6.21), a junction having many outgoing conductors (case d), and particularly, a solder hole in a ground plane (case e).

The appearance of the solder joint on the solder side depends on the shape of the pad and the way in which the conductors are connected to the pad. A fairly symmetrical configuration will always offer the best condition for the creation of a good and sound solder joint. In Fig. 6.22 some examples are shown of good and not-so-good types of pads. The following comments apply to these.

(a) An oblong pad gives an unsymmetrical solder joint.
(b) The molten solder flows away from the actual joint owing to the very wide conductor, resulting in a very unsymmetrical joint.
(c) Here, also, the solder joint has become very unsymmetrical.
(d) The solder joint in the middle will be all right but the solder joints at the extreme ends become unsymmetrical.
(e) Here, also, the solder joint in the middle is satisfactory, but as the solder flows away from the corners, the outer solder joints are inferior.
(f) In this case the solder flows towards the large hole.

6.5.8 Conductor Width

Existing official standards prescribe minimum widths; these must be judged very conservative in relation to the requirements of modern printed circuit boards with very complex circuitry and with so high a packaging density that they can only be designed if conductors of considerably smaller width are accepted. We shall therefore choose the width of the conductors without reference to these specifications. We shall not, however, permit a width less than 0.25 mm (0.01 in), as this is considered the smallest width which can be produced without undue difficulty by a competent manufacturer but even so it is advisable to secure the manufacturer's agreement.

6.5.8a Signal-carrying Conductors

Normally, the currents are so small that the width of the conductors is determined by practical considerations only regarding the manufacturing processes. As has been previously mentioned, the nominal width cannot be made arbitrarily small, i.e. less than 0.25 mm (0.01 in), as this would cause some production problems which, in turn, would be reflected in a higher price.

For the same reasons as given in Section 6.5.4b concerning the diameter of the solder pad, it is desirable to use as few conductor widths as possible. Standard widths should, therefore, be introduced as stated in the design rule given below. It should be

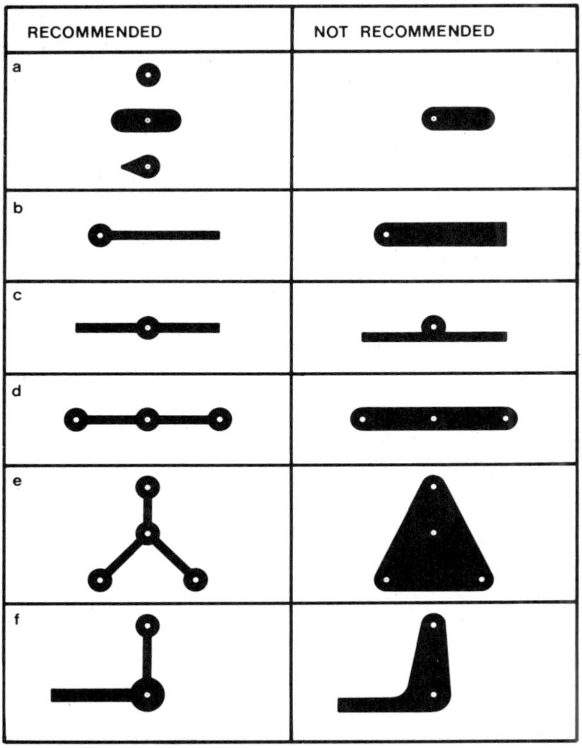

RECOMMENDED	NOT RECOMMENDED
a	
b	
c	
d	
e	
f	

Fig. 6.22 Design of solder pads. See text for comments.

noted that the design rule specifies only a range of narrow conductor widths to be used for signal-carrying conductors in complex boards.

Design Rule Number 9

	4:1	1:1
Preferred conductor width:	1.57 mm (0.062 in)	0.39 mm (0.015 in)
If Design Rule Number 2 cannot be fulfilled:	1.27 mm (0.050 in)	0.32 mm (0.013 in)
As an exception:	1.17 mm (0.046 in)	0.29 mm (0.011 in)
If distance to adjacent conductors is large:	2.03 mm (0.080 in)	0.51 mm (0.020 in)

When assessing the minimum width of the conductors of the finished board, a maximum shrinkage of approximately 0.13 mm owing to the manufacturing processes (cf. Section 6.5.1) should be allowed for. The effective width of the narrowest of the above conductors therefore becomes $0.29 - 0.13 = 0.16$ mm

(0.0063 in). Obviously, this is a very narrow conductor, and for the sake of a reasonable adhesion to the board, no narrower conductors should ever be specified.

6.5.8b Current-carrying Conductors

The current in a current-carrying conductor is usually of quite another order of magnitude than the current in a signal-carrying conductor. For this reason, the current-carrying conductor must be designed with due attention to the temperature rise caused by the power dissipated in the conductor, and to the voltage drop caused by the resistance of the conductor.

6.5.8b1 Temperature Rise

The maximum permissible temperature rise of a conductor is the difference between the maximum operating temperature of the base material and the temperature inside the equipment. As has been mentioned in Section 2.5.4, epoxy glass, type FR4, can withstand temperatures up to 130°C. Owing to the weakening and discolouration caused by this high temperature, the maximum operating temperature for design purposes is limited to 105°C. If the maximum ambient temperature is 35°C and the internal temperature rise in the equipment is 10°C, the maximum allowable temperature rise of the conductor is 60°C. It should, however, be considered that solder pads connected to a hot conductor will attain approximately the same temperature rise by conduction. Heat-sensitive components soldered to such pads can be affected and it is impossible for small heat-dissipating components to use the pads and the conductor as a heat sink. In addition, the component side of the board receives local heating which, in a corresponding manner, could affect the components. The permissible temperature rise, is, therefore, limited to 20°C and the width of the conductor is increased as necessary to achieve this. On the other hand, there are practical limits for conductor widths. For instance, the available space on the board may be limited and large copper areas can cause production difficulties in automatic soldering. We therefore limit the conductor width to 10–12 mm (0.39–0.47 in) and when necessary, a thicker copper foil can be used, for example 70 μm (2 oz) in order to limit the width of the conductors. Some PCB manufacturers dislike using so thick a copper foil owing to the greater undercut caused by etching, so they prefer to start with a thinner foil, for example 35 μm (1 oz) and deposit the necessary copper by plating. Normally, the thickness of the conductors is not particularly critical and, if a greater thickness than normal is required to give greater current-carrying capacity, the copper thickness should be clearly stated on the engineering drawing.

For ready assessment of the temperature rise in the conductors Fig. 6.23 gives a family of curves showing temperature rise as a function of the current and the conductor width and thickness. The curves should be regarded as approximate only as some more or less uncertain factors are involved: undercut, copper thickness due to plating tolerances, cooling conditions and heating from adjacent conductors or from heat-dissipating components.

The temperature curves apply to uncovered copper surfaces. In the case of plated

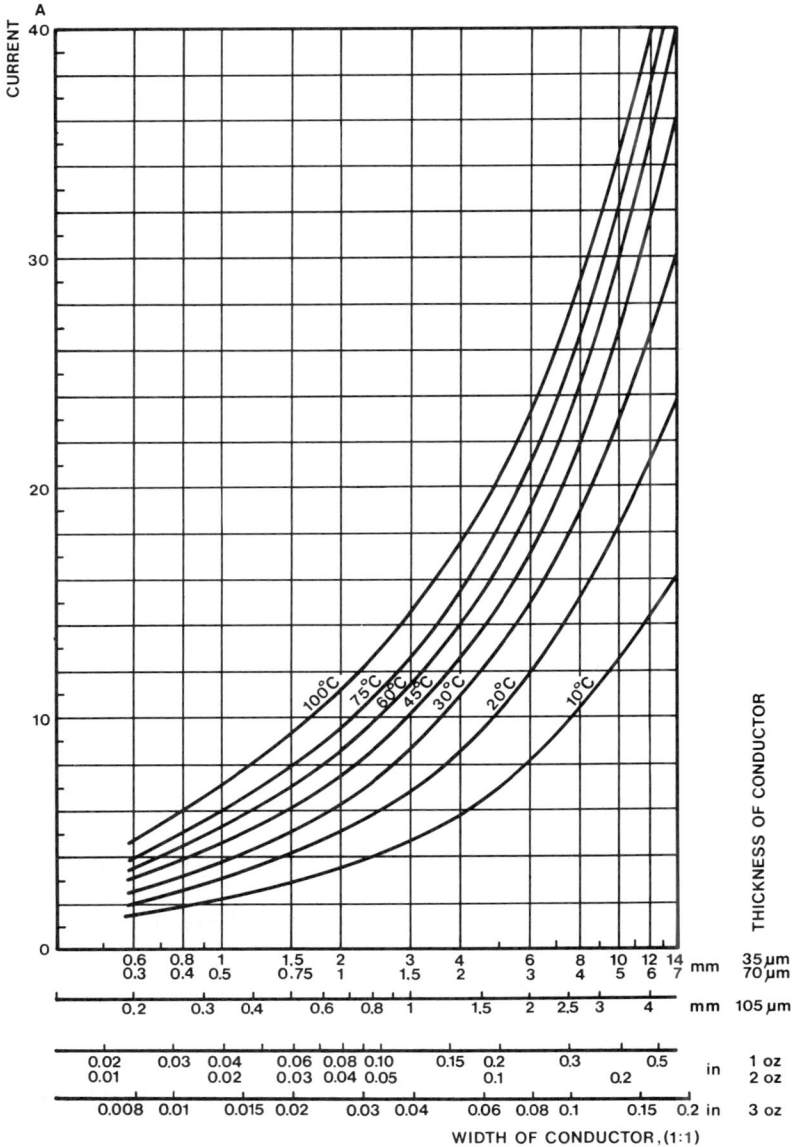

Fig. 6.23 Temperature rise in conductor as a function of the conductor cross-section (width × total thickness of copper) and current. (MIL-STD-275C)

boards, the copper thickness is, of course, to be understood as the total copper thickness which can vary between quite wide limits due to the plating tolerance. As this tolerance frequently is −0%, + 100% it is recommended to assess it fairly low, for example +50%. If the copper surface is covered, it is necessary to compensate for a change in the cooling conditions. In some standard specifications, for example, MIL-STD-275C, a certain de-rating is indicated. The PCB designer's starting point is normally the current in the conductor, so it seems more expedient to enter the curves with a current adjusted by the applicable de-rating factor, i.e. the current is multiplied by a factor of $100/100 - a$, where a is the de-rating percentage.

For a tin/lead surface or after automatic soldering of the board, the de-rating is stated as 30%, which gives a current factor of 1.42.

If the board is provided with an insulating layer or a solder mask, the de-rating is stated as 15%, which gives a current factor of 1.18.

For a copper thickness of 105 μm (3 oz) or above, a further de-rating of 15% is incorporated, giving an additional current factor of 1.18.

If the thickness of the base material is 0.8 mm (0.031 in) or less, the de-rating is 15%, giving a current factor of 1.18.

An example below shows how to use the curves:

The board is based on 35 μm (1 oz) copper foil and plated with 25 μm (0.001 in) copper plus 10 μm (0.0004 in) tin/lead. The thickness of the board is 1.6 mm (0.063 in), and the board is soldered automatically.

The total copper thickness is 35 μm + (25 μm + 50%) $\simeq$ 70 μm (0.0027 in). Note that if the copper thickness exceeds 105 μm (3 oz or 0.0041 in) an additional current factor is to be used. Therefore, the maximum plating tolerance of $+ 100\%$ should be used when checking that the copper thickness is below the 105 μm (3 oz or 0.041 in) limit. In our case, the maximum copper thickness is 35 μm + (25 μm + 100%) = 85 μm (0.0033 in) < 105 μm (0.0041 in or 3 oz). The current factor is therefore 1.42 and not 1.42 $\times$ 1.18 = 1.68.

In order to find a conductor width which gives a temperature rise of maximum 20°C at a current of 10 A, the designer enters the curves of Fig. 6.23 with a current of 10 $\times$ 1.42 $\simeq$ 14 A. The conductor width corresponding to a copper thickness of 70 μm (0.027 in) is found to be 3.8 mm (0.15 in).

If a solder mask is applied, the current factor is determined as 1.42 $\times$ 1.18 = 1.68. Now the conductor width is 4.5 mm (0.177 in).

If the conductor width is only a few times larger than the possible shrinkage of 0.13 mm (0.005 in), it is, of course, necessary to compensate for the reduced width by adding 0.13 mm (0.005 in) to the width found.

It is seen from the curves that if a temperature rise of 20°C is obtained, the temperature rise will be around 100°C if the current is doubled. The ratio varies between 1.9 and 2.3, but it gives an impression of the margin.

If there is a risk that a component, for example, an electrolytic capacitor, can fail so a conductor for a period of time carries a high short-circuit current, the conductor may burn out. Critical spots should be considered with due regard to possible short-circuit current, and the width of the involved conductors be designed for a maximum temperature rise of approximately 100°C. Normally, the fuse of the power supply will blow after so short a time that the conductors can stand the current without being damaged.

6.5.8b2 Conductor Resistance

In Section 5.5.4 it was mentioned that common ground conductors can be very troublesome because of the false signals introduced by the various voltage drops along the conductor in question. Similar effects are, of course, found in the voltage supply system. It is therefore very important that the ohmic resistances of current-carrying conductors are kept so low, that detrimental couplings between circuits connected to the power supply are minimized. The circuit designer will normally be able to specify the maximum acceptable resistance, and the PCB designer's job is to determine the conductor width. The resistance can be calculated using the formula:

$$R = 0.0172 \frac{L}{A}$$

where the resistance R is in milliohms (mΩ), the length L in millimetres, and the cross-sectional area A in square millimetres.

It is, however, easier to use the curves in Fig. 6.24. These display the width of the conductor as a function of the resistance of a 100-mm (4-in) length for various copper thicknesses. The designer knows the limiting value of the resistance and the physical length of the conductor from which he readily derives the permissible resistance per 100 mm (4 in). Entering the curve for the appropriate thickness with this value, the minimum width of conductor can be extracted. If the conductor width is only some few times larger than the maximum possible shrinkage of 0.13 mm (0.005 in) it is, of course, necessary to compensate for the reduced width by adding 0.13 mm (0.005 in) to the value found. The PCB designer must enter the curves with the total thickness of copper, i.e. the thickness of the copper foil and the plated copper, the latter being taken as 25 μm + 50% $\simeq$ 35 μm (0.0014 in). The curves are plotted for intervals of 35 μm (0.0014 in) $\simeq$ 1 oz corresponding to the intervals between the standard foil thicknesses available to the industry. The thickness of tin/lead plating is ignored as the conductivity of tin/lead is considerably lower than that of copper.

Since we are mainly concerned with current-carrying conductors it is assumed that they will show a slight temperature rise of 20°C. With an ambient temperature of 25°C and a temperature rise within the equipment of 10°C, the current-carrying conductors will reach a temperature of 55°C. The curves are therefore based on a copper temperature of 55°C. Temperature variations of ±30°C, however, introduce resistance changes of only ±12%.

From the above we derive the following design rule:

Design Rule Number 10

The width of current-carrying conductors shall first be determined for a maximum temperature rise of 20°C.

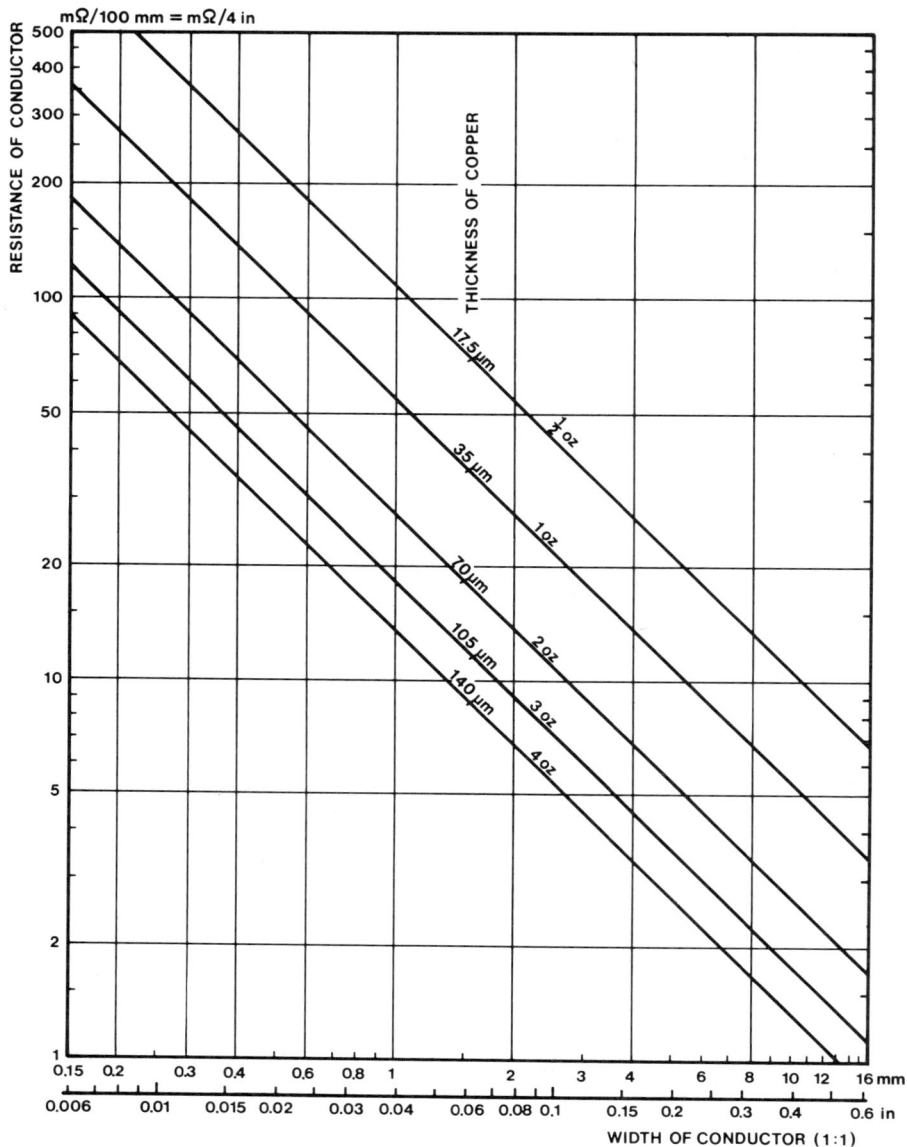

Fig. 6.24 Resistance per 100 mm (4 in) conductor length as a function of the conductor width and thickness.

If a maximum resistance is required, the width is determined in accordance with this requirement.

The larger of these shall be taken after correction for possible shrinkage in manufacture as the conductor width.

6.6 Pattern Details

There are many details to take into consideration when designing a circuit pattern which affect the ease of manufacture. There are narrow gaps through which conductors have to be taped while maintaining minimum spacing, and there are other specific details in connection with the conductor routing about which it is an advantage to know. These, and in addition, the 'equalization' or balancing of the pattern, special measures in connection with large ground and shield areas, and safety zones, will be dealt with in this section.

6.6.1 Conductors Through a Narrow Gap

In order to facilitate the manufacture, the conductors must be laid down as simply as possible, which means that they must be routed as directly as possible while avoiding narrow gaps between pads if another routing is possible. In order to utilize a narrow gap fully, the conductors shall be laid perpendicularly with respect to the centre-to-centre line of the pair of pads as shown in Fig. 6.25.

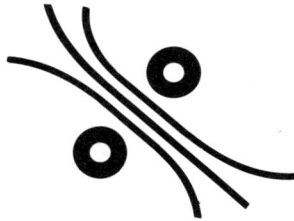

Fig. 6.25 Routing of conductors through a gap.

When two and more conductors are to be run through a narrow gap, it can be very difficult for the PCB designer to survey the conductor-to-conductor spacings as well as the conductor-to-pad spacings. It should be possible to tape the artwork without having to correct the layout sketch which should, therefore, be made in so realistic a manner that the prescribed minimum spacings are obtainable in taping. Figure 6.26 serves to facilitate the PCB designer's job in this respect as it indicates the mutual dependence of the design parameters given below:

(a) the centre-to-centre distance of a pair of pads, expressed as the number of 0.1 in modules in the X- and Y-directions;
(b) the number of conductors it is required to run through the gap;
(c) the width of the conductors;
(d) the maximum size of the two pads which form the gap, expressed by means of the combinations of pad sizes possible.

The following comments relate to the diagram which is designed for a nominal spacing of 0.40 mm (0.016 in) in the 1:1 scale. The gap size is defined as the centre-to-centre distance of the pads forming the gap, expressed by the number of 0.1 in modules in the vertical and horizontal directions. If the layout has been prepared in

NUMBER OF CONDUCTORS	2				3				4				5			
TAPE SIZE 4:1 in	0.046	0.050	0.062	0.080	0.046	0.050	0.062	0.080	0.046	0.050	0.062	0.080	0.046	0.050	0.062	0.080
mm	1.17	1.27	1.57	2.03	1.17	1.27	1.57	2.03	1.17	1.27	1.57	2.03	1.17	1.27	1.57	2.03
1 × 0	—	—	—	—	—	—	—	—	—	—	—	—	—	—	—	—
1 × ½	—	—	—	—	—	—	—	—	—	—	—	—	—	—	—	—
1 × 1	B-C	B-C	C-C	—	—	—	—	—	—	—	—	—	—	—	—	—
1½ × 0	B-B / A-C	B-B / A-C	B-C	C-C	C-C	—	—	—	—	—	—	—	—	—	—	—
1½ × ½	A-B	B-B / A-C	B-B / A-C	B-C	C-C	C-C	—	—	—	—	—	—	—	—	—	—
1½ × 1	A-A	A-A	A-A	A-B	B-B / A-C	B-B / A-C	B-C	C-C	C-C	C-C	—	—	—	—	—	—
1½ × 1½				A-A	A-A	A-A	A-A	A-B	A-B	B-B / A-C	B-C	C-C	—	—	—	—
2 × 0							A-B	B-B / A-C	B-B / A-C	B-C	C-C	—	—	—	—	—
2 × ½							A-B	B-B / A-C	B-B / A-C	B-B / A-C	B-C	—	C-C	—	—	—
2 × 1					A-A	A-A	A-A	A-B	B-B / A-C	B-C	B-C	B-C	—	—		
2 × 1½										A-A	A-A	A-B	A-A	A-B	B-B / A-C	C-C
2 × 2												A-A	A-A	A-A	A-A	A-B
2½ × 0													A-B	A-B	B-B / A-C	C-C
2½ × ½													A-A	A-A	B-B / A-C	C-C
2½ × 1															A-A	B-B / A-C
2½ × 1½																A-A
TAPE SIZE 1:1 in	0.0115	0.0125	0.0155	0.0200	0.0115	0.0125	0.0155	0.0200	0.0115	0.0125	0.0155	0.0200	0.0115	0.0125	0.0155	0.0200
mm	0.29	0.32	0.39	0.51	0.29	0.32	0.39	0.51	0.29	0.32	0.39	0.51	0.29	0.32	0.39	0.51

Fig. 6.26 Connection between number of conductors, width of tape, centre-to-centre distance between solder pads, and pad sizes. Centre-to-centre distance expressed by number of modules (0.1 in = 2.54 mm) in the X-and Y-axis directions.

the 4:1 scale, one module is 0.4 in (10.16 mm). In Fig. 6.13, three sizes of pads were defined, namely the A-, B- and C-pads. The space between pads in which conductors can be laid depends on which pads are used and enables the designer to trade-off pad diameter against conductor spacing in laying out the pattern.

The possible pad combinations are arranged in a column below, which gives more and more space in the direction of the arrow.

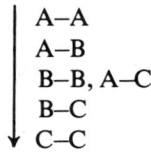

A–A
A–B
B–B, A–C
B–C
C–C

Figure 6.26 determines the largest size of pads which can be used and still permit the conductors to pass through the available space at the specified clearance. Following this, it is at the discretion of the designer whether he uses these or any smaller pads that are available.

An example shows the use of the diagram. The layout sketch has two pads located as shown in Fig. 6.27, i.e. with a centre-to-centre distance corresponding to 2 × 1 modules. The desired size of the pads is A and A, and 4 conductors are to be run through the gap. The width of the conductors is 0.32 mm (0.013 in) for which reason a tape width of 1.27 mm (0.05 in) is used.

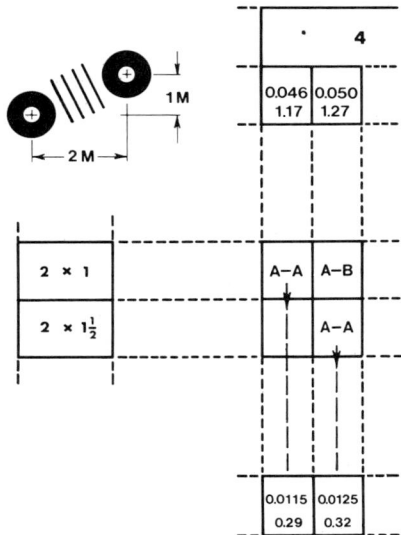

Fig. 6.27 Application of Fig 6.26.

It is easily seen from the diagram that this combination cannot be realized, and that A–B is the largest pad pair which can be used. The diagram gives three alternative suggestions for a solution:

1. One of the pads is changed from A to B.
2. A change in the conductor width to 0.29 mm (0.011 in) in the gap.
3. One of the pads is displaced so that centre-to-centre distance becomes 2 × ½ modules.

If solution 2 is chosen, a smooth transition from the wide tape to the narrow tape should be made as shown in Fig. 6.28, by tapering the wide tape.

Fig. 6.28 Transition between two different tape widths.

6.6.2 Conductor Routing

Conductors meeting or bent at an angle (less than 60°) should be avoided. This acute angle forms a sort of pocket which tends to retain some of the plating resist resulting in incomplete etching, and it provides a point where the conductor is most easily lifted from the board by handling in assembly or repair or as a result of thermal stressing of the board. Figure 6.29 shows some good and some bad ways of routing the conductors. The first examples are self-explanatory but there are good reasons for commenting on case (f). The suggested T-configuration must clearly restrict the routing of conductors in both directions of the board and should be avoided in the layout stage. It can be fairly difficult to survey alternative possibilities if one conductor should be transferred to the other side of the board. In the taping stage it is always easy to go from the V-configuration to the T-configuration. The examples in (g) and (h) show both how acute angles can be avoided, and how the total length of the conductors can be reduced.

If a conductor carries high voltages there is a risk of a voltage flash-over to adjacent parts of the pattern where pointed projections or acute angles occur. Two examples are shown in Fig. 6.30. In such cases, the draughtsman should remove all sharp points and round all corners. Narrow tapes of the crêpe type can reasonably easily be laid in fairly sharp curves, but the wider the tape, the more difficult it is to bend the tape on an edge. A dangerous point is found where a conductor runs through a narrow gap and has to make a bend just beyond the gap as shown in Fig. 6.31. Even when the tape is laid very carefully, i.e. without stretching the outer edge very much, some tension must be created in the tape. In the course of time, a slight displacement of the tape may occur in the narrow gap, whereby the minimum spacing is compromised. This effect can be remedied in several ways as shown in Fig. 6.32.

1. The draughtsman can use bends, the so-called elbows, of which an adequate selection exists. Angles of 90°, 60°, 45° and 30° are available.
2. The tape is laid at a suitable angle and the point is cut off by means of a short piece of tape laid across.
3. The tape is cut three-quarters through and bent, the flaps being laid over the already taped pieces.

Finally, it must be stressed again that manufacturing problems can arise when conductors run closely in parallel and, in these cases, the designer and the draughtsman must always try to space the conductors as widely as possible.

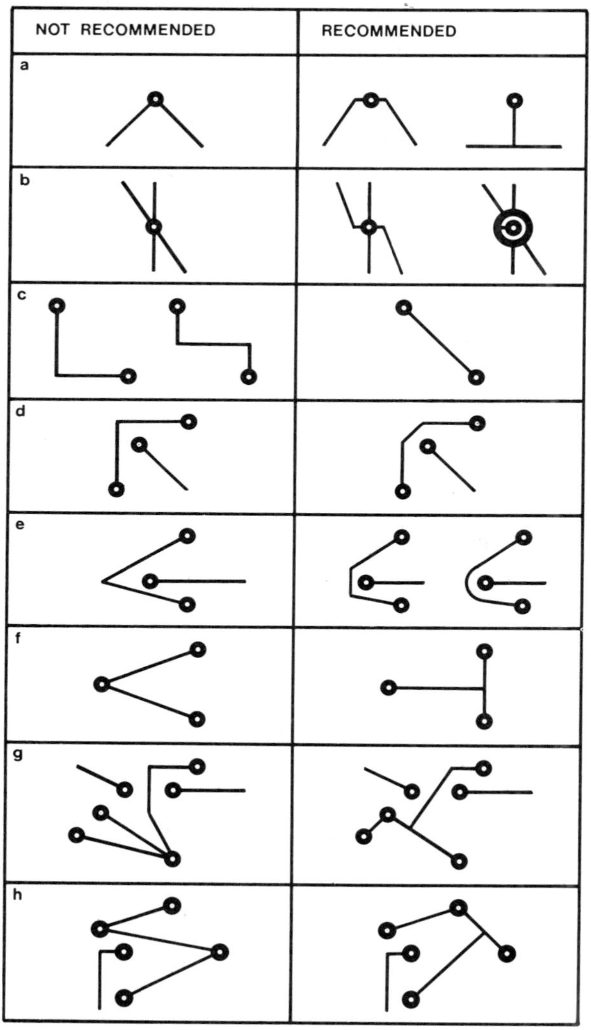

Fig. 6.29 Routing of conductors.

6.6.3 Equalization of Pattern Distribution

Equalization, or balancing of the pattern, which can considerably ease the manufacture, is not so often used as might be thought desirable although it serves two important purposes:

1. to protect isolated parts of the pattern from over-plating;
2. to create a fair balance between the pattern areas of the solder and component sides.

190

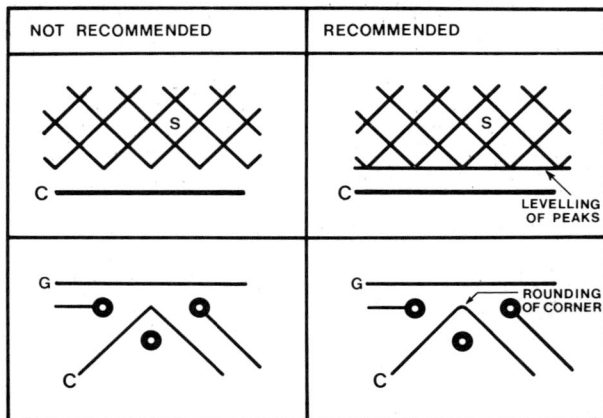

Fig. 6.30 Voltage flash-over from pointed projections. G, Ground conductor; S, shield or ground area; C, conductor carrying high voltage.

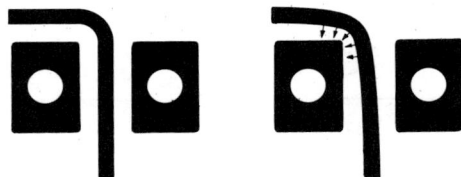

Fig. 6.31 Conductor 'creep'.

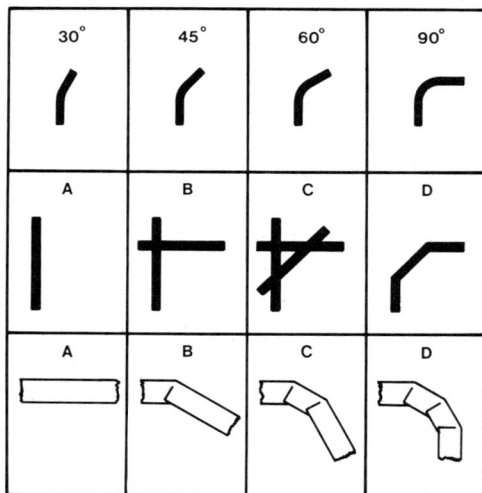

Fig. 6.32 Measures against conductor 'creep'.
Top: Various types of elbows.
Middle: Truncating the angle.
Bottom: 3/4-cutting and overlaying of tape.

The former, which has been mentioned in Section 4.5.3, ensures that the electrolytic deposition automatically becomes evenly distributed all over the area of the board. Better control of the process and thus higher quality is achieved. The latter ensures that the electrolytic deposition is of equal thickness on the two sides of the board, facilitating the adjustment of the plating process.

In addition, balancing is found to enable closer tolerances to be maintained on the finished hole diameters of plated through boards. The equalization or balancing is performed by introducing suitable dummy areas (non-functional areas) so that the density of the pattern is the same all over the board. The balancing cannot be expected to produce complete uniformity and its application is largely a matter for the judgement of the draughtsman. Figure 6.33 shows a typical example of a balanced board.

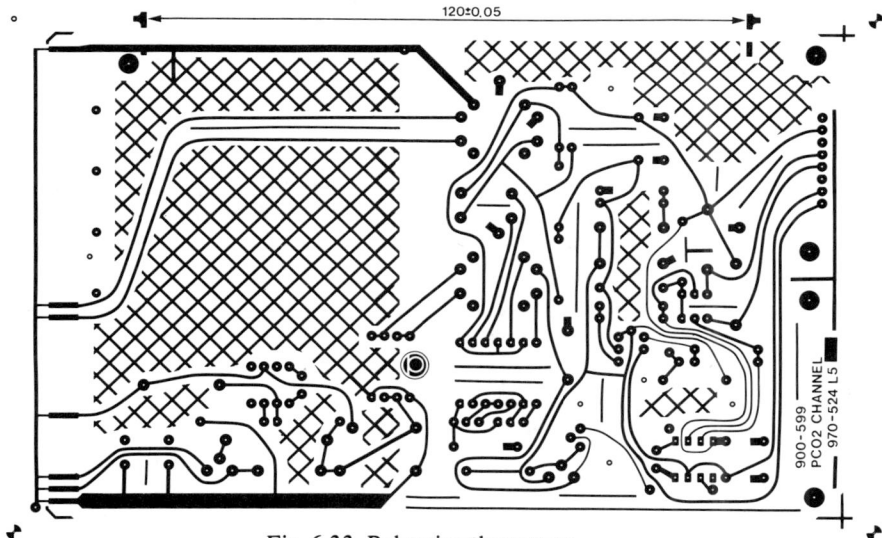

Fig. 6.33 Balancing the pattern.

Larger open areas are balanced by means of a cross-hatched, self-adhesive film. Smaller, but still open areas, are balanced by means of small pieces of tape. It is important that such non-functional areas visually distinguish themselves from the functional parts of the pattern in order to avoid unnecessary touch-up work on parts of the pattern which are of no consequence. Figure 6.34 shows two different types of balancing networks. The first type is unbroken and involves a risk of capacitive couplings; the other type is split into insulated bits and does not suffer from this disadvantage. When the board is passed through the soldering machine, the unbroken network is more or less liable to close the meshes unless a solder mask is applied.

6.6.4 Large Ground and Shield Areas

In certain boards, a large ground or shield area is needed for electrical reasons. Such areas, however, cause manufacturing difficulties: the board is liable to warp owing to

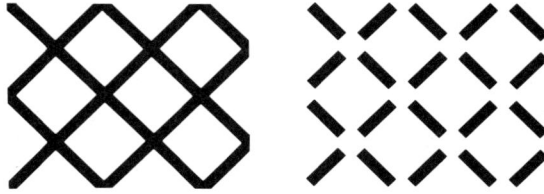

Fig. 6.34 Balancing networks.

the difference in the mechanical tension of the two sides of the board, and infrared reflowing is rendered difficult by the unequal temperature distribution that follows the unequal area distribution. Furthermore, problems common to both PCB manufacturer and the customer might occur due to the large areas: after reflowing or automatic soldering, the boards sometimes show signs of false dewetting, which is difficult to distinguish from the real, and harmful, dewetting, and therefore makes inspection inconclusive. Finally, there is a risk that excessive heat may cause the copper foil to blister and work loose.

The problems described above can be remedied by breaking up the large areas so that no unbroken areas larger than 100–250 mm (0.15–0.40 sq. in) remain. The ground or shield area is often produced by applying a red transparent self-adhesive film to the artwork and cutting the contour. In order to break the area, circular holes can be cut by means of bow compasses fitted with a cutting knife, or rectangular holes can be cut by means of a scalpel. Some examples are shown in Fig. 6.35. Such cutting operations require some manual work, and instead, a cross-hatched area can be produced conveniently by using a balancing network of the unbroken type shown in Fig. 6.34, possibly provided with a taped edge for the contour. A closer mesh can, if desired, be obtained by applying two networks displaced a little with respect to each other.

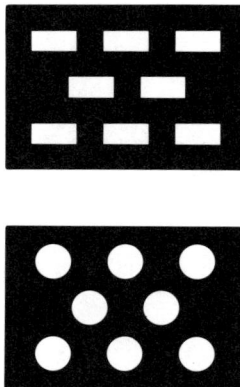

Fig. 6.35 Breaking up large areas.

6.6.5 Safety Zones

Most printed circuit boards are provided with a number of holes used for the mounting of larger components and mechanical sub-assemblies, or for fastening the board to the chassis. When taping the artwork, the draughtsman should take care that there is sufficient distance from the conductors and the pads to the outer edge of parts involved in the mounting, for example a screw head, a nut, or a mounting bracket. For this reason it is most practical if the draughtsman, at a fairly early stage, i.e. immediately after the mechanical restrictions and mountings have been determined, sketches on the artwork the 'forbidden' areas within which no conductor or pad can be permitted. A 'forbidden' area is to be perceived as a safety zone taking into consideration all tolerances and voltages and maintaining adequate clearances in accordance with these. The assessment of the size of the 'forbidden' area is based on several factors which are listed below for the 1:1 scale.

Radial displacements on board caused by:

0.5 mm (0.020 in) oversize of mounting holes	0.25 mm (0.010 in)
0.1 mm (0.004 in) diameter tolerance	0.05 mm (0.002 in)
positional tolerance of hole	0.08 mm (0.003 in)
side-to-side misregistration	0.05 mm (0.002 in)
growth along one edge of conductor	0.07 mm (0.003 in)
a: total radial displacement	0.50 mm (0.020 in)

Tolerance on edge of mounting elements:

b1: screws and nuts (estimated)	0.10 mm (0.004 in)
b2: mounting brackets, etc., originating from oversize of mounting hole plus tolerance on contour (estimated)	0.50 mm (0.020 in)

Minimum spacing:

c1: agreed minimum spacing, valid for d.c. voltages $\leqslant 20$ V (see Fig. 6.11 for other voltages)	0.35 mm (0.014 in)
c2: agreed minimum spacing, valid for a.c. mains voltages $\leqslant 250$ V	3.00 mm (0.118 in)

Now it is possible to find the safety distances measured from the centre of the mounting hole to the nearest conductor or solder pad for various mounting elements:

Screw heads or nuts:

(D = diameter of screw head or width of nut measured across corners)

D.C. voltage $\leqslant 20$ V: $\frac{1}{2}D + a + b1 + c1 = \frac{1}{2}D + 0.95$ mm (0.038 in)
A.C. voltage $\leqslant 250$ V: $\frac{1}{2}D + a + b1 + c2 = \frac{1}{2}D + 3.60$ mm (0.142 in)

Mechanical sub-unit, sub-assembly or mounting element:

(A = distance from centre of mounting hole to contour)

D.C. voltage $\leqslant$ 20 V: A + a + b2 + c1 = A + 1.35 mm (0.053 in)
A.C. voltage $\leqslant$ 250 V: A + a + b2 + c2 = A + 4.00 mm (0.158 in)

On the basis of the size of the standardized and commonly used mounting elements such as screws, nuts, etc., it is possible to set up a table which indicates the safety distance necessary. Should the safety distances be given in correspondence with the artwork scale used, for example 4:1, the values are rounded off to a convenient higher even value. As the difference between the diameter of a screw head and the measure across the corners of the matching nut is usually fairly small, it is possible to simplify the table by using the larger safety distance. The table must be worked out on the basis of the screw and nut system actually used. Table 6.4 shows the safety distance given as the radius of the 'forbidden' area for standard metric screws and nuts.

Table 6.4 Dimensions of 'forbidden' areas around metric screw heads and nuts. (Artwork scale 4:1.)

Thread size	Radius of 'forbidden' area (mm)	
	$\leqslant$20 V d.c.	220 V a.c.
M 2	13	24
M 2.5	15	27
M 3	17	29
M 4	20	31

6.7 Edge Connectors

Whether based on two-piece male and female connectors or on an edge connector and a receptacle, the connector system shall be perceived as consisting of precision components involving certain mechanical constraints if the contacts are to mate correctly. It is important that at least one part of the contact system must be free to flex or travel sufficiently to compensate for all mating misalignments without overloading mechanically the contact springs or the housing. If the travel is too small, the practical solution is to use a 'floating' mounting of one of the connector parts.

For plug-in boards the following systems are commonly used:

1. A back panel in which the receptacle is rigidly mounted and the contacts connected by means of soldered or wire-wrapped wires. The individual contacts will normally be located more or less loose in the connector housing so that they can easily align themselves with respect to the mating contacts. Whether these are mounted in the other part of a two-piece connector having the contacts soldered to the board, or take the form of an edge connector, the individual contacts will not be damaged mechanically.
2. A motherboard in which the individual contacts of the receptacle are seated and soldered. In this case a floating mounting is not practicable. Reliable mating

requires, therefore, that the contacts within the receptacle have sufficiently large travel and that the plug-in boards have sufficient clearance in the card guides of the equipment to permit correct alignment between the board and the receptacle.

This section deals with the artwork for edge connector, and the above considerations serve to illustrate that the contact system must be carefully designed.

The fixed receptacle for the edge connector is frequently produced on a modular basis, i.e. it can be purchased in the desired length. A certain standardization is, however, desirable in order to cut down the number of types carried in stock; in many cases it must, therefore, be accepted that a certain number of contacts will remain unused, especially in the case of a standard range of plug-in boards where standardization of the receptacle is very common.

Figure 6.36 shows two examples of edge connectors. In Fig. 6.36A the board is provided with a tongue which fits into the receptacle and guides the contact fingers

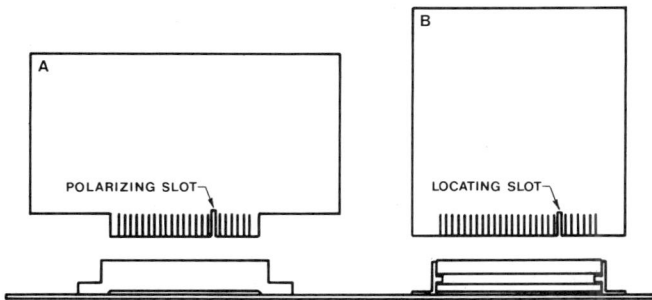

Fig. 6.36 Examples of edge connectors.

with respect to the contacts of the receptacle. A polarizing plug is inserted into the receptacle and engages with a corresponding slot in the board. In this case, where the position of the board is governed by the contact tongue and the housing of the receptacle, the function of the polarizing slot and plug is to ensure that only the correct type of board is inserted. The tolerance on the width of the polarizing slot can therefore be fairly wide, whereas the tolerance on the width of the tongue has to ensure that the mating contacts come correctly into line.

In other cases, the board is not governed by a tongue, and this happens when the receptacle has open ends. This is illustrated in Fig. 6.36B where the board is governed sidewise by means of a locating plug and a matching slot in the board. The slot requires much closer tolerances on its position and width since it has to ensure that on insertion the mating contacts are in correct alignment. As in the former case, the slot also serves to ensure that the correct type of board is inserted.

The width of the contact fingers of the edge connector must, in both cases mentioned, be adapted to accommodate for a certain lateral displacement of the board originating from a play in the guiding system, and from any lateral displacement of the contacts of the receptacles.

In the case of edge connectors based on a 0.1-in module, realistic tolerances of the machining of the slot or the tongue would be a symmetry tolerance with respect to the edge contacts of ± 0.1 mm (± 0.004 in) and a width tolerance of ± 0.1 mm (± 0.004 in).

A very common problem for a PCB manufacturer is to ensure the correct location of the slot or the edges of the contact tongue. Experience has shown that the corner marks defining the profile do not provide accurate reference. Introduction of a reference system is therefore recommended to very precisely determine the position of the slot or the tongue in relation to the adjacent contact fingers. This is described in detail in Section 8.9.3.

The receptacles can have many different spacings, for example, 0.2, 0.156, 0.125, 0.1, and 0.05 in. There is no advantage to be gained by choosing a particularly small spacing if this demands so high a concentration of conductors on the board that more board areas are lost than gained. The connector spacing must, in general, be consistent with the pattern density of the boards. A common solution is to use a 0.1 in module receptacle with double contacts, i.e. there are two independent rows of contacts. In the case of a group of boards employing the same contact arrangements, a row of via holes can be located on a level with the contact spacings. In this way, it is always possible to get from a contact finger on one side of the board to a conductor on the other side. This is clearly illustrated in Fig. 3.17.

Since the contact fingers are normally gold plated, the artwork must include a connection from each contact finger to a plating bar (see Section 3.7.4 and Fig. 3.17). As the edge of the board is to be chamfered at an angle of 45° (cf. Section 4.9.4 and Fig. 4.13) the actual contact area should preferably end a little before the chamfering as shown in Fig. 6.37. In this way, there is little likelihood of damaging the contact fingers during chamfering as well as during insertion in the receptacle. Both ends of the contact fingers should be designed with rounded ends in order to counteract any tendency to lifting the copper foil.

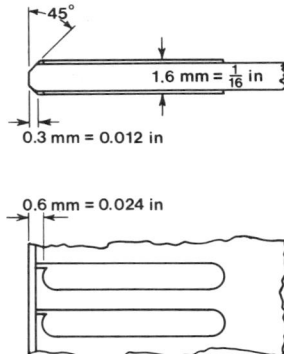

Fig. 6.37 Chamfering of edge connector. Note the shape and location of the contact finger.

Edge-connector patterns are commercially available in the following modules: 0.05, 0.1, 0.125, 0.150, 0.156, and 0.200 in. If other modules or types are desired it is necessary to prepare a pattern as it is too inconvenient and also too inaccurate to tape the pattern, contact finger by contact finger. Based on a single contact finger which is prepared oversize, for example 16 times, a photographer can produce a film with the necessary number of contact fingers and the desired spacing by a step-and-repeat process. The film is fixed to the artwork by means of transparent tape.

6.8 Marking Drawing

6.8.1 Screen-printed Component Identification (Notation)

Larger amounts of text and reference designations, for example for indicating the component location or the polarity of certain components, are normally screen printed on the component side of the board. While this gives a direct assembly instruction, the value in repair work is limited since the assembled components frequently cover the reference designations due to the high packaging density. This problem can be solved by inserting an assembly drawing of the component side in the service manual. The assembly drawing can easily be derived from the notation drawing, and, if desired, be contact printed together with the solder side pattern of the board.

Practice has shown that it is easier and faster to write the characters by means of a lettering template than to use the transfer type of letters. Figure 6.38 illustrates a

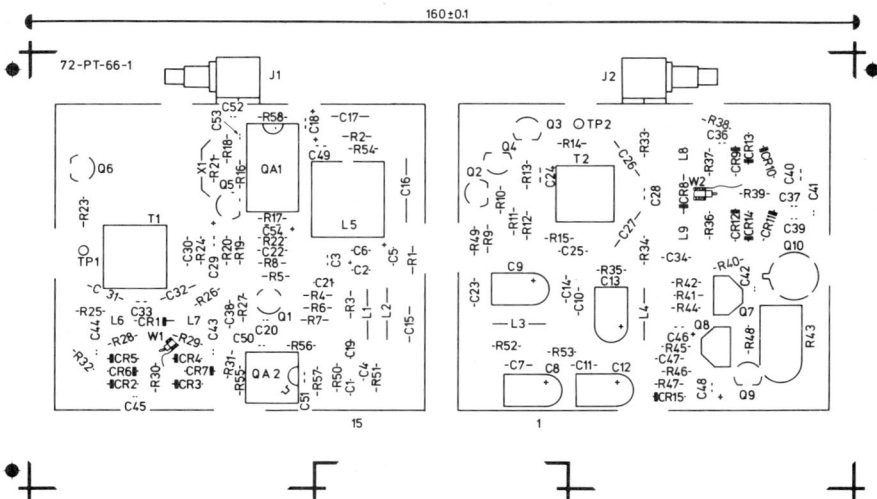

Fig. 6.38 Component identification (notation) drawing.

marking drawing where the component location is indicated by the reference designation, e.g. R17, and two leader lines indicates the solder holes to be used. Figure 6.39 shows another example where the component location is indicated by means of a frame corresponding to the contour of the component. Both methods are commonly used, but the author prefers the former as it indicates directly the solder holes to be used, and is also easier to draw. It is very important that the leader lines as well as the characters are kept well away from the solder holes so that the printing ink does not

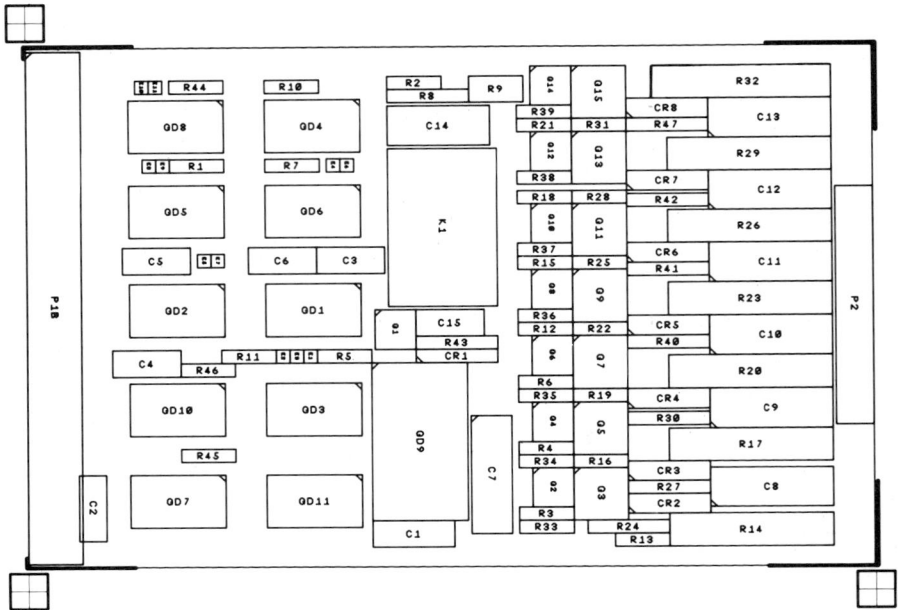

Fig. 6.39 Component identification (notation) drawing (CAD).

run down into the holes. This could cause some gassing at automatic soldering which, in turn, could cause blowholes.

Although the characters should, as far as possible, not be placed across the conductors since legibility thereby suffers, due to the high packaging density of present-day printed circuit boards, this will frequently be unavoidable and the inconvenience will not usually be too serious.

A more significant fault is that the lettering is too small, or the lettering width too thin, with consequent difficulty in producing sharp and legible prints. There are different opinions of the acceptable minimum size of lettering as various specifications cover the following ranges valid for the 1:1 scale:

> Lettering height: 1.5–5 mm   (0.059–0.20 in)
> Lettering thickness: 0.3–0.8 mm (0.012–0.031 in)

A compromise is 2.5 mm (0.1 in) and 0.3–0.4 mm (0.012–0.016 in) for the lettering height and thickness, respectively.

Screen printing of the marking drawing represents a modest part of the total price of the board, frequently just 1–2%. It shows a great advantage by enabling faster and more reliable assembly and it would therefore be wrong to discard it to save this small cost.

6.8.2 Etched Notation

It is an advantage to plate and etch certain notations in the same way as the pattern. The great advantage is that the notation with certainty is identified with the pattern. This is especially important in the case of the designation, code or type number and issue number of the board. Since good and clear edge definition is required, transfer characters such as Letraset (trademark) are normally used.

It is convenient to employ the same letter size used for the screen-printed component identification. Care must be taken that the notation observes the same minimum spacing as the pattern, and, furthermore, the draughtsman must take special care to avoid any reduction of creepage distances over which high voltages or mains voltages are present.

A small problem is found where inscriptions are placed in the area around a gold-plated edge connector (see Fig. 6.40). If the PCB manufacturer does not take care to

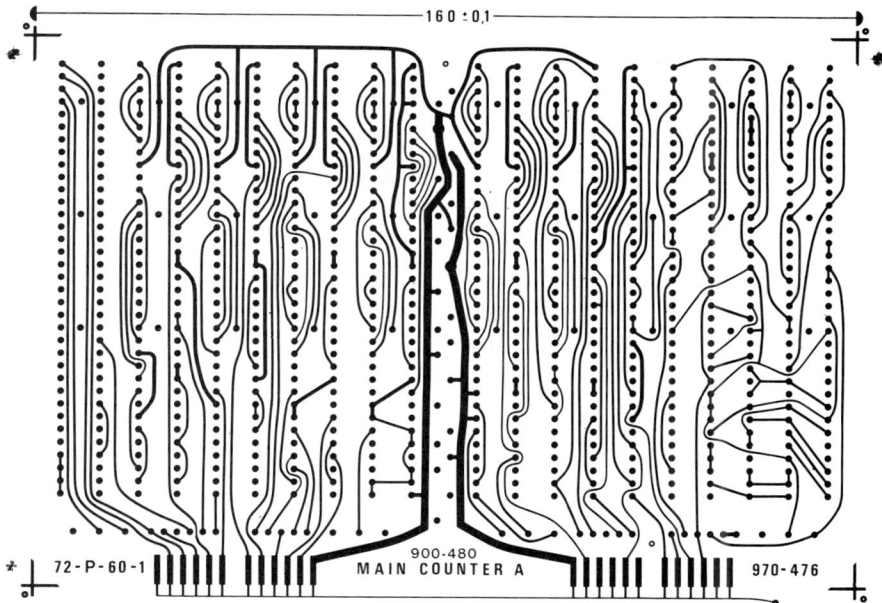

Fig. 6.40 Inscriptions close to gold plated edge connector. Dimension in millimetres.

cover the inscription with masking tape, the tin/lead layer will be stripped simultaneously with the stripping of the tin/lead layer of the contact fingers. As the inscription, contrary to the contact fingers, is insulated from the current source of the plating bath, it will not be gold plated and therefore appears as bare copper which, in the course of a short time, becomes oxidized. The attention of the manufacturer must be drawn to the above case.

Boards with very few items to be screen printed or boards being so small that the screen-printing process becomes a significant item of cost, can often have the text plated and etched instead of screen printed.

In a system comprising many boards inserted in a rack as shown in Figs. 5.5 and 5.6, a standardized location of the board designation and code number should be adopted, the location preferably being near the visible end of the board. Special care should be taken to ensure that the board designations and code numbers of the different boards turn the same way. This also applies to the screen-printed marking drawing. If boards are withdrawn from the rack and connected via extension boards, the lettering of all the boards should turn the same way and this requires a uniform procedure.

Finally, it should be noted that a PCB manufacturer should be required to mark his products with his company mark and the date of manufacture. If the location is not organized beforehand, it might be placed at random and therefore, difficult to find. For this reason a frame which indicates where the company mark and date of manufacture are to be located, can be included in the artwork and appears on the master film (see Fig. 6.41). In that case, the manufacturer introduces his company mark in the master film, which therefore is plated and etched together with the frame as part of the pattern. The date of manufacture is stamped on the finished board after having passed the outgoing inspection.

Fig. 6.41 Board marking.

6.9 Solder Mask

6.9.1 Introduction

When a board is soldered automatically in a soldering machine, solder bridging can easily occur, i.e. short-circuits between conductors and solder pads. The reasons for solder bridging can vary: e.g., faulty operation of the soldering machine, poor design of the pattern or careless assembly.

Solder bridging can be very difficult to find, and in the case of complex boards an average repair time of 15 min per fault is frequently found. This includes the identification and exchange of components damaged as a result of the solder bridging. Many soldering faults are, of course, found by the visual soldering inspection, but the remaining faults are so difficult to discover that they are not found until the testing phase. Therefore, quite a lot of money can be saved by preventing the occurrence of solder bridging, and the solution is the use of a solder mask.

A screen-printed solder mask consists of a heat-resistant ink film which is applied to the board in such a way that there are apertures corresponding to all pads which are to be soldered. The apertures are preferably a little larger than the pads so that the film does not encroach on the soldering area of the pads.

6.9.2 Generating Artwork for Solder Masks

The artwork can be prepared in several ways: The most complicated method is to use a cut-and-strip film, in which all the apertures of the mask are cut by means of bow compasses. As long as the apertures are circular it is a passable way but it is not readily applicable to rectangular apertures. The other method is based on a 'reverse' taping, which means a negative artwork showing all the apertures as pads. When the artwork is scaled down photographically, the negative film produced will turn the pattern to a positive one, i.e. all pads become apertures. The method is illustrated in Fig. 6.42 which shows a limited section of a board. Figure 6.42B shows the normal

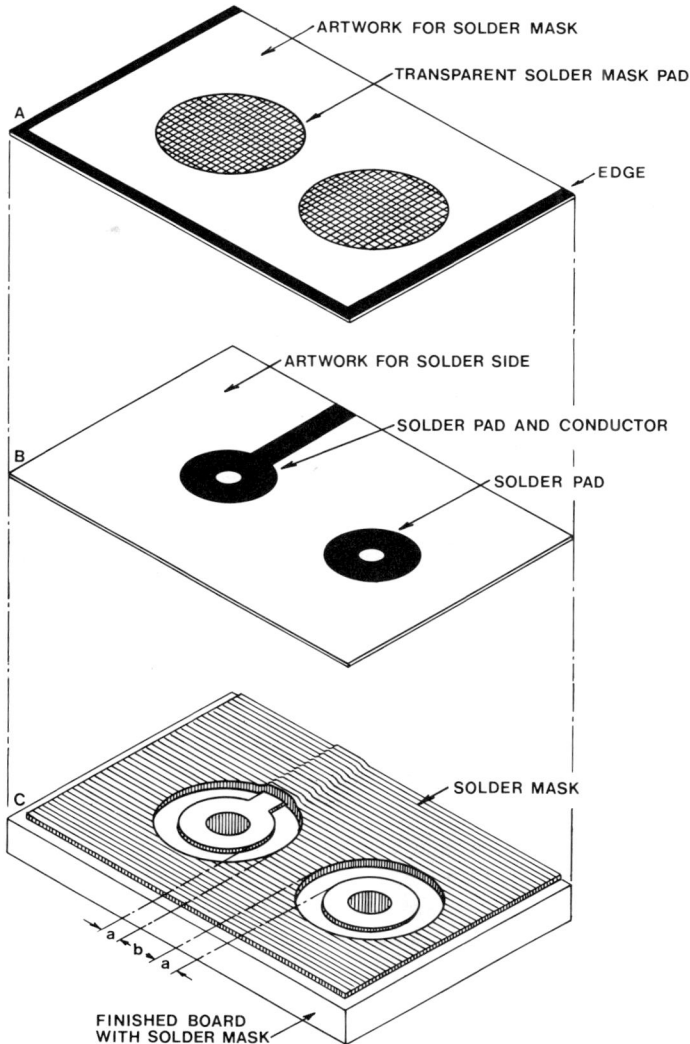

Fig. 6.42 Preparing the artwork for a solder mask.

artwork, and Fig. 6.42A shows the artwork for the solder mask. This is prepared by placing a polyester drafting film on top of the artwork for the solder side, whereafter the pads (shown cross-hatched) are taped concentrically with the solder pads. It is an advantage to use red, transparent, self-adhesive pads as it is easier to locate the pads correctly owing to the transparency. According to the principle, the solder-mask pads shall be solid, i.e. without any centre hole. If such pads cannot be procured, it is always possible to close the hole afterwards, preferably by a touch-up undertaken on the rear side of the artwork. Figure 6.42A is now a negative image of the solder mask, and by photography, the image is reversed so that the pads become clear apertures whereas the originally clear area becomes opaque. This film is used for the preparation of the screen for the screen-printing machine so that the printing ink can be applied all over the board with the exception of the solder pad areas. This is shown in Fig. 6.42C which, for easier understanding, illustrates a board in the same scale as the artwork. In the beginning it might be felt a little difficult to see everything in the 'negative' way, but, as soon as the draughtsman gets himself accustomed, the method will be found very easy and fast.

Before the artwork for the solder mask can be prepared, a compromise must be reached between the limitations of the screen-printing process and the clearance necessary between the solder pads and the solder mask apertures. Below we shall try to determine the clearance with due regard to both of the constraints, and we shall, therefore, take the starting point in our old example in Section 6.4.3. This is illustrated in Fig. 6.43. In order to achieve the desired protection against solder bridging, it is

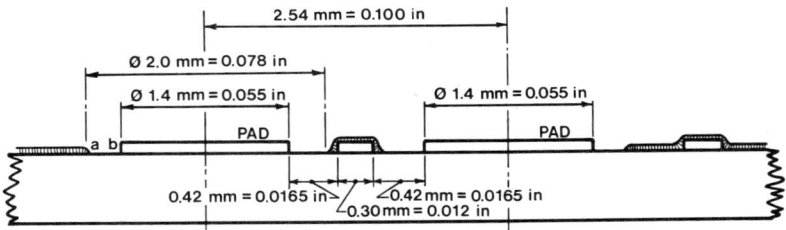

Fig. 6.43 Cross-section of board with solder mask.

required that the conductor which is running between the solder pads must be covered by the solder mask. The example shows one of the most critical situations, and it will be shown later in this section that it is nearly impossible to achieve full security for coverage of the conductor. In spite of this, solder masks do help to counteract the tendency to solder bridging and are therefore strongly recommended. The ideal condition would be to have a clearance of 0.20 mm (0.008 in) as the rim of the aperture of the solder mask would then fall exactly in the middle of the spacing between the solder pad and the conductor. The PCB manufacturer knows, however, that a clearance of 0.20 mm (0.008 in) is a little too small, for which reason he would claim a nominal clearance of minimum 0.30 mm (0.012 in). The main problems are that the screen always stretches a little, and that the surface to be screen-printed is not quite flat owing to the etched pattern. If the screen is stretched, for example, 0.15 mm (0.006 in), the edge a of the solder mask (see Fig. 6.43) will move towards the rim b of the pad. As the screen is not sufficiently elastic to get in touch with the base material in the bottom of the space a–b, it is very likely that so much ink flows from the screen that the pad becomes more or less flooded.

The above description shows that it is important for the manufacturer to find the correct tension for the screen. A very tight tension gives a high printing accuracy, but, at the same time, the reduced elasticity gives some problems in controlling the ink at the edges of the apertures. On the other hand, a less tightly stretched screen causes reduced printing accuracy, but the screen is in a better position to restrain the flow of the printing ink. This is, obviously, a difficult compromise to reach, and it is very easy to understand that the manufacturer wants the largest permissible clearance. The draughtsman should, therefore, bear these problems in mind and allow for a clearance of minimum 0.50 mm (0.020 in) at all non-critical locations.

Later we shall estimate the tolerances involved but, at present, it can be said that a solder mask, if it is to succeed at the critical places, must be taped accurately. Obviously, the necessary artwork should be made in a large scale, preferably 4:1.

We can now state the following design rule:

Design Rule Number 11

Critical places	4:1	1:1
Design clearance:	1.2 mm (0.047 in)	0.30 mm (0.012 in)
Accepted clearance:	1.0 mm (0.039 in)	0.25 mm (0.010 in)
Non-critical places		
Minimum clearance:	2.0 mm (0.079 in)	0.50 mm (0.020 in)

By the accepted clearance is understood the value actually measured by means of a magnifying glass (minimum ×7). Unfortunately, it is a recurrent problem that some manufacturers do not fully understand the very demanding conditions described above, for which reason, they are liable to treat the solder mask with insufficient care. Therefore, it is the customer's duty, in his own interests, to point out to such manufacturers that a solder mask requires at least the same degree of accuracy as does the image transfer.

6.9.3 Dimensions of Solder Mask Pads (Screen-printed Solder Masks)

Below we shall determine some mask pad sizes which match the standard range of pads (A-, B-, and C-pads). Unless special solder mask pads are custom designed in the correct size, it is necessary to select the most suitable from those available on the market. Selection made in this way is set out in Table 6.5.

It is seen that the pads fall into three groups of clearances, namely, approximately 0.25 mm (0.01 in), 0.30 mm (0.012 in), and 0.50–0.60 mm (0.020–0.024 in). Before

Table 6.5 Selection of solder pads suitable for use as solder-mask pads. (Dimensions in millimetres.)

Standard solder pad			Solder-mask pads					
			Pad	Clearance	Pad	Clearance	Pad	Clearance
Type	4:1	1:1	4:1	1:1	4:1	1:1	4:1	1:1
A	10.16	2.54	12.09	0.24	12.70	0.32	14.27	0.51
B	7.92	1.98	9.98	0.26	10.16	0.28	12.70	0.60
C	5.54	1.39	7.62	0.26	7.92	0.30	10.16	0.58

making the final choice we shall evaluate the tolerance chain from the artwork to the finished solder mask, and compare the result with the positional tolerance of the pads.

The solder mask is taped with the solder side as an underlay, and the solder mask pads are, as far as possible, applied concentrically with the pads of the solder side. The displacement between the solder pads and the solder mask pads of the finished board will be due to the manufacturing tolerances, and, to a less degree, the artwork tolerances.

The maximum radial displacement of the solder pad is now determined:

0.5 × possible change in width (Section 6.5.1; Table 6.2)	0.065 mm (0.0026 in)
Accuracy of master film	0.05 mm (0.002 in)
Misregistration	0.05 mm (0.002 in)
	0.16 mm (0.0066 in)

The maximum radial displacement of the solder mask pads (apertures) is determined below:

Positional accuracy
Discriminating ability of the eye / Parallax — 0.15 mm (0.006 in) (scale 4:1) — 0.04 mm (0.0016 in)
Accuracy of master film — 0.05 mm (0.002 in)

Screen-printing accuracy
Misregistration — 0.05 mm (0.002 in)
Stretching — 0.05 mm (0.002 in) — 0.10 mm (0.004 in)

The maximum displacement under worst-case conditions is 0.16 + 0.19 = 0.35 mm (0.014 in). If the clearance is the same size as the displacement, the rim of the apertures of the solder mask will just touch the rim of the solder pads of the finished board. In practice, it will not be that bad. It is evident, that the manufacturer wants the clearance as big as possible, and it seems reasonable to choose a clearance of 0.30 mm (0.012 in) as the design value (cf. Design Rule Number 11). The reason for an accepted clearance of 0.25 mm (0.01 in) is the fact that the contribution 0.04 mm (0.0016 in) can be ignored, so by a slight rounding-off the value becomes 0.25 mm (0.01 in).

We have now been consistent with Design Rule Number 11, but it must be emphasized that, although the artwork, in some cases, can easily be prepared with smaller clearances, it will be of no use, if, owing to the manufacturing tolerances a clearance of 0.30 mm (0.012 in) is demanded. We have, therefore, to ignore the pads of the first group and use the pads of the second and third groups for critical and non-critical places, respectively.

Regarding the IC symbols, the choice is fairly easy as only one commercially available type fits the special pad size discussed in Section 6.5.5. By comparing pad type number 1 and number 5 stated in Table 6.3, clearances of 0.27 mm (0.011 in) and 0.31 mm (0.012 in) in the longitudinal and transverse directions, respectively, are found.

Now we can determine the solder mask pads as stated in Table 6.6. *Note*: If the selected type of solder mask pad causes too large a clearance—for example, if a conductor runs very close to a solder pad—there are two possibilities. The solder mask pad can either be displaced a little or it can be trimmed slightly (see Fig. 6.44).

Table 6.6 Standard sizes of solder-mask pads. (Artwork scale 4:1.)

Standard solder pads		Standard solder-mask pads			
		Critical areas		Non-critical areas	
A	10.16 mm	A1	12.70 mm	A2	14.27 mm
B	7.92 mm	B1	10.16 mm	B2	12.70 mm
C	5.54 mm	C1	7.92 mm	C2	10.16 mm
−	5.6 × 8.0 mm	−	8.12 × 10.16 mm	−	8.12 × 10.16 mm

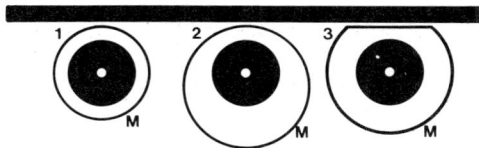

Fig. 6.44 Application of solder mask pads.
1. Concentric application, symmetrical minimum clearance.
2. Eccentric application, minimum clearance at conductor.
3. Trimming of pad, minimum clearance at conductor.

The above calculation shows, however, that it is impossible to screen print a 100% perfect solder mask. Figure 6.43 shows that the spacing between the conductor and the pads is 0.42 mm (0.0165 in). The nominal clearance between the solder mask apertures and the pads is 0.30 mm (0.012 in), leaving a 0.12-mm (0.0047-in) margin for coverage of the conductor. Above we have shown that the positional accuracy of the screen-printed solder mask is about 0.19 mm (0.0075 in), which means that theoretically 0.07 mm (0.0028 in) of the conductor side will not be covered by the solder mask. Only if the PCB manufacturers could be persuaded to accept 0.20-mm (0.008-in) clearances between the pads and the apertures, and to warrant a positional accuracy of the solder mask better than 0.15 mm (0.006 in) could full coverage always be achieved. This is, however, not likely, and the present situation must be accepted until dry-film solder masks break through (see Section 6.9.5).

To recapitulate: There are two main problems connected with screen-printed solder masks. The most serious and difficult one is to achieve full coverage of a conductor running in the gap between two IC pads, which at present seems more or less inextricable. The other problem is to avoid overprinting of the solder pads by the solder mask as this could imply soldering problems. This is solved by following Design Rule Number 11, the clearances stated being in agreement with the conditions imposed by the PCB manufacturers.

The two problems are highly interrelated, and a solution of the first one could be achieved by an adaptation of the requirements embodied in the latter. This is dealt with in Section 6.9.4.

A solder mask can also serve as an insulating layer which prevents dust or other particles from causing short-circuits between conductors. If equipment is to be used under particularly demanding conditions, a solder mask should also be applied to the component side. Normally, it will not be necessary to prepare a special artwork if a solder mask artwork already is available. In order to facilitate the screen printing, larger clearances should be provided as there is no risk of solder bridging on this side.

The application of a solder mask costs, normally, less than 3–5% of the price of the board, and, in the case of boards which are to be soldered in a soldering machine, a considerable saving is obtained. A greater flexibility in the layout is achieved since it is not necessary to pay special attention to the soldering process when routing the conductor tapes. For the above reasons, solder masks will, undoubtedly, find more general use in the future. It is, therefore, to be hoped that the suppliers of drafting aids will produce correctly dimensioned solder mask pads.

6.9.4 Requirements Adapted to the Solder Mask

The preceding account is based on the principle that the solder mask is not allowed to touch the solder pads as this can cause problems in soldering. The screen-printing method, however, has some imperfections which makes it difficult to work on the uneven surface of a plated and etched printed circuit board. The two contradictory requirements are: the need for a tightly suspended screen in order to achieve a sufficiently close positional tolerance and the need for sufficient elasticity of the screen to make it possible to print the solder mask with acceptable definition. There will always be areas between two closely located parts of the pattern where the solder mask will prove unsatisfactory, although the artwork has been prepared with great care and the manufacturer has spared no pains.

It should, therefore, be considered whether it would be better to accept a certain degree of overprinting of the rim of the pads if this implies that adjacent conductors really become covered by the solder mask. Overprinting can take place along the entire rim of the pad or just where the conditions are most critical. In this connection, it must be remembered that the annular ring can be as narrow as 0.3 mm (0.012 in). For this reason the overprinting should be limited to maximum 0.15 mm (0.006 in). In order to reduce the risk of printing ink running down into the solder holes, overprinting should be allowed only at critical places and not along the entire rim of the pads, especially if the annular ring is as narrow as 0.3 mm (0.012 in). In the case of pads having wider annular rings—which would be the case with single-sided, non-plated through boards—overprinting along the entire rim of the pads can be allowed. This results in the pads being better 'locked' to the board and, consequently, able if necessary to stand several soldering and desoldering operations.

The above considerations lead to an adapted design rule:

Design Rule Number 12

	4:1	1:1
At special critical places the design clearance can be reduced to:	0.6 mm (0.024 in)	0.15 mm (0.006 in)

The artwork is prepared by applying the solder mask pads stated in Table 6.6 and the reduced design clearance can be achieved by a displaced application of the solder mask pads or by trimming the pads as shown in Fig. 6.44.

6.9.5 Dry-film Solder Masks

The preceding discussion leaves no room for doubt that it is nearly impossible to screen-print solder masks on complex, high-density boards with the accuracy needed.

A new type of solder mask has just appeared on the market. Owing to the high price of the necessary equipment and of the material itself compared with the price of ordinary screen printing, only very few PCB manufacturers have until now taken a positive interest in the method. The new type of solder mask is based on a dry film which works in fundamentally the same way as the dry film used for the image transfer to the board (cf. Section 3.2.2). The procedure is that the dry film is applied to the finished board by means of a special laminator, exposed by ultraviolet light, developed and cured. It is inherent that the solder mask has a positional accuracy corresponding very closely to that of the pattern itself and with a resolution which ensures an effective protection against the formation of solder bridging.

In principle, it should be possible to design the solder mask apertures (pads) with exactly the same size as the solder pads and this would facilitate the work of the drawing office enormously. A further facility and a higher degree of accuracy could be achieved by producing the artwork for the solder mask on the basis of a diazo contact print of the common pad master (cf. Sections 6.3.8, 8.4.2b2, and 8.6.2).

As dry-film solder masks seem likely to give better results than screen-printed solder mask, following future development attentively is recommended. Finally, a word of warning: Dry-film solder masks appear to be far more expensive than ordinary screen-print solder masks, in many cases prohibitively expensive. It seems, however, that thinner and therefore cheaper dry-film material will be introduced on to the market in order to bring dry-film solder masks into a competitive position.

7

Contact Printing on Diazo
Film. Theory

7.1 Introduction

In Section 6.3 we treated the general methods of preparing the artwork, and in Section 6.3.8 particularly, it was shown how the problem of achieving a perfect registration between the solder pads of the two sides was solved in an easy and elegant manner by contact printing on diazo-coated polyester films. In the author's experience this is the fastest, cheapest and most reliable method for the preparation of artwork for double-sided boards and also for multilayer boards. As the method plays an important part in improving efficiency in the PCB drawing office, this chapter is devoted to a detailed description of the systems and the necessary equipment, and to an analysis of the restrictions which are necessary for its successful application. For optimum use of the method, the user must be familiar with its advantages and limitations so that he can adapt it to his special requirements. Section 7.3 which contains the analysis may seem somewhat tedious but it is necessary for a thorough grasp of the system and should be carefully studied in conjunction with the perspective drawings illustrating the various ways in which the contact printing can be carried out. The diazo-printing process corresponds to the process commonly used in making black-and-white prints in the drawing office. This process produces a positive copy of the drawing which is being reproduced. In order to achieve maximum accuracy it is, however, necessary to perform the contact printing under flat conditions which require special printing equipment.

7.2 Equipment for Contact Printing

The contact printing equipment consists of a light source and a vacuum exposure frame including a vacuum pump. One arrangement of a diazo copying-room is shown in Fig. 7.1. In addition, an ammonia developer is required and in many drawing offices this will already be available. Exposure of the sensitized diazo film requires a light

210

Fig. 7.1 Arrangement of a contact-printing room.

source with a wavelength in the range 350–450 nm, i.e. blue–violet/ultraviolet light, which can, very conveniently, be a metal halide lamp. Other types of lamps can be used but, as the spectral distribution is not optimum for the diazo film, longer exposure is required. For example, a xenon lamp takes 3–4 times as long. A short, intensive exposure gives a higher definition than one which is longer but less intensive. This is another reason for using the metal halide lamp. In order to reduce the parallax error occurring when printing, the lamp must be located at a reasonable distance from the exposure frame, for example, 1.5 m (~1.5 yd). As a rule, the distance should not be less than the length of the diagonal of the exposure frame. This gives relatively even light distribution over the entire area of the frame and enables good quality prints to be produced even from larger artwork. Exposure time depends on the lamp power and at 1.5 m from the exposure frame, a 5-kW metal halide lamp requires approximately 3 minutes for full exposure.

This type of lamp improves in efficiency as it warms up in use, and even if preheating is arranged in the stand-by condition, the early prints in a run will be underexposed in relation to the later ones unless a compensating adjustment is made to the exposure time. Ideally, a control unit is used which senses the light intensity and adjusts the exposure time to compensate for lamp ageing, temperature and mains voltage variations.

The printing frame is a shallow box with a flexible sheet-rubber base and a closely fitting plate-glass lid. A diazo film and the artwork are placed on the rubber base, the lid is closed, sealing the box which is then evacuated by the vacuum pump so that the rubber base creates intimate contact between the two films by pressing them against the glass lid. If there is a tendency for air pockets to be trapped between the films, a string run along the edge of the film (as shown in Fig. 7.2) will provide a channel to facilitate the elimination of such pockets over the whole area of the films.

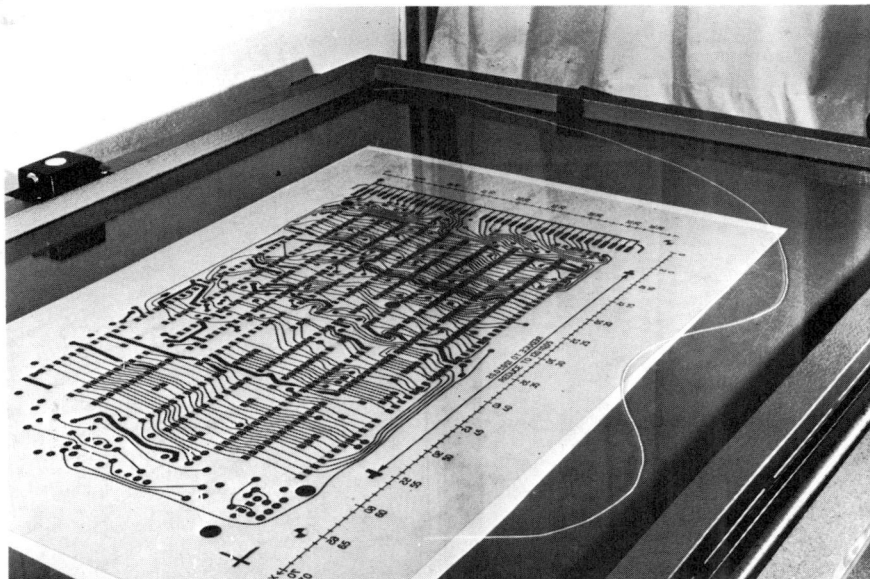

Fig. 7.2 Elimination of air-pockets.

In order to achieve good prints it is important to keep the glass plate completely clean. Stains and dust cause black spots which require subsequent re-touching.

Contact printing should be performed only in a flat vacuum frame of the type described. The conventional printing machine is unsuitable because it passes the original and the diazo film around a cylindrical exposure chamber. This is liable to cause a slight displacement between the original and the diazo film, and, furthermore, some distortion occurs owing to the different radii of the diazo film and the original. In addition, if taped artwork is being printed some of the conductors may be displaced or work loose when the artwork is bent around the cylinder.

7.3 How to Make Contact Prints

The work flow of the PCB drawing office was dealt with in Section 6.3.8 and il-lustrated in Fig. 6.6. We shall now study the application of contact printing to the production of permanent masters and secondary copies. The artwork for the solder side and the component side must show the pattern facing the same way as on the finished board which means that the two sides should be turned opposite to each other. The contact printing can be performed in three different sequences and below we shall discuss the advantages and disadvantages of these. The handling of the master and the diazo films in these sequences is somewhat complicated but the perspective drawings shown in Figs. 7.3 and 7.5–7.9 make it easy to understand these methods. In the drawings, the solder pads are represented by an 'L' which can easily be recognized when reversed owing to its unsymmetrical structure. In accordance with the practical conditions, the 'L' is shown lying on top of the drafting film whereas on the diazo film it is shown lying in the diazo layer.

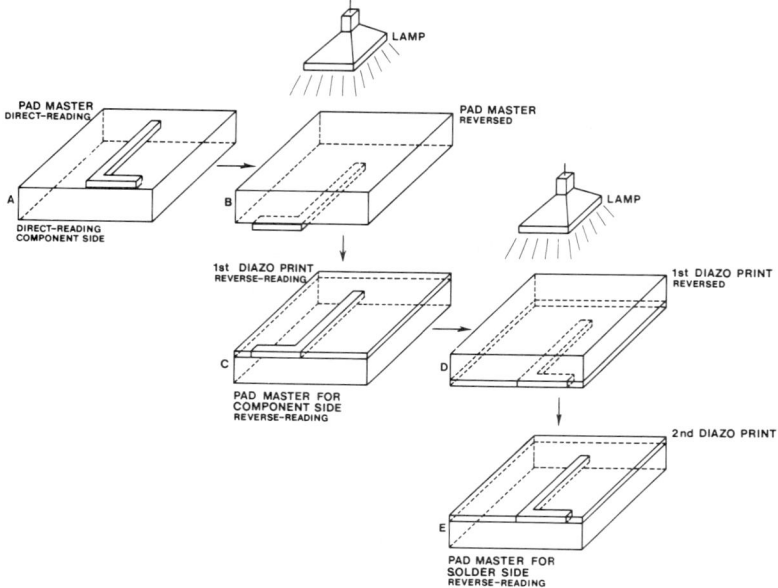

Fig. 7.3 Contact-printing method A.

It should be noted that the sequences described for printing on diazo films apply also to contact printing on photographic films (Section 6.3.7). Photographic reproductions must usually be performed by a photographer outside the company. It is difficult, therefore, to ensure that the correct sequence is followed so as to result in correctly facing masters for both sides of the board. The discussion below shows that these conditions are so confusing that it is not always easy to ensure that the photographer has complete understanding. An in-house facility, with a well-established routine, will ensure far better results.

7.3.1 Method A

The contact-printing method A, which is illustrated in Fig. 7.3, is the most accurate of the three sequences described. The layout sketch and the pad master (A) will usually show the component side as seen when looking on that side of the board. A first contact print is made with the pad master reversed, i.e. the taped pads face downwards in contact with the diazo layer (B). The print will of course be inverted, as seen by the reversed 'L' (C). In order to get a direct-reading copy which is printed with tight contact between the pads and the diazo film, the print (C) is turned over and used as a new master (D) for the next contact print (E). The first print (C) is used for the component side, but as it is reading reverse, the layout sketch has to be reversed in order to match the pad location. The print (E) is used for the solder side but, in this case, it directly matches the layout sketch which is drawn as seen from the component side. The conductors are taped directly on the diazo layer of prints (C) and (E) for the solder side and component side, respectively. The two prints are reverse-reading, and for making secondary copies they are turned over and printed with direct contact between the pads/conductors and the diazo layer. This means that the secondary copies will be direct-reading.

Without going into details here, the author refers to Fig. 7.8 which shows how the contact printing of the secondary copies is performed. A detailed description is given in Section 7.3.5. Here we shall ignore the 'F' in Fig. 7.8 and let the 'L' represent the solder pads and the conductors.

From a technical point of view, the contact-printing method A is best since the tight contact between the pads of the master and the diazo layer gives the sharpest image. There are, however, some drawbacks to this method:

1. In practice, it is unsatisfactory to tape the conductors on the same side as the solder pads because it is easy to damage the pads when cutting the tape (see Fig. 7.4).

Fig. 7.4 Damage to a pad.

Furthermore, imperfect contact printing can occur if the pads do not achieve tight contact with the diazo layer due to the conductors which are taped over the pads.

2. All inscriptions and markings have to be reverse-reading requiring a mirror-image number and alphabet kit. The alternative is to reverse the film for application of the inscriptions to make it direct-reading but this may cause displacement of the taped conductors owing to the additional handling.

3. The second print (E) is derived as a copy of the first print (C), and any difference in density which occurs between these will be repeated in subsequent copies. Although it is not possible to find any difference of practical importance between the first and fourth prints shown in Fig. 6.5, there will be a progressive, though slight, deterioration in both definition and density. This is, of course, of special interest if the file, as mentioned in Section 6.3.8, is based on secondary copies (see the comments in Section 7.3.5).

4. If modifications and rectifications are introduced after the artwork is finished and extra pads are required, it will be necessary to check the registration of these pads. In order to avoid parallax errors between the registration marks of the two

214

masters, it is necessary to put them together with the diazo layers facing each other. This is unsatisfactory as conductors of one master may easily engage with conductors of the other master, resulting in possible displacement of some of the conductors. Frequently, such displacements are not noticed until the boards have been assembled and fail when tested. It is a firm rule, therefore, that the two taped artwork sides should never be put together.

5. The method requires sequential contact printing. If the developer is in another department, the draughtsman will have to walk to and fro twice reducing his efficiency.

7.3.2 Method B

Contact-printing method B as illustrated in Fig. 7.5 results from a minor modification of method A. The first print (C) and the second print (E) are made as before, but when the conductors are to be taped, the prints are turned upside down (F) and (G) so that the conductors are taped on the sides which are not diazo-coated, i.e. on the sides opposite to the pads. The first print (C) is, as before, a reversed-component side but when

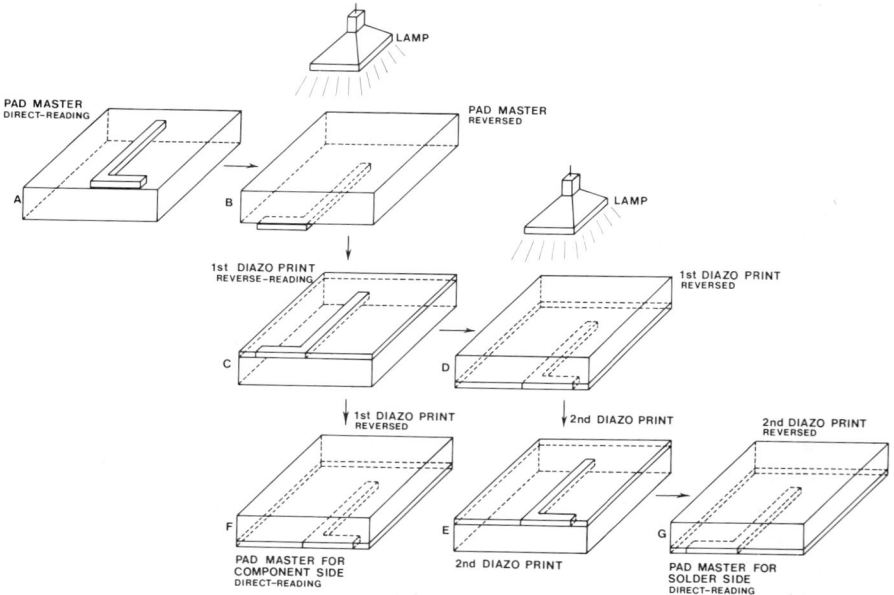

Fig. 7.5 Contact-printing method B.

it is turned over it becomes direct-reading (F) so that the layout sketch can remain direct-reading. The other print (E), however, is a reversed-solder side and when it is turned upside down it becomes direct-reading (G), so that in this case, the layout sketch has to be reversed. In this way, disadvantages 1, 2, and 4 (Section 7.3.1) are removed. However, another disadvantage is introduced, namely, that there is no close contact between the conductors and the diazo layer when the secondary copies are made. For this reason, a less sharp image of the conductors could be expected. In practice, this effect is small and is negligible when the artwork is scaled down.

7.3.3 Method C

Contact-printing method C shown in Fig. 7.6 differs fundamentally from methods A and B in deriving the two prints from the same pad master. The procedure is that the pad master is placed in the vacuum frame, and then reversed (B). The print (C) is used

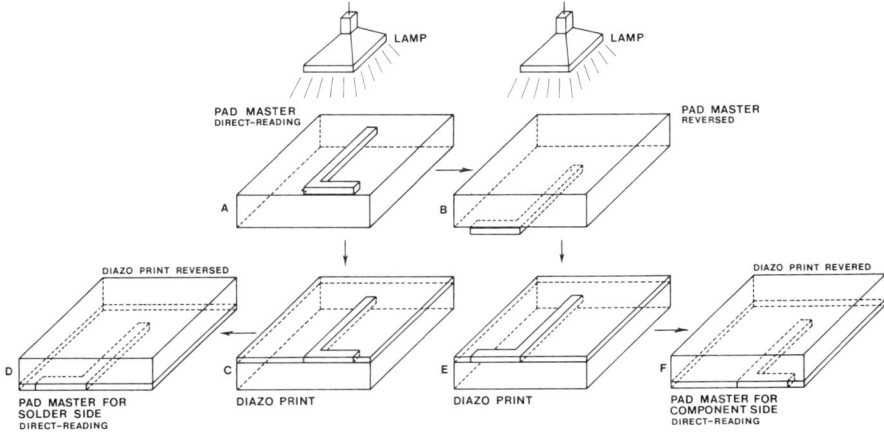

Fig. 7.6 Contact-printing method C.

for the solder side, so that, in order to show a direct-reading solder side, it has to be turned over (D). The other print (E) is used for the component side, and, in order to be direct-reading, it also has to be turned over (F). The conductors are now taped on the uncoated side, i.e. opposite from the solder pads. We have now eliminated the five disadvantages listed in Section 7.3.1, but, in turn, have introduced another disadvantage, the importance of which has to be evaluated. When the first print (C) is made, we do not have close contact between the pads and the diazo layer because they are separated by the drafting film. The less sharp image is, however, of no practical importance if the lamp is suspended at a suitable distance from the vacuum frame. What is more important is that the pattern of the print (C) is slightly enlarged with respect to the pattern of the common pad master (A). This appears in Fig. 7.7 which shows a cross-section through the exposure lamp, the pad master and the diazo film as placed in the vacuum frame. The trigonometric relations give as a close approximation:

$$\Delta L \simeq (t + T)\,\frac{L}{H}$$

Where

ΔL: Displacement of the pad, measured on the diazo print
H: Height of lamp above the pad master
L: Distance from the centre line through the lamp to the solder pad
     (this is equal to half the length of the artwork)
t: Thickness of the pad symbol
T: Thickness of the drafting film used for the pad master.

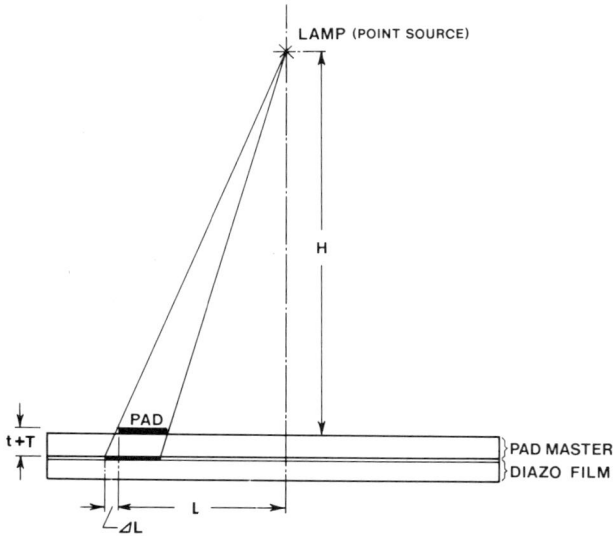

Fig. 7.7 Unavoidable enlargement when using contact-printing method C.

The total increase in length of the print is $2\Delta L$ and the problem is now to keep this as small as possible. In practice, it is only possible to change two of the quantities, namely t and T. It is not possible to change the height H arbitrarily since the available space is limited and the exposure time varies as the square of the distance H, i.e. doubling the distance quadruples the exposure time. L is the half length of the artwork and cannot be changed unless another artwork scale is chosen.

We shall now examine a couple of examples. A practical compromise for the height H is 1500 mm (59 in) which implies an exposure time of $2\tfrac{1}{2}$–3 min, if a 5 kW metal halide lamp is used. The length of the artwork is 800 mm (31.5 in), so $L = 400$ mm (15.75 in).

Example 1

The symbol as well as the drafting film are fairly thick, i.e. $t = 0.12$ mm (0.0047 in) and $T = 0.18$ mm (0.007 in). Therefore, we find:

$$\Delta L = (0.12 + 0.18)\,\frac{400}{1500} = 0.08 \text{ mm (0.0031 in)}$$

The total growth in the length of the print is therefore $2 \times 0.08 = 0.16$ mm (0.0063 in). If the photographer uses a fixed camera adjustment corresponding to a reduction of exactly four times, or, if he uses the same adjustment as found for the print (E) (Fig. 7.6) which was printed with the close contact, the master film becomes 0.04 mm (0.0016 in) too long. Compared with the normal ± 0.05 mm ($\pm$ 0.002 in) tolerance of the photo reduction this enlargement of the print is unacceptable.

Example 2

Here we are using the thinnest symbols available and a very thin drafting film for the pad master, i.e. $t = 0.01$ mm (0.0004 in) and $T = 0.076$ mm (0.003 in). Therefore we find:

$$\Delta L = (0.01 + 0.076) \frac{400}{1500} = 0.022 \text{ mm } (0.0086 \text{ in})$$

The total growth in length of the print is therefore only $2 \times 0.022 = 0.044$ mm (0.0017 in) and under the same conditions as above the master film is only 0.01 mm (0.0004 in) too long, which is considered quite satisfactory. It appears from Fig. 7.6 that only one of the prints is enlarged as the other has close contact. By contact printing according to method C, we have created a small difference in the dimensions of the two artwork sides, and, in order to avoid registration problems for the manufacturer, we have to keep the difference as small as possible. It is interesting to compare these changes in length with those caused by changes in the temperature and the humidity (see Section 6.3.3). So long as the two prints are always subjected to the same environment, they will vary always to the same extent and there will be no registration problems. A small difference between the actual length and the nominal length of a board is normally of no importance for the assembly of the board except in the case of automatic component insertion.

Examples 1 and 2 showed that it is important to use pad symbols and drafting films as thin as possible. Symbols of the adhesive type are normally fairly thick, 0.06–0.12 mm (0.0024–0.0047 in), whereas transfer symbols are very thin, frequently just 0.01 mm (0.0004 in) thick. For this reason, transfers are preferred for the contact-printing method C. When choosing the type of drafting film, the draughtsman may be tempted to use too thick a film, for example, 0.18 mm (0.007 in) thick on account of his wish to achieve the highest possible stability. In the case of the common pad master, a high long-term stability is, however, less important since the pad master is only once used for the copying process and thereafter discarded. A thickness of 0.076 mm (0.003 in) is considered to meet the requirements of sufficient mechanical strength and a reasonable short-term stability.

7.3.4 Comparisons between Methods A, B, and C

In order to compare the above methods of contact printing we have to study the enlargement which occurs when the printing is performed under ideal conditions, i.e. the pads are in close contact with the diazo coating. Therefore $T = 0$, and according to Fig. 7.7 we find for the thick-pad symbols—where $t = 0.12$ mm (0.0047 in)—that $2\Delta L = 0.064$ mm (0.0025 in) whereas we find for the thin-pad symbols—where $t = 0.01$ mm (0.0004 in)—that $2\Delta L = 0.005$ mm (0.0002 in). After reduction by a factor of four, the master film is too large by 0.016 mm (0.00063 in) and 0.001 mm (0.00004 in), respectively. It should be noted, however, that only the left side of the pad of Fig. 7.7 moves when $T = 0$, i.e. the pad shape is distorted depending on its position on the master. For the sake of simplicity, this effect is ignored in the comparison. Obviously, the thick-pad symbols cause an 'enlargement' a little greater than that which occurs when the thin-pad symbols were used in method C. It is also seen that methods A and B give a perfect result when the thin-pad symbols are used.

Contact-printing method A has the five disadvantages listed in Section 7.3.1. Taping directly on the diazo layer implies disadvantages in the form of a risk of damaging the pads and a possibility of 'enlarging' the secondary copies derived from the finished artwork. A special problem arises where the taped conductors are connected to the pads. The tapes usually cover part of the pads, for which reason such pads will be lifted a little from the diazo layer and therefore give rise to a little displacement on the print. Pads having no conductors connected will, of course, be in close contact, and, consequently, no displacement will be found. The commonly used crêpe tape has a thickness of 0.12 mm (0.0047 in) which causes pads in a distance of 800 mm (31.5 in) to be displaced by 0.064 mm (0.0025 in), or 0.016 mm (0.00063 in) on the reduced master film. This is about 50% more than caused by method C when the thin-pad symbols were used. For all the above reasons, we shall discard contact-printing method A.

In the case of contact-printing methods B and C, the prints are taped with the conductors on the uncoated side of the diazo masters. Unlike the drafting film used for the pad master, the diazo master should be fairly thick, preferably 0.18 mm (0.007 in) for reasons of long-term stability as well as reduction of warping due to tensions in the taped conductors. When contact printing secondary copies, the conductors will be displaced a little, whereas the pads which have a direct, close contact to the diazo layer, will not be displaced (see Fig. 7.8). For $t = 0.12$ mm (0.0047 in) and $T = 0.18$ mm (0.007 in) the maximum displacement of a conductor is:

$$\Delta L = (0.12 + 0.18)\ \frac{400}{1500} = 0.08 \text{ mm } (0.0031 \text{ in})$$

measured at the edge of an 800-mm (31.5-in) long artwork, and gradually decreasing to zero at the centre of the artwork. On the reduced master film, the displacement is only 0.02 mm (0.0008 in) which, in comparison with the accepted minimum conductor spacing of 0.35 mm (0.014 in), is acceptable (see Section 6.5.1, Design Rule Number 1).

The choice between the contact-printing methods B and C depends on whether a small difference in the size of the two artwork sides can be accepted. On the reduced master film, the difference is about 0.01 mm (0.0004 in) over 200 mm (8 in), and, furthermore, this difference has to be related to the ± 0.05 mm (± 0.002 in) tolerance of the photo reduction.

In the author's experience, gained from several hundred artworks, this difference in size is of no importance in practice, and contact-printing method C is accordingly recommended. It is, however, a condition that the pad symbols are thin ($t = 0.01$ mm or 0.0004 in) and the drafting film for the common pad master also is thin ($T = 0.076$ mm or 0.003 in).

7.3.5 How to Make and Utilize Secondary Copies

In Section 6.3.8 the use of secondary artworks for filing and despatching to the PCB manufacturer was mentioned. Below, we shall deal with these topics in detail and describe how to make the secondary prints.

The main reason for sending the artwork to the PCB manufacturer is to leave the responsibility of the total manufacturing process, including the preparation of the master film, in his hands. Every little fault appearing on the master film will be transferred to the board, and, in the case of a customer having supplied the master films, unpleasant disputes as to who is responsible for such faults may easily arise. This can be avoided by allowing the manufacturer to be in charge of the photographic reduction, and, as mentioned in Section 6.3.8, the best and safest procedure is to provide him with secondary artwork in the form of diazo prints.

It is very convenient to arrange for the manufacturer to send positive films to the customer, not only of the first issue of the artwork but also of any subsequent issues. These films are used by the customer for Incoming Goods' Inspection and for preparation of assembly instructions.

We shall now see how contact printing of the secondary copies is performed. Figure 7.6 shows how the diazo copies for the solder side (D) and the component side (F) are printed, and Fig. 7.8 shows the subsequent taping of the conductors on the un-

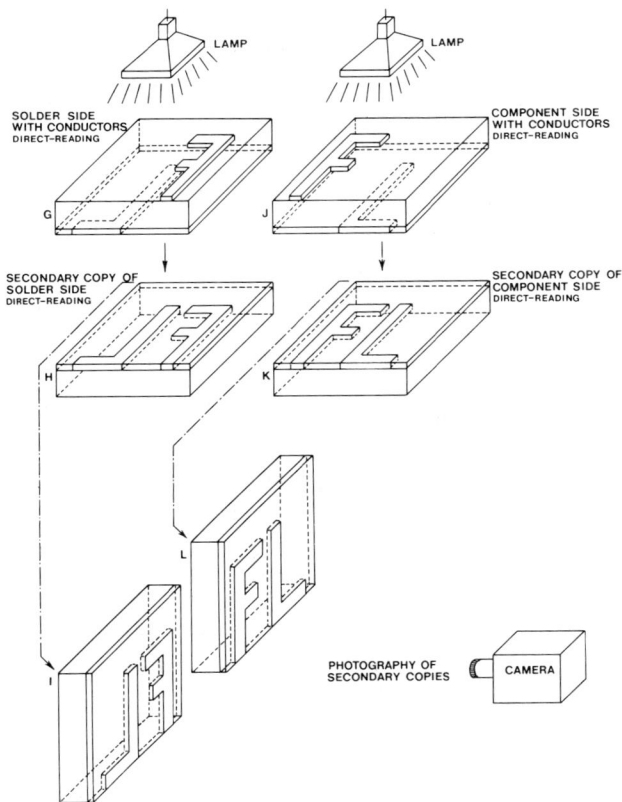

Fig. 7.8 Contact printing and photography of secondary copies.

coated side of the diazo copies (G) and (J), the conductors being represented by an 'F'. The secondary copies (H) and (K) are printed by contact printing, and it is seen that direct, close contact is obtained between the pads represented by the 'L' and the diazo layer, for which reason no displacement of the pads occurs on the copies. Furthermore, it is seen that the secondary copies are direct-reading so that they can be photographed when the diazo layer faces the camera.

Filing of secondary artwork instead of the original artwork offers indisputable advantages. The conductors cannot creep or work loose, and there is no adhesive to flow out along the tape edges whereby the image of the conductors can become blurred, either due to the presence of the adhesive alone or because of adhesion of dust particles. The greatest disadvantage is that amendments can be a little more troublesome to perform on the diazo prints than on the original artwork. As has been mentioned earlier, it may be advantageous to delay a little before converting to the secondary copies so that the majority of rectifications can be performed on the original artwork. Sooner or later, however, it will be necessary to make an alteration if, for no other reason, certain components have become obsolete and therefore have to be replaced by others of different sizes.

The rectification of secondary copies is performed by removing the unwanted pattern, either by scraping with a knife, or more easily, by means of a diazo erasing fluid. The pattern is taped with the new pads on the diazo side (in order to achieve close contact during the subsequent printings) and the new tapes on the opposite side (see (M), Fig. 7.9).

If the rectification is not extensive, filing of the rectified secondary copies can be accepted, but, for large rectifications it may be advantageous to produce 'clean' third-generation copies for filing. Some caution should be observed to avoid too many successive printings between the original pad master and the final artwork.

The printing of third-generation copies by contact printing of the rectified secondary copies is illustrated in Fig. 7.9 which, for the sake of clarity, shows the component side only. The rectified component side (M) which is derived from (K) (Fig. 7.8) is direct-reading and has to be turned over (N) in order to obtain close contact between both the 'diazo pads' and the new pads and the diazo film. The third-generation copy (O) which is filed, is consequently reverse-reading and must be seen through the film in order to be direct-reading. If the new copy is to be sent to the PCB manufacturer, it is necessary to turn the third-generation copy (O) over (P) in order to obtain a close contact between the diazo layers. The print (Q) is direct-reading and corresponds to the secondary copy (K) (Fig. 7.8).

After this rather elaborate description of various ways of contact printing, the author will, with great peace of mind, leave it to the reader to find out how to make fourth-generation prints. The secret is very simple: Always take care that the pads are in direct, close contact with the diazo layer, and preferably use a diazo film which is glossy on both sides (see (Q), Fig. 7.9).

7.3.6 Materials Used for Contact Printing

Below we shall discuss the materials used.

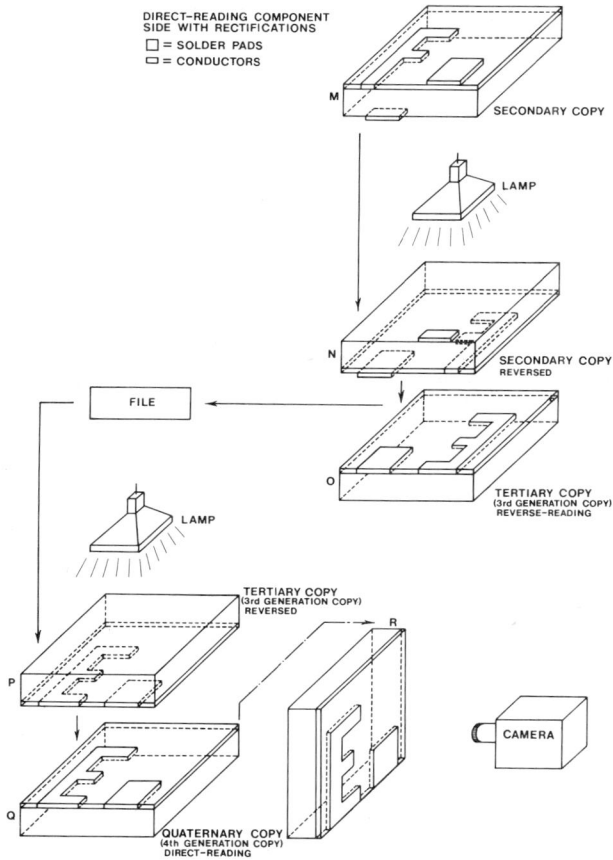

Fig 7.9. Contact printing of 3rd and 4th generation copies.

a. Pad Master

As has been previously mentioned, the pad master is prepared on a thin polyester drafting film of thickness 0.076 mm (0.003 in). The drafting film should preferably be glossy on both sides as this gives the least light-scattering effect when printing the component side (A) (Fig. 7.6). If a sheet is taken from a roll it should, if possible, be cut to size the day before and stored flat to allow for settling out of any strains due to rolling. Ready-cut sheets will, obviously, be an advantage.

b. Solder Side and Component Side

In order to achieve good contact-printing quality, the diazo film used for the solder and the component sides should preferably be glossy on both sides. The use of matt diazo films is discouraged as light scattering occurs at the edges of the pattern when contact printing secondary copies (see (G) and (J), Fig. 7.8). To avoid warping after taping the conductors, the diazo film should not be too thin—never less than 0.1 mm (0.004 in)—and preferably 0.18 mm (0.007 in) thick.

Diazo films can be purchased in rolls and in sheets of standard sizes. It is advisable to use sheets which have been stored flat, as the material does not need time for stabilization prior to contact printing. The shelf life is normally stated as six months under dry and not too warm conditions, and the diazo film should not, of course, be exposed to light and to ammonia fumes during storage.

c. Copies for the Supplier

The advantage of being able to send diazo prints to the PCB manufacturer has been mentioned previously in Sections 6.3.8 and 7.3.5. This type of print does not require extremely high, long-term stability and, as no conductors will be taped on the film, a rather thin diazo film of 0.076 mm (0.003 in) will suffice. Consequently, the weight will be less—of importance when the films are being sent by post (see Section 7.3.8). Depending on the method used for the image transfer (screen printing or photo-resist printing) the master film should be direct- or reverse-reading when looking at the emulsion side. For this reason, it is advantageous to use a glossy diazo film as the photographer can turn the secondary copies in the reverse-reading position without inconvenient light-scattering effects.

d. Copies for the File

These copies require good long-term stability and freedom for warping when additional conductors are taped in case of amendments. A rather thick film of 0.18 mm (0.007 in) is therefore appropriate, preferably glossy on both sides.

7.3.7 Economic Comparison of Diazo and Conventional Methods

In the preceding sections we have studied only the technical advantages of using two contact prints derived from a common pad master instead of taping separate pad masters for the two sides. There is also a cost advantage and considerable time saving in using two contact prints derived from a common pad master as is shown by the following comparative analysis of the costs. Below we shall analyse the cost of an artwork for a board of intermediate size, by way of example, 100×180 mm (4×7 in), prepared on a scale of 4:1. The labour cost is taken as £3.50 per hour including overhead costs.

a. Taping of pad master and contact printing on two diazo films

1 piece of drafting film, 600×800 mm $\times 0.076$ mm	£ 1.00
($23.6 \times 31.5 \times 0.003$ in)	
1 set of symbols	£ 5.00
2 pieces of diazo film, 600×800 mm $\times 0.18$ mm	£ 5.00
($23.6 \times 31.5 \times 0.007$ in)	

Taping (1 hour)	£ 3.50
Time for contact printing (approximately 12 min)	£ 0.70
Total	£15.20

b. Taping of two separate artwork sheets

2 pieces of drafting film (600 × 800 mm × 0.18 mm)	£ 3.00
(23.6 × 31.5 × 0.007 in)	
2 sets of symbols	£10.00
Taping (2 hours)	£ 7.00
Total	£20.00

It is seen that a saving of material as well as time is achieved. If the 'product mix' is known, it is possible to estimate the annual saving to see whether an investment of approximately £1000 for the equipment is worthwhile. To this may be added the saving gained by a cheaper method of despatching artwork to the manufacturer and this is described in the following section.

The decision regarding the purchase of the diazo contact-printing equipment is, however, not exclusively a question of basic cost. The most weighty factor must be the higher technical level of the artwork which can be produced by the diazo contact-printing method. The higher quality can be reflected in lower prices, and, although the extent of this is hard to estimate, a saving of 3–5% due to the perfect registration between the two artwork sides seems to be possible. In addition, the higher quality enables the PCB manufacturer to avoid delays in delivery, a factor of great importance to the customer.

7.3.8 Despatching Secondary Copies to the PCB Manufacturer

The importance of a fast and cheap method of sending secondary artworks to a manufacturer has been discussed in Section 6.3.8, and we shall now conclude this section by discussing the most practical way of despatch. Normally it will be fastest to send the artwork by letter post. It is possible to send a cardboard tube as a letter, provided certain limitations of weight and size are observed. If secondary artwork is to be sent abroad, the internationally accepted weight and size of the tube is:

Length (L): 900 mm (35.4 in) maximum
Diameter (D): $D = 0.5 \times (1040 - L)$ mm, maximum
Weight: 2 kg maximum

For internal despatches the reader is advised to consult the regulations of the post office of his own country.

A cardboard tube having an outer diameter of 100 mm (4 in) and a length of 750 mm (29.5 in) will be accepted in most countries. This size of tube will accommodate most artworks. For reasons of strength, the thickness of the tube wall should be 2.5 mm (0.1 in).

The manufacturers of cardboard tubes frequently suggest that the plastic lids of the cardboard tube are sufficiently retained by frictional forces, but experience has shown that a heavy roll of diazo prints can, at times, push the lid off. The easiest way of securing the lids is to use staples which, by means of a heavy-duty stapler, are pressed into the lids through the cardboard tube, preferably with the ends of the staples inside the tubes so that no one tears their fingers (see Fig. 7.10). To accommodate the bead of the lids it may be necessary to grind a matching groove in the stapler.

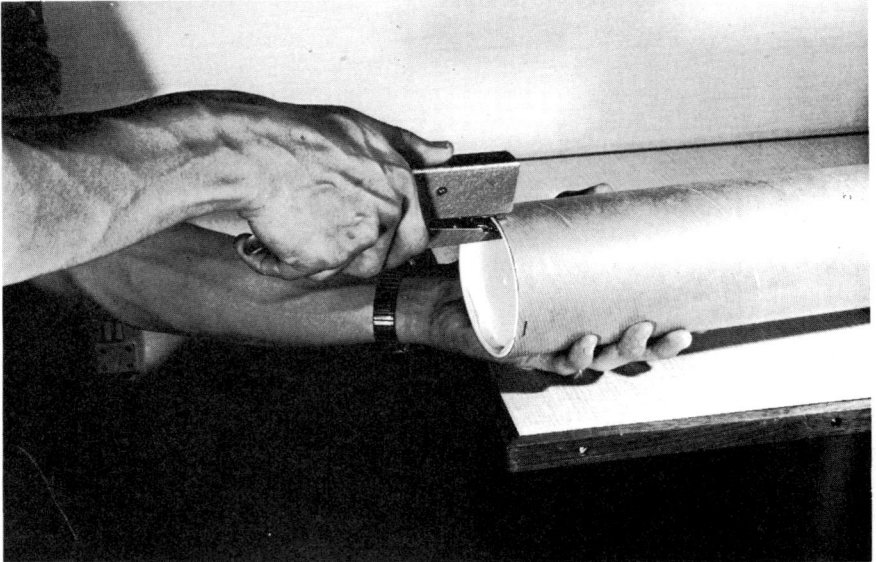

Fig. 7.10 Securing lids by means of a stapler.

The diazo prints are rolled and fastened by means of tape to prevent unrolling in the cardboard tube, since, otherwise, difficulty may be experienced in removing the prints from the tube. A pad of foam plastic is placed in each end of the tube in order to prevent the roll from moving backwards and forwards and damaging the edges of the prints.

The cardboard tube can be provided with the following markings: 'Letter', to ensure that it is sent by letter post; 'Registered', to ensure safe delivery; 'Express', to ensure fast delivery. Despatching artwork to foreign manufacturers this way is the easiest and cheapest method. There are no despatch documents nor customs formalities, and packing in cardboard tubes is convenient and cheap. In the author's experience, a despatch in Europe reaches its destination within 2–3 days which is very fast, the cost being about £3.00 for artwork necessary for four or five models.

8

Preparation of Precision Artwork. Practice

8.1 Introdution

In Chapter 6 we discussed the theoretical aspects which should be taken into consideration when preparing the artwork, but these are far removed from the practical situation. In this chapter we will consider, in a fairly detailed manner, the methods which, in the author's experience, bring artwork quality up to the level which the PCB manufacturer is entitled to expect.

The treatment is biased towards the method described in Section 6.3.8, i.e. the taping of a common pad master and the subsequent contact printing on diazo film of the two copies. Since, however, the methods which are described—apart from the contact printing—are universally applicable, the treatment will also cover most of the methods described in Chapter 6.

We shall also deal with the practical preparation of the engineering drawing; that is, the drilling or dimensional drawing. In connection with the master drawing, we shall consider a registration system which is easily administered.

8.2 Equipment

A principal precondition for the preparation of an accurate and professional artwork is that the PCB drawing office is adequately equipped. Perfect artwork greatly facilitates manufacture of the boards, put it can only be produced if high-quality drawing tools and equipment are available. We shall, therefore, deal in some detail with the equipment which is necessary in the drawing office—apart from the contact-printing equipment which was dealt with in Chapter 7—and try to lay down certain quality criteria.

8.2.1 Light Table

A light table large enough to accommodate the board size in the chosen work scale, is indispensable in the preparation of the artwork (see Section 5.2.3a and 6.4). The light table should be so constructed that the work surface can be tilted to allow the draughtsman to perform his work in a sitting position (see Fig. 8.1). It can be extremely tiring to work at a light table where the artwork is so placed that it is necessary to stand up in order to reach the whole of the work area (see Fig. 8.2). A sitting position,

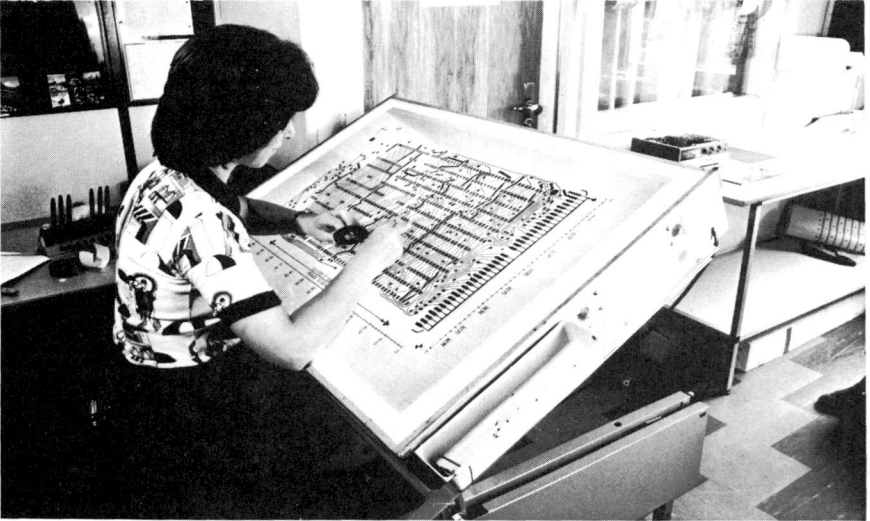

Fig. 8.1 Light table. Work surface can be tilted from level to 75°.

Fig. 8.2 Light table. Work surface level.

together with an adjustable work surface should result in better quality work because the draughtsman suffers from less fatigue. Also, the parallax error will be decreased as it is easier to view the work surface perpendicularly (cf. Section 6.4.1). The optimum solution is to mount the light table on a column, thereby permitting it to be tilted and adjusted for height (see Fig. 8.3).

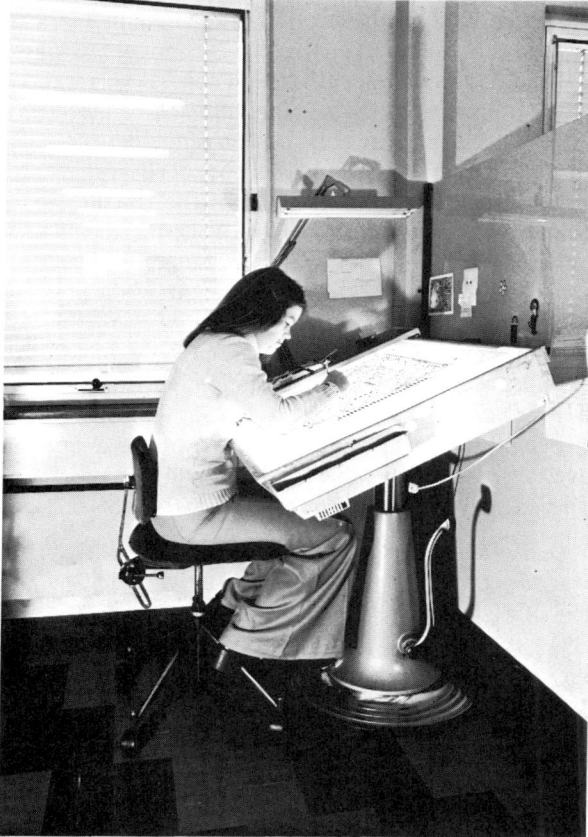

Fig. 8.3 Light table mounted on a column.

The work plate must not be flexible. In many light tables an acrylic plate, which is rendered opalescent in order to eliminate dark spots, is used. For cost reasons, the plate is made fairly thin, often just 4–5 mm (0.16–0.2 in) so that it flexes when loaded, making it difficult to perform precision measurements using a glass scale (see Section 8.2.4). In order to overcome this defect and also to protect the needle of the drop bow compasses, a 6-mm (0.25-in) glass plate should be placed under the acrylic plate. The additional thickness implies that the rabbet of the frame must be deepened and adjustment made to the balance of the light table.

Often the light table is provided with fluorescent tubes with a total power of 100 W which can easily increase the temperature of the top side of the acrylic plate by up to

228

20°C. The result can be that the grid which is used as a permanent underlay, expands by 0.2 mm over 600 mm (0.008 in over 24 in). An electronic light dimmer, whereby the power is reduced, will lower the surface temperature and ensure that the draughtsman's eyes are not unnecessarily fatigued by strong light.

It is possible to modify existing light tables using commercially available dimmers. A separate filament transformer is required for each fluorescent tube which should be the 'quick-start' type. Ignition can be effected with a longitudinal electrode printed externally on the glass tube with conductive ink. It is not always possible to obtain these special fluorescent tubes in all lengths but, if necessary, a standard type can be used by winding a piece of wire around the glass tube and connecting it to the voltage source through a 1-MΩ resistor. Such installations should, of course, be made by a qualified electrician who observes accepted safety procedures.

Ventilation can be provided by holes in the sides of the light table and the increase in temperature will limit the temperature rise of the work surface. In order to facilitate ventilation when the light table is tilted, the holes should be located at the front and rear of the light table. Built-in fans are not recommended as the noise and the draught can be distracting to the draughtsman. The best solution is, therefore, to tilt the light table and turn down the light. If the light table is tilted through more than 20–30°, the friction between the drawing tools and other material and the work plate will be insufficient to hold them on the plate. It is not practical to use a table positioned beside the light table or behind the draughtsman. The solution is a small utility carriage which can be moved across the light table as demonstrated in Fig. 8.4. The carriage is provided with a special high-friction plastic foil so that the tools remain on the surface even when the light table is tilted 70–80°.

Fig. 8.4 Movable utility carriage on light table.

It is not difficult to make light tables for special purposes. Figure 8.5 shows a light table, which is used for reasonably large artwork, mounted on the base of an old drawing board. Figure 8.6 shows another light table which is used when taping long

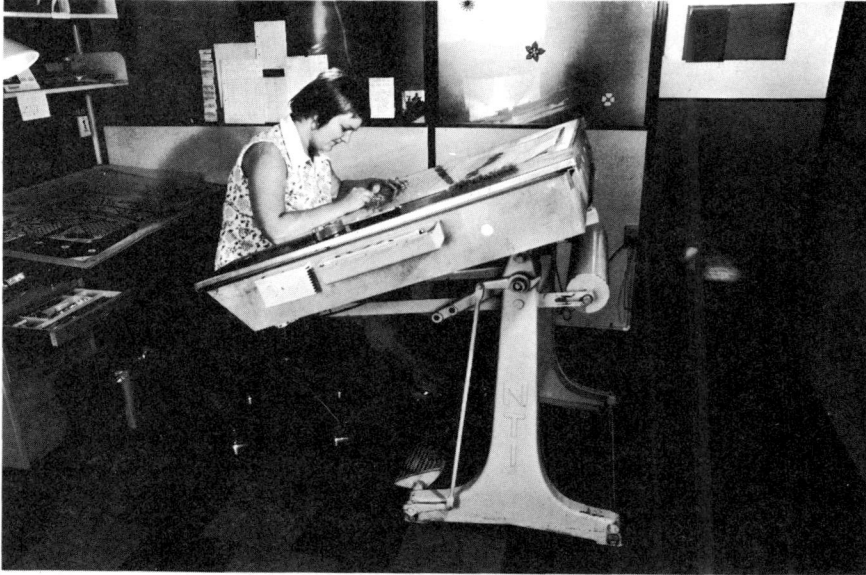

Fig. 8.5 Home-made light table mounted on base of old drawing board.

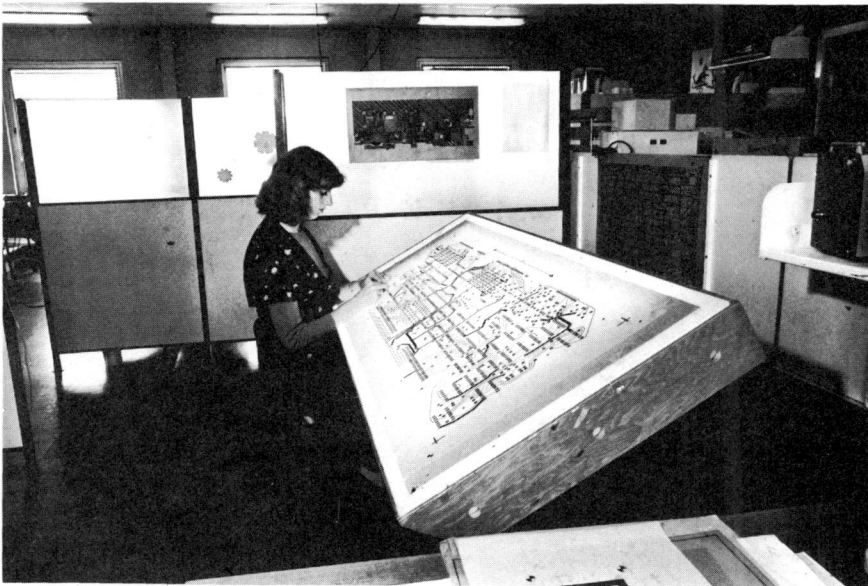

Fig. 8.6 Home-made light table based on an available table frame.

and narrow artwork, for example, for motherboards. The light box is hinged to the front of the table frame. A very simple bar system, by means of which the work plate can be tilted at 3 different angles, is located at the rear. Details are given in Figure 8.7.

Fig. 8.7 Details of the light table shown in Fig. 8.6.

8.2.2 Other Facilities

An auxiliary table should be located near the light table so that the draughtsman can place the sketches and other artwork belonging to the set within easy reach.

A drawer cabinet, mounted under either the light table or the auxiliary table is indispensable; the draughtsman needs somewhere to keep his drawing tools and his personal supply of the commonly used tapes and symbols.

Finally, the PCB drawing office needs a file for artwork which is finished but not yet released. In order to avoid damage, the artwork should be stored horizontally in a drawer file.

8.2.3 Personal Equipment

Figure 8.8 shows the most important tools which should be given to the draughtsman as personal drawing tools. Since nothing is so frustrating to the draughtsman and

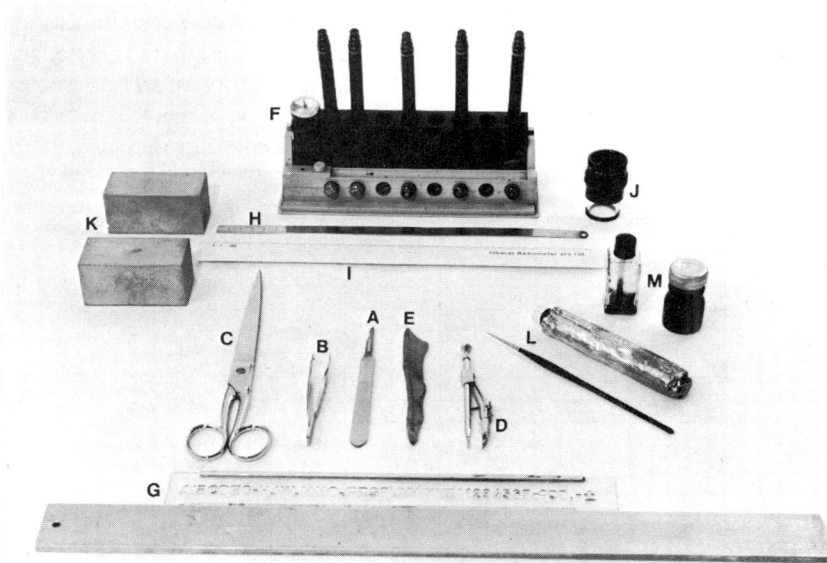

Fig. 8.8 The draughtsman's personal equipment. (See text.)

damaging to the quality of the finished product as poor drawing tools, he should be provided with professional instruments. The tool kit contains the following items:

A. A scalpel with exchangeable blade
B. A pair of pointed tweezers
C. Large paper shears
D. Stable drop bow compasses with a cutting blade mounted in the lead holder
E. A small wooden stick for fixing transfer symbols
F. A Rapidograph fountain-pen set
G. A lettering template with guiding rail
H. A thin steel measuring scale
I. A 1:4 measuring scale, calibrated in 0.1-in modules
J. A magnifying glass, minimum magnification ×7, with graticule for precision measurement
K. Some weights for clamping the drawings
L. Touching-up ink and a pointed touching-up brush
M. Erasing fluid for diazo and a brush

In addition to the above list, he should have standard draughting tools such as pencils, triangles, erasers, etc.

232

The light table should be provided with a precision inch grid which is a first-generation copy of a precision glass master. For a new grid, at a temperature of 20°C and a relative humidity of 50%, the cumulative error over a distance of 1000 mm (40 in) should not exceed ±0.05 mm (±0.002 in) and the line spacing should not vary by more than ±0.015 mm (±0.0006 in). The grid lines must be reasonably thin, preferably less than 0.10–0.15 mm (0.004–0.006 in) for the accent lines, and 0.05–0.075 mm (0.002–0.003 in) for the standard lines. In the author's experience, it is preferable to use a grid which matches the artwork scale, i.e. for an artwork scale of 4:1, the grid should be accented on every fourth line, as shown in Fig. 8.9, in order to facilitate the counting of modules, especially if the draughtsman does not have the 1:4 measuring scale mentioned under I in the above list.

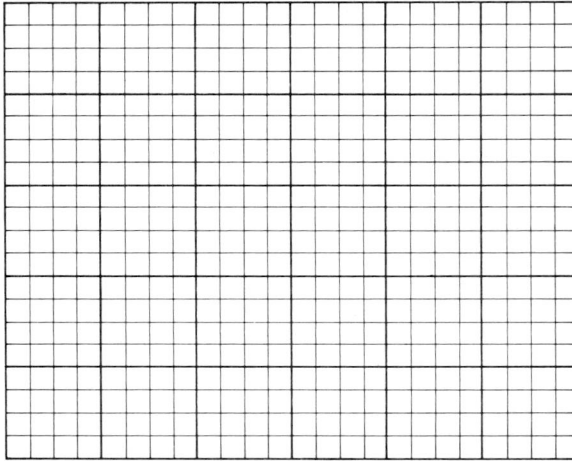

Fig. 8.9 Section of a grid.

The grid will expand and contract with rise and fall in temperature of the light table (see Section 6.3.7) and eventually there will be a progressive deterioration in accuracy of the grid to the extent that the error becomes unacceptable. In order to check this, a precision glass scale, of the type mentioned in Section 8.2.4 below, should be used to ascertain the error. Any deviation greater than 0.1 mm (0.004 in) over a length of 1000 mm (40 in) is unacceptable.

8.2.4 Precision Measuring Equipment

In addition to the draughting equipment mentioned in Section 8.2.3, the draughtsman needs, on certain occasions, precision measuring equipment. There is no need for each draughtsman to have a set of this equipment for his personal use and one set will normally be adequate even for a very large PCB drawing office. A precision steel scale for marking the various mechanical dimensions will be indispensable, especially if metric dimensions are to be used and the grid is an inch grid. The steel scale, as shown in Fig. 8.10, should have a bevelled, calibrated edge which counteracts the parallax

Fig. 8.10 Precision steel scale with chamfered edge.

error which could otherwise be introduced with scales of this type. An indication of the level of accuracy of such scales can be obtained from the German standard DIN18701 where the accuracy at 20°C is given as:

$$\Delta L = \pm\,(0.02 + 2L \times 10^{-5})\,\text{mm}$$

where L indicates the length of the scale in mm, and ΔL indicates the deviation from the nominal length. For a 1000-mm (40-in) long steel scale the accuracy is ± 0.04 mm (± 0.0016 in). The steel scale should, preferably, be a little longer than the largest dimension to be marked or checked, since some accuracy will be lost if the measurement has to be taken in two parts.

The steel scale is not adequate for accurate measurement of the photo-reduction dimensions, partly because of its inherent inaccuracy and partly because the necessary interpolation will be too inaccurate. For this reason, the purchase of a precision glass scale is recommended, and this should be regarded as the primary measuring standard of the PCB drawing office. An example of a precision glass scale is shown in Fig. 8.11. Several makes are found on the market but common to all of them is that the precision glass scale is held firmly in a rack which carries a carriage provided with a microscope offering a magnification of about ×50. The glass scale is aligned by means of the microscope so that its zero-point corresponds very accurately to the start-point of the dimension to be measured. Then the carriage is moved to the extreme point, and the glass scale is read. A commonly used glass scale is 500 mm long and has principal divisions of 2 mm. The scale of the microscope sub-divides the 2 mm divisions into 0.02 mm sub-divisions with a possibility of interpolation down to 0.005 mm. The accuracy of the precision glass scale is approximately ± 0.01 mm at 20°C. Similar glass scales are available calibrated in inches.

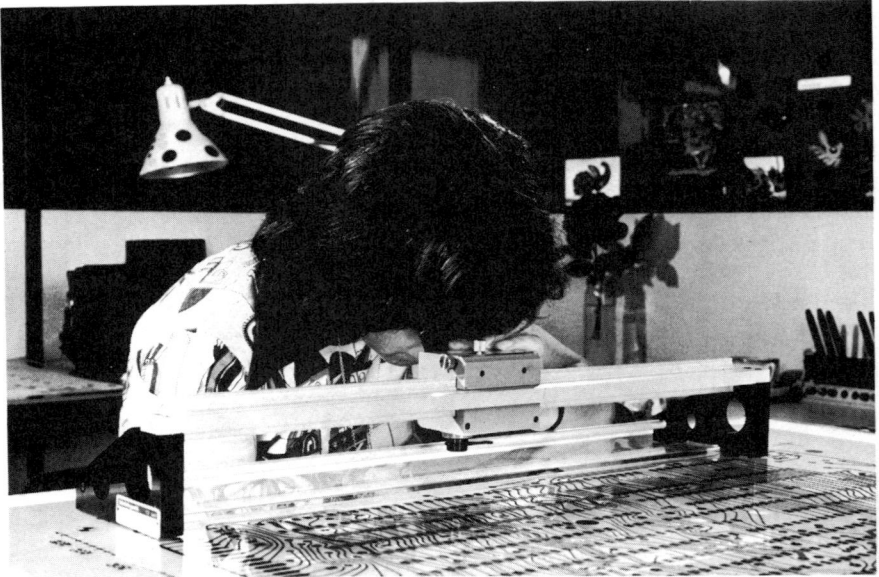

Fig. 8.11 Precision glass scale.

Obviously, having gone to so much expense to ensure accurate measurements, it would be unwise not to ensure that the work plate was flat and did not, therefore, introduce a further error. The point made in Section 8.2.1 regarding a glass plate placed under the acrylic work plate is clearly important.

8.3 Fundamental Disciplines of Artwork Preparation

8.3.1 Introduction

In order to produce accurate taping—a prerequisite for precision artwork—the draughtsman must be competent in the various techniques for the application of pads and tapes. In this section we will break down the working procedure into its individual processes, which then can be combined in accordance with the present demand. It is assumed that the draughtsman is fully conversant with the contents of Chapter 6 including the various design rules.

8.3.2 Applying Pads and Other Symbols

The symbols are available in two fundamentally different types:

A. Pressure-sensitive self-adhesive symbols mounted on backing paper
B. Transfer symbols

Both types comprise individual symbols such as pads and multi-pad configurations, for example all the pads for a dual-in-line IC. An IC configuration can, of course, always be made by applying individual pads, but it is easier and considerably more accurate to use multi-pad configurations.

8.3.2a Self-adhesive Symbols

The self-adhesive symbols can be divided into two groups. The first group comprises symbols which are removed from the backing paper by slipping a knife blade under the edge of the symbol and peeling it off. It is recommended that the symbols be held against the blade with a finger, but when loose from the backing paper, the symbol sticks to the blade and can easily be positioned on the drafting film (see Fig. 8.12). It is important that the symbol is not pressed down so hard initially that it cannot be re-positioned easily. When the exact position is determined the symbol can then be firmly affixed to the drafting film.

Fig. 8.12 Applying self-adhesive symbols.

The other group comprises self-adhesive symbols, also in multi-pad con-figurations, which are mounted on a transparent carrier strip, the adhesive side away from the carrier strip and covered with backing paper. When a multi-pad symbol is to be applied, the backing paper is removed. Due to its transparency, the carrier strip can then be easily positioned. Again, when the exact position has been determined, the symbol can then be pressed down on to the drafting film.

8.3.2b Transfer Symbols

These symbols are printed on a very thin film (thickness 0.01 mm or 0.0004 in) which, again, is mounted on a transparent carrier strip. The symbols are transferred from the carrier strip to the drafting film by pressure, for example, by rubbing the reverse side of the carrier with a small wooden stick. The carrier strip can then be lifted from the drafting film (see Fig. 8.13). In order to improve the adhesion, the separating paper strip,

Fig. 8.13 Applying transfer symbols.

which protects the 'adhesive' side, is now placed on top of the symbol and rubbed with the wooden stick. Because the symbol does not adhere until it is rubbed it can be laid directly on the surface of the drafting film without any risk of its sticking. This fact, together with negligible parallax, makes it easy to locate the symbol very precisely. However, since the material from which the symbol is made, is very thin, transfer symbols are rather delicate and, therefore, easily damaged. On the other hand, transfer symbols are very suitable when using the diazo contact-printing method because they are so thin (see Examples 1 and 2 in Section 7.3.3) and also because they are required to last only until the common pad master has been contact printed.

8.3.2c Location of Pads

Normally the draughtsman tries to apply the pad so that the centre hole is located symmetrically about the grid intersection (see Fig. 8.14) taking into account the

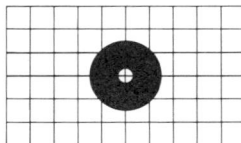

Fig. 8.14 Location of pad.

parallax error described in Section 6.4.1. The centre hole must not be too small or else it will be too difficult to locate the pad accurately. Pads come with many different standard sizes of centre hole, but a centre hole diameter of 1.57 mm (0.062 in) is most suitable for use with the 4:1 scale. This is also important when making prototype boards in the laboratory. After etching, the centre hole will be about 0.4 mm (0.016 in) which is a good size for locating the drill point.

Multi-pad symbols are frequently provided with two small crosses which are used for location purposes. These crosses, which are very thin, will disappear when the artwork is photographically reduced.

Since the pads are located in accordance with the layout sketch but adjusted to coincide with the grid of the light table, it is practical to place the layout sketch under the drafting film and to align it to the grid (cf. Section 5.22.2). If puppets or semi-sticky symbols have been used in the layout, their thickness plus that of the drafting film may introduce a significant parallax error (see Section 6.4.1). In order to minimize this, a thin contact print of the component layer should be produced so that the thickness of the component layer can be reduced.

The size of the pads has been dealt with in detail earlier (Sections 6.5.4b, 6.5.5 and 6.5.6) and will not be further discussed here. Incidentally, the size of the pads should have been determined at the layout phase in order to facilitate realistic conductor routing.

8.3.2d Trimming of Pads

During the taping of conductors it might occur that the minimum requirements for spacing between conductors and pads cannot be maintained using the conventional symbols. If this occurs, an obvious solution is to trim the pads (see Fig. 8.15). In practice, this is a relatively delicate task since the draughtsman must ensure that he does

Fig. 8.15 Trimming of pads.

not reduce the width of the annular ring to less than 0.3 mm (0.012 in) in a 1:1 scale. Alternatively, he can control the width of the annular ring by using a smaller pad or even an oblong pad, the width of which corresponds to the desired trimming. The draughtsman must be very careful when applying the new pad to ensure that it is absolutely concentric with the pad on the reverse side, thereby avoiding a reduction of the annular ring on the finished board or, in the worst case, a hole breakout. If it turns out that the requirements of minimum width of the annular ring and minimum spacing cannot be met, the only solution is to modify the affected area of the layout.

8.3.2e Removal of Pads

A self-adhesive pad is removed using a scalpel in the same way as described for its removal from the backing paper. It is more difficult to lift a pad from drafting film than from backing paper as the latter is impregnated in order to facilitate the release. Therefore, there is a risk that a pad removed from drafting film will be deformed and, for this reason, a pad, once removed must not be re-used (cf. Section 6.4.1).

A transfer pad is most easily removed using a piece of ordinary adhesive tape which is applied to the pad and then quickly peeled off. It is also possible to use a scalpel, either to lift the pad or simply to scrape it off.

A diazo pad is removed by diazo erasing fluid. The fluid is applied to the pads with a small brush and the diazo layer is dissolved quite quickly. It can then be wiped off with a paper napkin. The procedure is shown in Figs. 8.16A–8.16C. A scalpel can be used but the blade quickly becomes blunt. Apart from the irritating nature of the task, the method has the additional disadvantage that small chips are produced. These chips are liable to stick to any glue along the edges of the conductors, causing blurred conductor images on the master film.

8.3.2f Improving the Solderability

When the artwork for both the component and solder sides has been taped with the conductors, it may be necessary to adjust some pad sizes in order to improve the solderability. The object is to reduce the heat transfer away from the pads on the component side. These techniques have been discussed in Section 6.5.7 and illustrated in Figs. 6.21 and 6.22.

8.3.3 Taping Conductors

Black self-adhesive crêpe tape is normally used for conductor taping. This tape can be laid in soft curves, something which cannot be done with the red and blue tapes used for red/blue taping on one sheet of drafting film (cf. Section 6.3.6). The routing of the conductors was dealt with in Sections 6.6.1 and 6.6.2 and will not be repeated here. Instead, we shall consider the practicalities because a poor taping technique incurs a significant risk of creepage, of the order of magnitude of 0.2–0.3 mm (0.008–0.012 in), in the course of a few months. The most vulnerable places are where the tape is laid down in curves so that it is stretched. The tape will contract a little in the course of time, and in returning to its natural length, will change the radius of curvature. The most serious effect is that the spacing between adjacent conductors or between conductors and pads will be reduced and, in certain cases, the minimum spacing conditions will not be met. This will almost certainly complicate the manufacturing processes and probably make it impossible to produce satisfactory boards. Another effect is that the tape which has moved will leave some adhesive behind it making the conductors on the reduced master film blurred. In Fig. 8.17, the tape has crept 0.5–1 mm (0.02–0.04 in) and left a shadow of glue.

It is clear, from the foregoing, that we can profit from devoting some time in consideration of the details of the working methods. The use of special tape holders or special guiding devices, however, is not recommended since their working accuracy, in general, is inadequate.

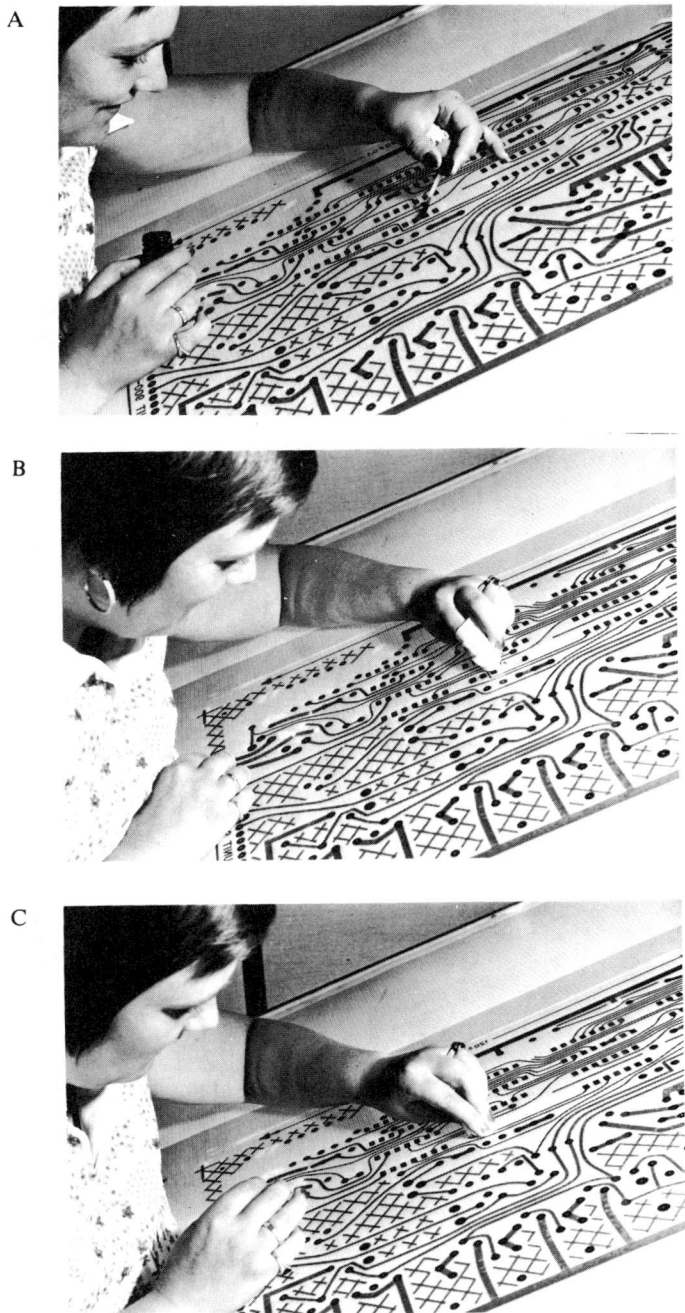

Fig. 8.16 Using diazo erasing fluid. A, application; B, dissolving; C, wiping off.

Fig. 8.17 Creeping of taped conductors.

8.3.3a Taping Conductors in Soft Curves

The most important rule for the taping of conductors is that the tape must never be stretched. In order to avoid contraction later, it is important, when the tape is laid down in a curve, not to stretch the outer edge (see Fig. 8.18). The tape should,

Fig. 8.18 Laying the tape without stretching.

therefore, be laid in such a way that the outer side maintains its natural length. Consequently, the inner edge has to be compressed. When laying conductors in soft curves the tape should be held fairly loosely and the curve produced by compressing the inner edge with the index finger, piece by piece, and turning the tape in the direction of the curve (see Fig. 8.19).

8.3.3b Taping Conductors in Sharp Curves

Obviously, the sharper the curve, the more difficult it is to avoid stretching the tape. The space will probably also be restricted since otherwise there would be no need to

Fig. 8.19 Taping of conductors in soft curves.

lay the conductor in a sharp curve—under these conditions any creepage is likely to become a problem. In order to avoid significant creepage, it is better to use pre-cut angles or to truncate the angle. Alternatively, the tape can be cut transversely three-quarters through, bent in the desired direction and the tape flap lifted over the tape which has already been laid. By repeating the procedure, it is possible to bend the tape rather sharply but this method is not practicable with the narrower tape widths. The three methods are illustrated in Fig. 6.32 (Section 6.6.2).

Pre-cut elbows as well as concentric rings which can be cut to any desired angle, are made both as self-adhesive and transfer types in black, red and blue. The transfer symbols tend to be too delicate and therefore the self-adhesive type is recommended. Black symbols usually are 0.10–0.12 mm (0.004–0.005 in) thick—approximately twice the thickness of the red and blue ones. This leads us to select coloured self-adhesive symbols and for photographic and contact-printing reasons the red symbols are preferred.

8.3.3c Taping Irregularly Shaped Conductors

Occasionally, a conductor must have an irregular shape as shown in Fig. 8.20. It is inconvenient to produce the conductor by laying down several tapes side by side with a small overlap. It is far better to choose a red, transparent, self-adhesive lithotape such that it accommodates for the total width of the irregularly shaped conductor. Lithotape can be obtained in widths up to 50 mm (2 in). The contour is trimmed with a scalpel, and the surplus lithotape peeled off. These conductors are, of course, produced before the start of the ordinary taping.

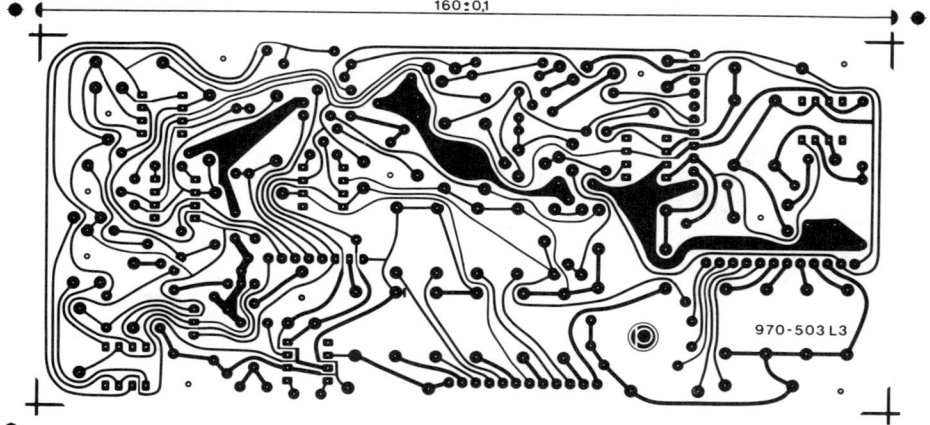

Fig. 8.20 Irregularly shaped conductors. (Dimension in millimetres.)

8.3.3d Taping Parallel Conductors

Previously we emphasized that parallel conductors can cause manufacturing problems (cf. Section 6.5.1). The draughtsman should, therefore, overcome his desire to produce neat taping in which the conductors run strictly parallel to one another, if it is possible to increase the spacing, even if it is only over a short distance, as shown in Fig. 8.21. On the other hand, there are many cases where it is necessary to lay a

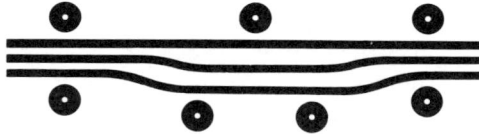

Fig. 8.21 Spacing parallel conductors.

number of conductors in parallel, separated only by the minimum spacing (see Design Rule Number 2, Section 6.5.1). An easy way of achieving even spacing is to use a spacer tape with a width corresponding to the desired spacing. When the first conductor tape is laid, the spacer tape is laid close to the conductor tape. Then the next conductor tape is laid close to the spacer tape and so on. When all the parallel conductors are taped, the spacer tapes are pulled up and a regular routing conforming to the spacing requirements is produced. The draughtsman should bear in mind that the spacing will be slightly larger than the width of the spacer tape, so, perhaps, it is wise to make a test taping in order to determine the correct width of the spacer tape before undertaking the final taping.

8.3.3e Taping Branching Conductors

If a conductor branches out from another conductor, the branch should preferentially be laid perpendicular to it (see Fig. 8.22). The branching conductor is laid over the

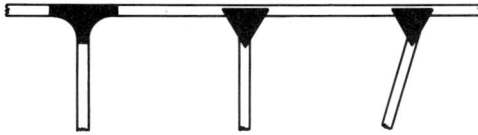

Fig. 8.22 Reinforcing the branch-point of a conductor.

trunk conductor but if the latter is narrow, for example only 1.27 mm (0.05 in) in the artwork scale 4:1, the overlapping is small. If the conductors are stretched during the taping operation there is a risk that the branching conductor will contract or the trunk conductor will creep away from it. The result is an interruption at the branch point. In such cases, a 'Tee', or a triangle, which can possibly be taped on the rear side, should be used. The triangle is preferred because it does not require to be located as precisely as the 'Tee'. In all cases, in order to avoid confusion, symbols, which could be mistaken for ordinary pads, should not be selected.

8.3.3f Taping Conductors through a Narrow Gap

If conductors are to be routed through a narrow gap between two pads, they should be so laid that they run perpendicular to the line joining the centres of the pads (see Fig. 6.25). In this way, the maximum conductor spacings will be achieved.

Figure 6.26 illustrates the relationship between the number of conductors, the width of the tape, the centre-to-centre distance between the pads, and the pad sizes. This diagram is of particular use to the PCB designer. In the draughtsman's case it is more a question of determining the spacing, for which purpose the curves shown in Figs. 8.23–8.26 are more suited. The curves are derived from the expression given below and apply to an artwork scale of 4:1

$$a = \frac{A - nb}{n + 1}$$

where a = the spacing; b = the width of the tape; A = the width of the gap; n = the number of conductors (of the same width). The curves serve first and foremost to give the draughtsman a feeling for the size of the spacing so that the first conductor can be laid down correctly.

8.3.3g Transition between Conductors

A conductor should, preferably, be taped in one piece although sometimes it is necessary to extend a conductor, for example, if modifications are introduced. Where this is necessary, the overlap should be significant, if possible not less than 10 mm (0.4 in). The overlap should be made very precisely in order to avoid increased width.

At certain points it may be necessary to reduce the width of the conductors, to neck down the conductors so that they can pass through a narrow gap. If so, the conductor can be tapered as shown in Fig. 6.28 but tapering is not absolutely necessary.

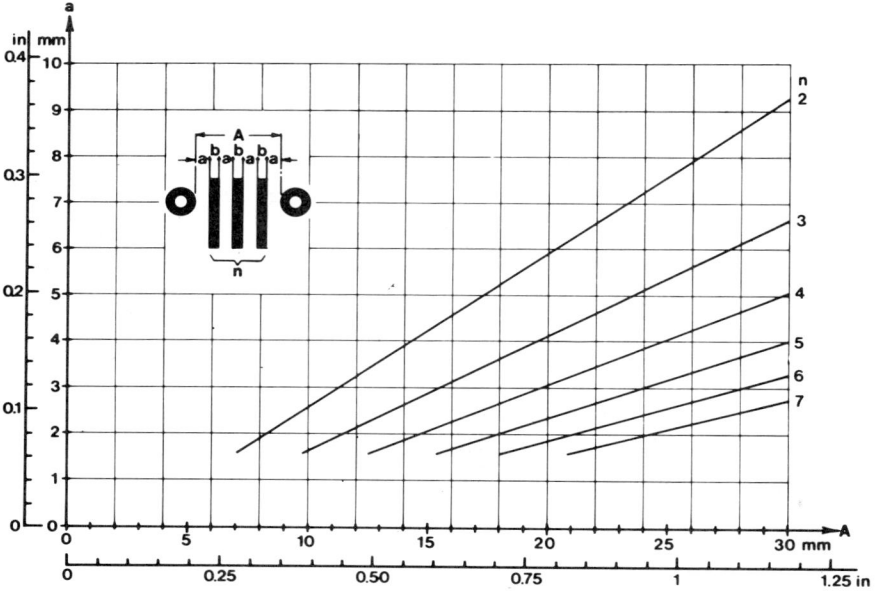

Fig. 8.23 Taping conductors through a narrow gap. Tape width b = 1.17 mm (0.046 in).

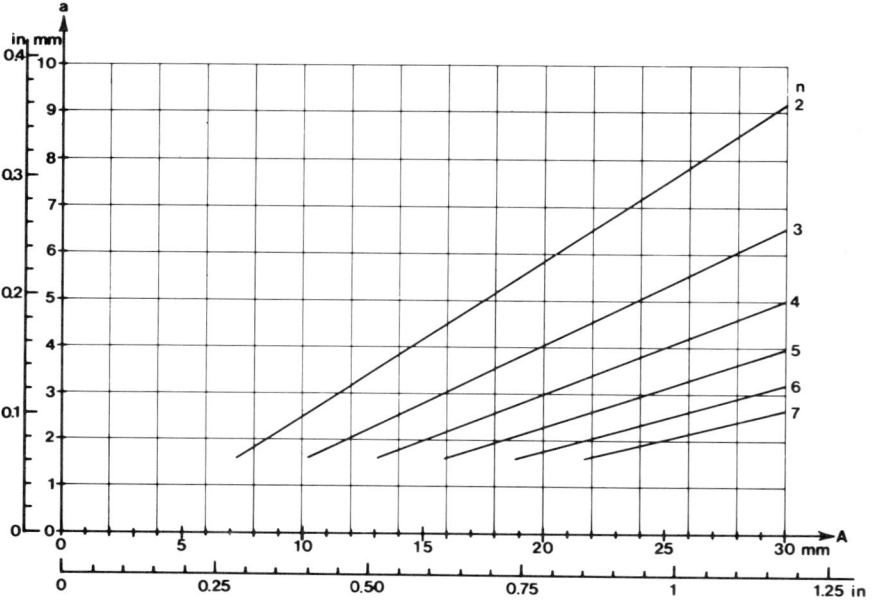

Fig. 8.24 Taping conductors through a narrow gap. Tape width b = 1.27 mm (0.050 in).

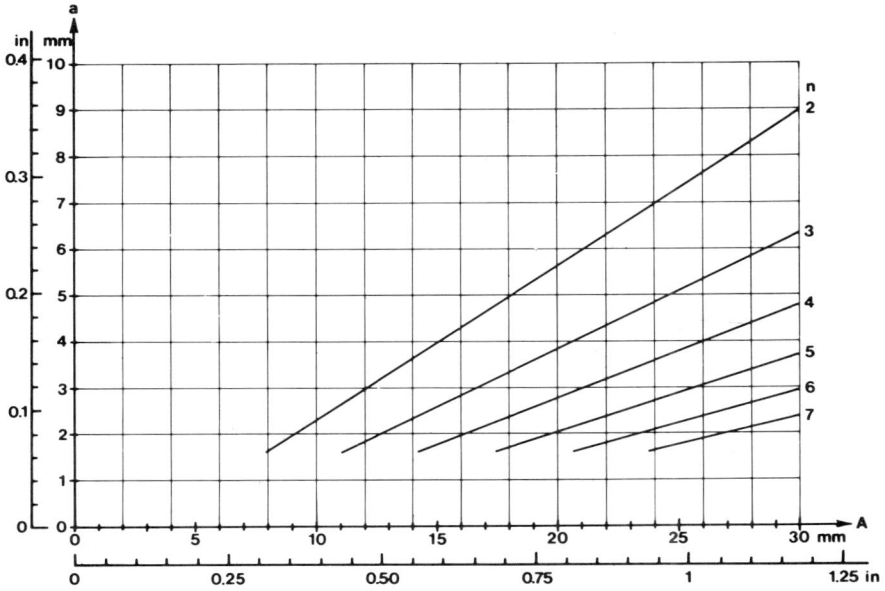

Fig. 8.25 Taping conductors through a narrow gap. Tape width $b = 1.57$ mm (0.062 in).

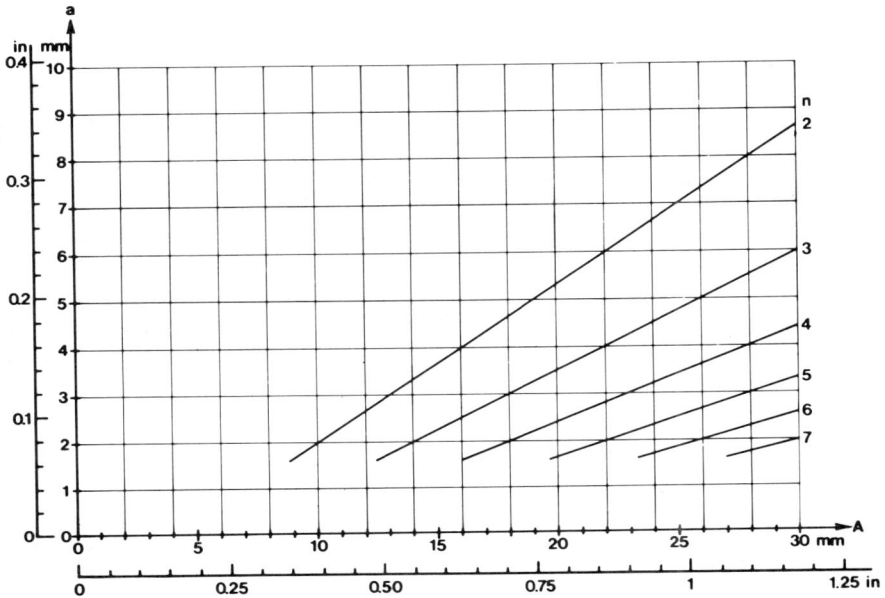

Fig. 8.26 Taping conductors through a narrow gap. Tape width $b = 2.03$ mm (0.080 in).

8.3.3h Terminating a Conductor on a Pad

A conductor must 'land' symmetrically on a pad as this produces the best condition for soldering. The tape should overlap the pad to such a degree that the connection cannot be broken if the tape contracts following stretching at the laying stage. On the other hand, the tape should not overlap such that the centre hole is covered as the PCB manufacturer uses the centre hole rather than the circumference of the pad when digitizing the board (see Fig. 8.27). 'Teardrop' pads produce a smooth transition

Fig. 8.27 Transition between conductor and solder pad.

between the pad and the conductor, but as mentioned in Section 6.5.6, the use of tear-drops is not always practical.

Wide conductors should be narrowed down to the pad dimension at the meeting-point (see Figs. 6.21a and 8.28). A conductor should not be connected to an oblong

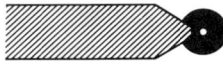

Fig. 8.28 Tapering of wide conductor.

pad where the width of the 'annular ring' is narrowest but to the end of the pad. This is particularly important in the case of an IC configuration where two adjacent pads are to be interconnected as shown in Fig. 8.29(A). Eccentric drilling of the holes will

Fig. 8.29 Interconnecting adjacent oblong pads.

reduce the reliability of the electrical connection between the conductor and the pad as a breakout of the hole reduces the area of the transition zone (B). Actually, the connection should be laid as shown in (C) in order to avoid the connections shown in (A) and (B) being taken for solder bridgings during the inspection of the soldered boards. However, the space around the IC pads will usually be fairly limited so that (C) is not so advantageous because the number of conductors it is possible to run longitudinally through an IC or between two ICs is reduced by one. Therefore the connection shown in (A) is to be preferred.

Tape is best cut using a scalpel. If the tape is laid on the same side of the drafting film as the pads, the cutting must be done with extreme care (cf. Fig. 7.4). The knife blade should be kept a little above the drafting film, and the tape tightened up against the knife edge simultaneously with sideways movement of the scalpel. If the tape is laid on the other side (which is more practical) tape cutting does not present a problem. The conductor is taped across the pad, and the cutting is performed by pressing the knife edge down against the tape and pulling the tape upwards.

8.3.3i Conductors Close to the Board Edge

As mentioned earlier, the tolerance of the contour of the board is estimated to be ± 0.2 mm (± 0.008 in). It is necessary, therefore, to keep the conductors away from the edges by at least 0.6 mm (~ 0.025 in) on the finished board, or minimum 1 module (~ 0.1 in or 2.54 mm) on artwork prepared on a 4:1 scale.

8.3.3j Finishing the Taping of Conductors

After the conductor tapes have been laid they are pressed against the drafting film using a finger nail or rounded metal point, in order to achieve positive adhesion.

Finally, a check should be made that the minimum spacing requirements are met. The draughtsman will soon learn to judge the spacings by eye but if he is in doubt he can check the spacing using a magnifying glass with a dimensioned graticule. All the critical areas must be inspected: conductors which may have crept and violated the minimum spacings, e.g. between a conductor and an adjacent pad, or between parallel conductors. Other critical areas could be branchings between thin conductors.

8.3.4 Photo-reduction Marks

The artwork must be provided with a photo-reduction dimension which is used by the photographer for adjusting the degree of reduction and for checking the size of the finished master film. The photo-reduction dimension is normally located lengthwise the board (see A, Fig. 8.30), but not necessarily tied to the overall dimension of the board. A length corresponding more or less to the largest dimension of the board is marked on the artwork. The dimension is stated as the desired length on the reduced master film. A tolerance of ± 0.05 mm (± 0.002 in) seems to produce a reasonable compromise for both the photographer and the PCB manufacturer.

When the diazo contact-printing method is used, it is practical to place the photo-reduction dimension on the common pad master in order to achieve exactly the same lengths on the contact prints. The dimension and the tolerance should not be given on the common pad master as the lettering will be reversed on one of the copies. On the other hand, if the common pad master is used as the basic standard for a series of standard boards, it might be practical to state the dimension and the tolerance both direct-reading and reverse-reading on the pad master.

As shown in Fig. 8.30, the photo-reduction dimension is indicated as the length between two marks, and in the author's experience it is very convenient to use 'Tee's. These symbols are normally provided with small alignment marks which facilitate their location. Precise marking of the photo-reduction dimension is not possible if it is based on small pencil marks located using a precision steel scale. It is better to use the grid for the precise location, and the steel scale for seeking out the correct grid lines. If the grid lines are all equally thick, the photo-reduction marks can be aligned with, say, the right edge of the grid lines. A particular problem arises if the dimension is to be given in metric measure, and an inch grid is being used. Fortunately, there are some

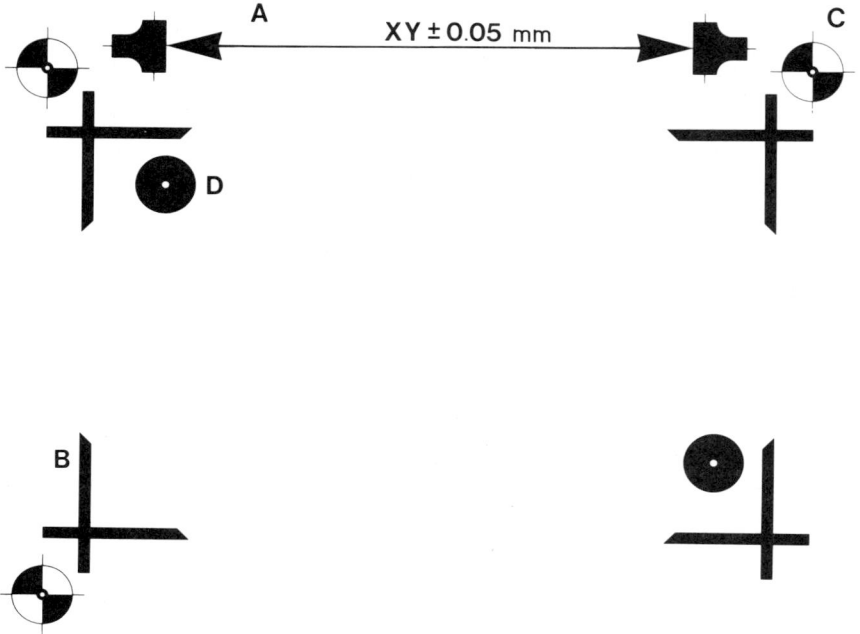

Fig. 8.30 Preparation of artwork.

module measures which can closely approximate to whole numbers of millimetres. These module measures are given in Table 8.1.

Table 8.1 Specially selected metric photo-reduction dimensions.

Photo-reduction dimension to be stated on artwork (mm)	Accurate dimension (not used) (mm)	Number of 0.1-in modules	Dimension for identifying correct 0.1-in module (mm)
23.5	23.495	37	94
40	40.005	63	160
63.5	63.500	100	254
80	80.010	126	320
120	120.015	189	480
160	160.020	252	640
207	207.010	326	828
254	254.000	400	1016

It is clear that the longer the photo-reduction dimension is the better. The limitations are the size of the drafting film, which must match the artwork size, and the length of the precision scale. It should be a matter of course to check a photo-reduction dimension but if this has to be done by adding two measurements using a glass scale which is too short, the overall accuracy must be affected.

8.3.5 Board Profile

The board profile is defined by means of corner marks (board delineation marks) as shown in B (Fig. 8.30). If the drafting film is matt it is possible to indicate the location of the corner marks by means of small pencil marks, but on a glossy drafting film this is rather difficult, so it is better to draw the profile on an underlay of matt drafting film. After the corner marks have been applied, the dimension should be checked using a precision steel scale.

In order to avoid copper remaining on the board after profiling, it is common practice to apply the corner marks so that their inner edges are flush with the edges of the board. It is unnecessary to tape the complete contour or profile. Inward corners and cut-outs should not be sharp or right-angled as this increases the cost of manufacture. If a rounded corner can be accepted, the draughtsman can indicate this by writing $0 < R \leqslant 3.2$ mm ($\frac{1}{8}$ in). In this case, the corner mark is omitted at the rounding; otherwise the inner part will remain as metal after profiling. The machining is carried on the master engineering drawing (see Section 8.9.3).

8.3.6 Register Marks

The register marks serve to facilitate accurate registration between the two artwork sides so that all the pairs of pads are located concentrically. When the two sides are derived by contact printing from a common pad master, registration problems do not arise. If, however, the two sides are taped individually, the draughtsman must take particular care when laying the register marks.

Normally, three unsymmetrically located marks are provided, as demonstrated in C (Fig. 8.30) so that the two artwork sides can be overlaid in only one way. These marks should be thin and have sharply defined edges.

8.3.7 Tooling Holes

Two tooling holes are provided within the contour of the board, as shown in D (Fig. 8.30) in order to make it possible to locate the board on a jig for profiling. The diameter of the tooling holes should be approximately 3 mm or $\frac{1}{8}$ in, and, if possible, of the same size as one of the sizes already used on the board. This avoids a further drill change at the drilling stage. If the board is not plated through, the hole centres can be marked by pads, whose diameter should be slightly less than the diameter of the finished tooling holes. This means that a very thin copper ring around the tooling holes can be avoided. The tooling holes in plated through boards may be plated along with the other holes. They can, therefore, be drilled in the same drilling operation, improving their positional accuracy. In this case, the pads should be slightly larger than the drilled holes, and it might be an advantage to add a note on the engineering drawing stating that scratches to the plating of these holes will be accepted. The tooling holes should be located to that the centre-to-centre distance is as large as possible, i.e. a diagonal location is preferred. The location should be slightly unsymmetrical so that if the board is incorrectly located in the profiling jig, it is immediately obvious. A very useful reference system is described in Section 8.9.3.

8.3.8 Mounting Holes

It is practical to mark all the mounting holes which are used for mounting the board in the chassis, and for the mounting of mechanical units on the board, at the profile marking stage. The mounting holes are most often non-plated through and they are indicated in the same way as the tooling holes.

8.3.9 Edge Connectors

An edge connector can, of course, be taped contact finger by contact finger but it is much faster and also more accurate to use finished symbols comprising a number of contact fingers. Such symbols, which can be of considerable length, should be applied very carefully in order to ensure that they are not stretched, thereby reducing their accuracy.

If a special edge connector cannot be laid using commercially available symbols, the symbols necessary can be custom-designed and delivered by a symbol supplier, or they can be reproduced on a photographic film which is applied to the artwork using transparent tape. The starting point is, in both cases, a single contact finger drawn oversize, for example four times the size of the finished symbol. As mentioned in Section 3.7.4 the individual contact fingers must have an electrical connection to a plating bar. The connections from the contact fingers are made 0.5 mm (0.02 in) wide, and the plating bar about 2 mm (0.08 in) wide, in the 1:1 scale. The correct shape of the contact finger is described in Section 6.7 and illustrated in Fig. 6.37. On the basis of a single contact finger (as shown in Fig. 8.31) it is possible for the supplier to produce the

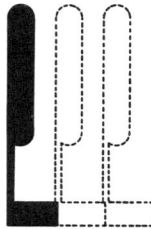

Fig. 8.31 Preparation of film of contact fingers.
————, Original contact finger;
– – – – –, 'stepped' contact fingers.

desired symbol. When only a small number of symbols is required and the tooling and set-up costs cannot be justified, the drawing is sent to a photographer who reduces the contact finger and, by means of a step-and-repeat process produces a film with the desired number of contact fingers spaced at the required module. If the draughtsman has diazo contact-printing equipment at his disposal he can always produce the symbols when the need arises so that just one positive film is required.

A polarizing slot, if required, can often be located through one of the contact fingers. Since the PCB manufacturer prefers that metal is not present where the slot is to be milled, the contact finger in question is removed from the artwork by cutting or erasing. The location of the slot is not to be marked on the artwork since this information is embodied in the master drawing (engineering drawing).

8.3.10 Ground Planes

The use of a red, transparent lithotape for the taping of irregularly-shaped conductors has been mentioned earlier. If a large ground plane is required it is possible to use a red, transparent and self-adhesive film which can be obtained from most suppliers of drafting aids in sizes up to 400 × 600 mm (16 × 24 in) or more. This film is applied to the artwork side on which the conductors are to be taped, i.e. opposite the pads. As the film is transparent, trimming of the contour is easy.

When a solder hole is required in the ground plane, a pad should be provided with a clearance from the ground plane and connected to it by two short lenghts of tape (see Section 6.5.7 and Fig. 6.21e). If the solder hole is not to be connected to the ground plane sufficient clearance must be allowed around the pad to ensure that solder bridging will not occur between the pad and the ground plane. This means that the clearance around the pads must be at least 0.6 mm (0.024 in) wide in the 1:1 scale. This also applies when the ground plane is on the component side since solder can come up through the hole, flow along the terminal lead and cause a solder bridging.

After the necessary clearances around the pads or along shielded or guarded conductors have been made, using drop bow compasses and scalpel, the ground plane must be relieved so that unbroken areas larger than 100–250 mm^2 (0.15–0.4 sq. in) are not found. This has been described in Section 6.6.4 and illustrated in Figs. 6.34 and 6.35. In high-frequency boards, however, the relieving of the ground plane ought not to be carried out without consulting the circuit designer since the high-frequency characteristics of the board may be affected.

8.3.11 Equalization of Pattern Distribution

The advantages of equalization or balancing of the pattern distribution has previously been discussed (see Section 6.6.3). Equalization, or balancing, is effected by introducing a number of isolated pattern areas which are not connected to any part of the electrical circuit (see Fig 6.33). So far as the author is aware, no self-adhesive balancing network films with suitable mesh sizes are available from drawing aids' suppliers. It is therefore necessary to tape a cross-hatched pattern on larger areas, unless balancing network films can be procured to special order from the suppliers of drawing aids. For this purpose, an accurate master is made for a small section of the network, usually in a scale of four times full size. Using a step-and-repeat process, the supplier can now produce the balancing network in the desired format. Two possible styles are shown in Fig. 6.34. The closed network should not be used on the solder side unless a solder mask is specified, as some of the meshes might otherwise be closed in the soldering process.

The balancing pattern is applied after the board has been taped with all the active conductors. Where large areas have to be filled, the balancing network is applied to the artwork side which does not carry the conductors, i.e. the diazo side. The contour is then trimmed to give adequate clearance from all pads and conductors. For correct contact printing, the balancing network should be not more than 0.05 mm (0.002 in) thick. Smaller areas are conveniently balanced by means of ordinary tape which is applied to the taped side of the artwork.

The extent to which balancing is used depends largely on the draughtsman's judgement of a reasonably equalized pattern. To ensure some uniformity in plating thickness on the two sides of the board, the two artwork sides should be balanced to show an approximately equal pattern density as judged by eye.

As mentioned in Section 6.6.3, the balancing areas should, if possible, differ in appearance from the functional parts of the pattern in order to avoid unnecessary touch-up work on non-functional areas.

8.4 Contact Printing on Diazo Film. Practice

In Chapter 7 the theoretical background of contact printing on diazo films was discussed, and in this section practical instructions for its application will be given. First of all we shall deal with the determination of the correct exposure time and then describe how to make the diazo contact prints which are used when taping the artwork.

8.4.1 Determination of Correct Exposure Time

It is important that the diazo film is exposed correctly. Too short an exposure gives more or less pronounced shadows across the surface, and too long exposure time results in insufficient density so that subsequent generations of contact prints will be of inferior quality.

Since the thickness of the diazo layer varies a little from batch to batch, a contact-printing test has to be made to establish the correct exposure. It is not necessary to use a large sheet of diazo film for the test, a strip having a width of 100 mm (4 in) being suitable. By means of a piece of cardboard, 5/6 of the strip is covered and exposed for half a minute. Then the cardboard is moved so that 4/6 of the strip is covered, and the strip is further exposed for another half minute. In this way, the test exposure is continued until the final 1/6 is exposed. If the normal exposure time is around 2 to 3 min, it would be advantageous to prolong the last exposure time to, for example, 1 min. In this way the test strip contains six partial tests with exposure times in increments of half a minute, ranging from 1 to $3\frac{1}{2}$ min. After the strip has been developed, it is very easy to select the section which has a high density, combined with a clear background and therefore the optimum exposure time.

The above test method assumes that the light intensity from the copying lamp is constant, i.e. that the lamp has been allowed to heat up and reach a stable light output before making the test. Alternatively automatic control equipment using an ultra-violet light meter to monitor the total quantity of light can be provided. In this case, the control box is set to a suitable quantity of light, roughly corresponding to an exposure time of 20–40 sec, whereafter the test is carried out as described above.

8.4.2 Contact Printing Diazo Prints for the Artwork

The description given below is based on contact-printing method C (cf. Section 7.3.3 and Fig. 7.6). Before contact printing is started, the glass plate of the vacuum frame must be completely clean on both sides. It is particularly important to check that no remnants of tape adhesive are present. A clean glass plate saves unproductive touch-up work later.

8.4.2a Pad Master for Solder Side

The diazo film (both sides glossy, thickness 0.007 in or 0.18 mm) is placed in the vacuum frame directly, i.e. symmetrically, under the copying lamp with the diazo layer upwards. So placed, unavoidable enlargement will be minimized. The common pad master is placed on top of the diazo film with the pads upwards. The vacuum frame is closed and the vacuum pump started. Any air pockets must be removed as, otherwise, the quality of the contact print will be inferior. A string should therefore be laid around the artwork as shown in Fig. 7.2. When the air is sucked out (in approximately 30 sec) the exposure can start. The required exposure time, or better, the quantity of light, is set on the control box, and the copying lamp is lit.

Warning: The operator must protect his eyes by not staying in the copying room during the exposure. The light intensity is very high and can damage the sight, or at least cause painful eyes. Certain types of copying lamps, such as xenon lamps produce rather large amounts of ozone which is noxious at high concentrations. Ventilation of the copying room is therefore necessary.

When the exposure is finished, the copying lamp is switched off by the control box. The vacuum pump is then stopped, the air vent opened and the pad master and diazo film removed from the frame.

In the case of boards with extremely high packaging densities, it is always prudent to check and rectify, if necessary, the component clearances, i.e. the pad locations, before the subsequent diazo masters are contact printed. In this way, rectifications of ready taped solder and component sides with consequent side-to-side misregistration of possible new pads can be avoided.

The pad master for the solder side is sent to the photographer who produces a 1:1 negative. A single-sided laboratory board having solder pads only, can then be etched and drilled. When the components are mounted on this 'board' it is easy to check the component clearances.

8.4.2b1 Pad Master for Component Side. A new diazo film (both sides glossy, thickness 0.007 in (0.18 mm) is placed in the vacuum frame with the diazo layer

upwards. The common pad master is placed on top of the diazo film with the pads downwards. The subsequent procedure follows the description given above for the solder side.

8.4.2b2 Master for Solder Mask. Sections 8.8.1–8.8.4 deal with a special method for the preparation of the solder mask artwork which increases the accuracy and reduces the work of preparation. The method uses an extra diazo print of the solder side but in this case the diazo pads have to turn upwards. This copy is made in exactly the same way as the master for the component side and is printed on the same type of diazo film (see above).

8.4.2c Master for Master Drawing

The preparation of the master drawing is described in Section 8.9.3. This drawing is based on a diazo print of the common pad master (both sides matt, thickness 0.003 in (0.076 mm). Matt material is selected to facilitate subsequent drawing and lettering. As the digitizing of the board is normally based on the master film of the solder side, this diazo print is an exact duplicate of the pad master for the solder side, i.e. the diazo pads are turned downwards. The remaining drawing details are added on the upper side which is of advantage when introducing changes since these can be done without interfering with the details on the other side of the drawing.

8.4.2d Developing Diazo Contact Prints

When the prints have been made they are developed in an ammonia developer, which can be part of an existing diazo printing machine (see Fig. 8.32). As over-development

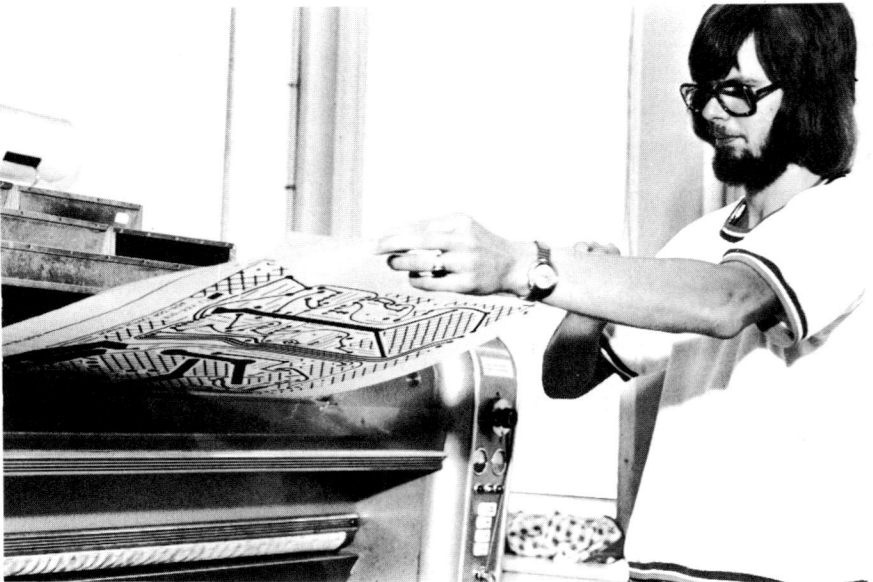

Fig. 8.32 Development of diazo contact prints.

cannot take place, it is an advantage to use a low speed through the machine. In order to avoid the diazo layer being scratched in the developer, it can be protected by a piece of tissue paper or nylon mesh when passed through the machine. After being developed the masters are ready for immediate use.

8.5 Identification and Marking of the Artwork

It is extremely important that the different artworks and issue levels can be easily identified. It is obvious that there can be as many systems of identification as there are manufacturers of electronic equipment. The essential requirement for any of these is that the PCB manufacturer must easily be able to understand exactly what he is being asked to make. One such system is described in this chapter and, naturally, it is the one favoured by the author. It is simple and yet comprehensive enough to be effective. It is described more fully in Section 8.9.

In this system a printed circuit board is identified by means of a six-figure code number, for example, 970–525 where the first three digits identify the group of printed circuit boards within the component classification system and the last three digits form the serial number of the model in question within the system. The solder-side artwork and the component-side artwork for the particular model are distinguished from one another as follows:

Solder side: 970–525 S3
Component side: 970–525 C3

The last part of the designation, for example S3, indicates that the artwork is the solder side (S) and the issue is number (3). When the artwork is up-issued S3 will be changed to S4. Transfer letters and numerals are used for these code numbers and other inscriptions (the name of the board, etc.). For the artwork scale 4:1 a height of 8 mm (0.3 in) and a line width of 1.5 mm (0.06 in) are recommended.

In a system of plug-in boards, all inscriptions should face in the same direction and preferably be similarly located on each board.

The PCB manufacturer's placing of his company mark and the date of manufacture has been mentioned in Section 6.8.2 and illustrated in Fig. 6.41. This marking is usually located on the component side so that the date of manufacture, which is usually added by stamping, will not be rendered illegible in the soldering operation.

8.6 Practical Procedure for Artwork Generation

In the preceding sections we have dealt with the various practices to be followed leading to a satisfactory quality of artwork. Below we shall combine these practices in a description of the procedure for preparing the master drawing and the artwork for the solder side, the component side, the component notation and the solder mask. An example of a documentation package is shown in Fig. 8.33–8.37.

8.6.1 Common Pad Master

The common pad master must contain all the details common to the solder side and

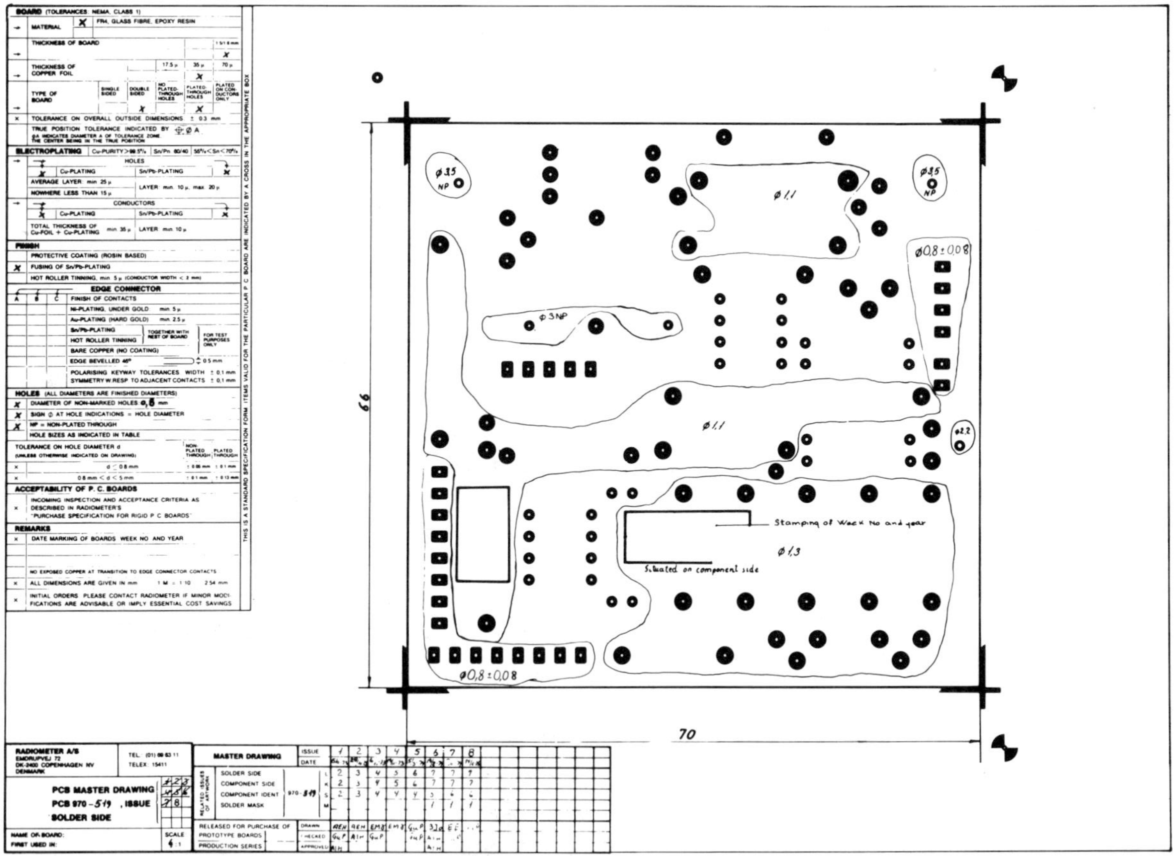

Fig. 8.33 Master drawings (dimension in millimetres).

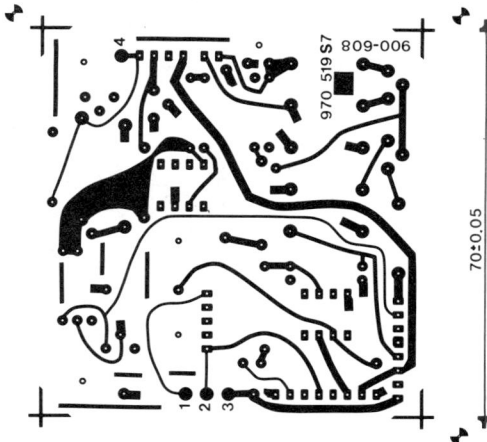

Fig. 8.34 Artwork for the solder side. Dimension in millimetres.

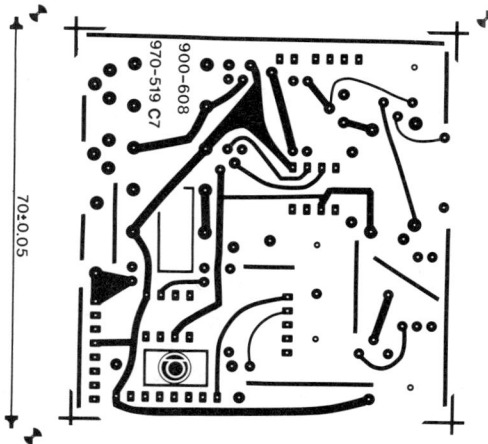

Fig. 8.35 Artwork for the component side. Dimension in millimetres.

the component side. By means of diazo contact printing, two identical pad masters are produced on which are taped the conductors and other details specific to the individual sides. The practical procedure is described in Table 8.2 where all the various operations are listed and provided with references to the describing text.

8.6.2 Diazo Contact Printing

When the common pad master has been completed it is contact printed on diazo film. The printing is performed as described in Table 8.3. It is assumed that the common pad master displays a direct-reading view of the component side and that contact-printing method C is followed. The table indicates the printing of a third and a fourth contact print which are used for the solder mask artwork and the master drawing.

258

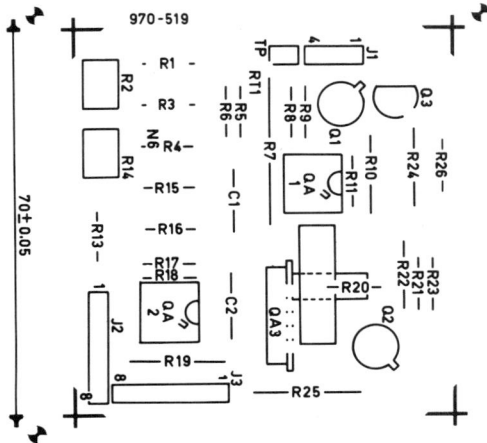

Fig. 8.36 Artwork for the component notation. Dimension in millimetres.

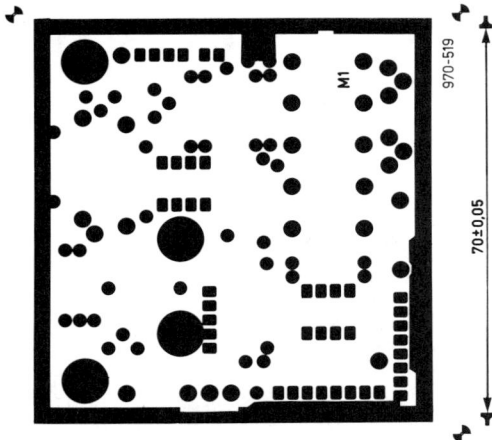

Fig. 8.37 Artwork for the solder mask. Dimension in millimetres.

8.6.3 Finishing the Artwork

The developed diazo prints are now ready for the taping of the conductors. The procedure has already been described and is shown in summary form in Table 8.4 and calls for no further comment. The artwork is completed by adding the title and revision blocks as shown in Fig. 8.38. Changes of issue are easily recorded by deleting the obsolete issue number and adding the new number. The issue number located on the artwork itself must also be changed. The use of transfer characters renders the rectification easy.

Table 8.2 Survey of the preparation of the common pad master.

Activities	Section
Accurate marking on matt drafting film of:	
Contour of board	8.3.5
2 tooling holes	8.3.7
Mounting holes	8.3.8
Edge connector	8.3.9
Placing of 0.076 mm (0.003 in) drafting film, glossy on both sides, on top of the marked drafting film	7.3.6
Taping of:	
Contour of board, by means of corner marks	8.3.5
2 tooling holes	8.3.7
Mounting holes	8.3.8
Taping of edge connector	8.3.9
Taping of:	
Register marks	8.3.6
Photo-reduction marks	8.3.4
Checking of:	
Photo-reduction dimension, by means of a glass scale	8.3.4
Mechanical dimensions, by means of a precision steel scale	8.3.5
Exchange of marked drafting film with layout sketch Alignment of layout sketch with respect to grid	8.3.2c
Accurate marking of safety zones Alignment of glossy drafting film with respect to pattern	6.6.5
Taping of pads common to both sides	8.3.2c

Table 8.3 Survey of diazo contact printing. (Contact printing method C.)

	Activities	Section
a	Contact printing of pad master for solder side	8.4.2a
	Diazo film: thickness 0.18 mm (0.007 in), glossy on both sides Placing of common pad master: pads turned upwards	
b1	Contact printing of pad master for component side	8.4.2b1
b2	Contact printing of master for solder mask Diazo film: thickness 0.18 mm (0.007 in), glossy on both sides Placing of common pad master: pads turned downwards	8.4.2b2 8.8.2
c	Contact printing of pad master for master drawing Diazo film: thickness 0.076 mm (0.003 in), matt on both sides Placing of common pad master: pads turned upwards	8.4.2c 8.9.3
d	Developing diazo contact prints in ammonium developer	8.4.2d

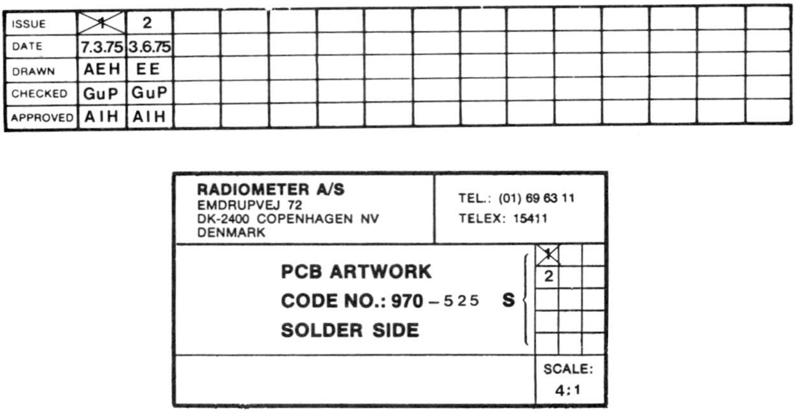

ISSUE	✗	2													
DATE	7.3.75	3.6.75													
DRAWN	AEH	EE													
CHECKED	GuP	GuP													
APPROVED	AIH	AIH													

RADIOMETER A/S EMDRUPVEJ 72 DK-2400 COPENHAGEN NV DENMARK	TEL.: (01) 69 63 11 TELEX: 15411
PCB ARTWORK **CODE NO.: 970 – 5 2 5** S **SOLDER SIDE**	✗ 2
	SCALE: 4:1

Fig. 8.38 Revision block and title block for the artwork.

Table 8.4 Survey of the finishing of the artwork.

Solder side	Component side	Activities	Section
×	×	Layout sketch reverse-reading Layout sketch direct-reading	7.3.3
×	×	Diazo print placed on top of layout sketch Diazo film to be aligned, diazo layer turned downwards	7.3.3
×	×	Taping conductors	8.3.3
×	×	Adjusting size of solder pads in order to improve the solderability	8.3.2f 6.5.7
×	×	Lettering of: Photo-reduction dimension	8.3.4
×	×	Part numbers	
	×	Name of board	8.5
	×	Frame for company mark and date of manufacture	6.8
×	×	Taping of ground plane	8.3.10
×	×	Balancing of pattern	8.3.11
×	×	Checking minimum spacings	6.5.1
×	×	Applying the title and revision blocks	8.6.3

8.7 Screen-printed Notation

The screen printing of the component notation has been described in Section 6.8.1, and some examples are shown in Figs. 6.38 and 6.39. The author prefers the form shown in Fig. 6.38 to which the following more detailed description applies.

The starting point is the layout sketch which provided the information for applying the pads on the common pad master (see Section 8.3.2c). Since this gives the location, the outline and the designation of the components, the easiest way of preparing the artwork for the component notation is to lay a matt drafting film over the layout sketch. The details of the notation are then drawn with pen and ink. Firstly, the three register marks are taped in close agreement with the register marks on the artwork for the component side. The contour of the board is not to be indicated on the notation artwork.

A flat surface is desirable for screen printing, for which reason printing over conductors and pads should be avoided wherever possible. It is therefore helpful to place the finished artwork for the component side (or a diazo print of the artwork) under the layout sketch so that the areas to be avoided can readily be seen. The minimum spacing between a pad and the screen-printed notation should be 0.5–1 mm (0.02–0.04 in) in order to avoid the risk of screen-printing ink entering solder holes.

Two-terminal components are indicated by a reference number and two leader lines indicating the solder holes to be used. Other components, e.g. large electrolytic capacitors, semiconductors and ICs, are indicated by an outline in addition to the reference number. Polarity, where applicable, must be shown to ensure correct insertion of the components.

In Section 5.22.4 was mentioned a special layout technique based on diazo contact printing of layout symbols which were sprayed with a semi-sticky adhesive. The same technique can be utilized for making outline symbols for larger components, semiconductors and ICs. Large numbers of ICs are frequently used in digital boards and drawing the outlines becomes very tedious. In order to eliminate this task, pre-printed semi-sticky outline symbols are applied to the notation artwork. These symbols are drawn so that the clearance from the solder pads is adequate. The procedure is as follows: All very simple contours are drawn and hand-lettered. The contour symbols are then applied but not lettered. After that a direct-reading contact print is made on matt diazo film. The type codes of the ICs can now be inserted on this film to complete the artwork. Changes can also be made on this film at a later date if required.

This technique requires that the correct exposure time is determined so that a clear background and a reasonable line density is obtained (see Section 8.4.1). It is important to apply only sufficient adhesive on the contour symbols as any excess could cause contact-printing problems. When the printing has been completed, the contour symbols are removed from the master and filed in the same way as the layout symbols. The diazo print is now used as the master and the drafting film can be discarded.

In addition to facilitating the work, the method has the great advantage that the outlines of identical components look alike, not only in the individual board, but also in all the boards in a group. When the contours are drawn individually they are bound to vary within the board and also when different draughtsmen are employed in the work.

Height and line width of the lettering must not be too small. A suitable choice for the artwork scale 4:1 is a height of approximately 10 mm (0.4 in) and a line width of approximately 1.5 mm (0.06 in). It is, of course, possible to use transfer letters and figures for the lettering, but it is considerably faster to use a lettering template, and the characters produced are quite satisfactory (see Fig. 8.39). A drawing ink with good adhesion and high density, Castell TG for example, should be used.

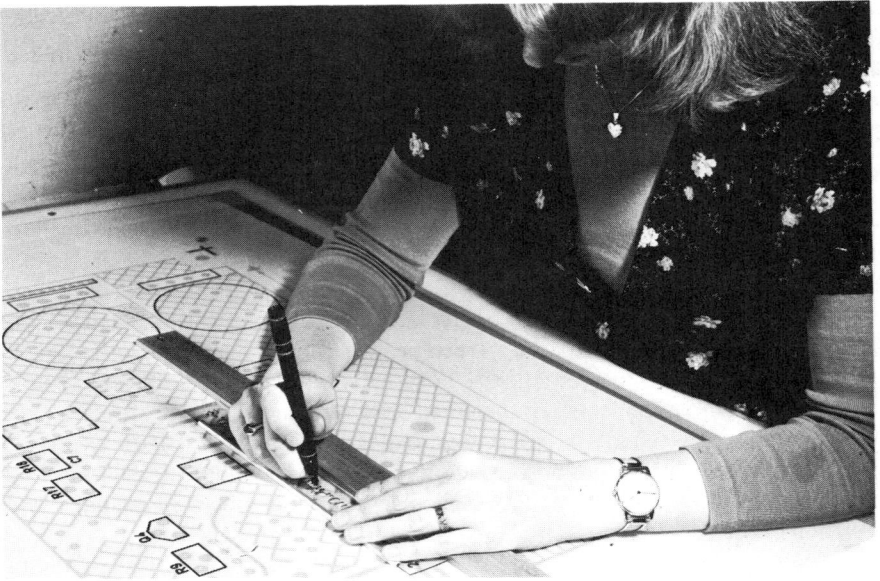

Fig. 8.39 Lettering using a template.

The artwork for the screen-printed component notation should be identified in the same way as those for the solder and component sides, and, in accordance with the system outlined in Section 8.5 the designation will be

970–525 N3

The 'N' indicates that the artwork is for the notation, and the figure indicates the issue.

It is unnecessary to use space for the code number 970–525 as it appears etched on the component side of the board. Therefore, the 'N3' is located on the artwork in such a manner that when screen printed it appears just after the etched code number including the side designation and the issue number. The component side of the finished board will therefore bear the following lettering:

970–525 C3 N3

The title block and the revision block are applied and filled in the same way as for the solder and component sides.

Sometimes the title block is omitted on the reduced master film and it may, therefore, be difficult to identify the master film. As a precautionary measure the code number can be written just outside the outline of the board.

8.8 Solder Mask

8.8.1 Introduction

The theoretical considerations which dictate the design of the solder mask have been detailed in Section 6.9 and we will, therefore, deal only with the practical preparation of the artwork. The 'reverse' taping method is used where the apertures in the mask are taped as solder mask pads, cf. Fig. 6.42. The work falls in the following three stages: the basic artwork; the outline and identification; the application of the solder mask pads.

8.8.2 Basic Artwork

The artwork for the solder side is placed direct-reading on the light table and aligned to the grid. The drafting film [thickness 0.14 mm (0.005 or 0.006 in)] is placed on top, and the photo-reduction dimension and the register marks are applied in close agreement with the solder side. The basic artwork is then ready for the next stage.

Alternatively, and with the advantage of higher accuracy, another contact print of the common pad master, as stated in Table 8.3, activity b2, may be used, with the diazo layer turned upwards (cf. Fig 7.6E). The advantage is a reduction of the parallax error which would otherwise occur when applying the solder-mask pads. In addition to the register marks and photo-reduction dimension, the print shows all the common pads of the pattern. These pads are to be left uncovered by the solder mask on the finished board, and owing to the reverse-taping method they have to be covered on the artwork. The presence of the pads on the top side of the diazo print facilitates the accurate location of the solder-mask pads. If the basic artwork is based on the drafting film instead of the diazo print the pads on the solder side artwork would be located below the surface of the drafting film by the thickness of the diazo film used for the solder side artwork 0.18 mm (0.007 in) plus the taped conductors which are found on the top side of the solder side artwork, 0.12 mm (0.05 in) plus the thickness of the drafting film 0.14 mm (approximately 0.005 in), in total 0.44 mm (0.017 in). This distance implies a significant risk of a parallax error of 0.10–0.15 mm (0.004–0.006 in), not least because the solder-mask pads have no centre hole so that they have to be aligned with respect to the rim of the solder pads. In some cases it may transpire that the diazo film and the red, transparent solder-mask pads are not fully compatible, i.e. the pads do not adhere sufficiently well to the diazo coating. The remedy is to make the diazo print in the same way as the pad master for the solder side (see Table 8.3, activity a), a little larger parallax being accepted when taping the solder-mask pads.

8.8.3 Outline and Identification

Special attention has to be given to the outline of the solder mask along the edges of the board and where the outline cuts through etched inscriptions on the solder side, as described below and illustrated in Fig. 8.40. The outline is marked by means of a red, transparent self-adhesive lithotape which is applied so that it covers about two modules (0.2 in or 5 mm) inside the edge of the board in the artwork scale 4:1 (see A). If the lithotape tends to cover certain parts of the pattern which have to be covered by the mask, the lithotape can be moved a little closer to the edge so that it only covers one module (0.1 in or 2.5 mm) inside the edge (B). For the sake of legibility, inscriptions such as code numbers should not be cut through by the outline of the solder mask. The inscriptions must either be covered completely by the lithotape (C), i.e. no solder mask

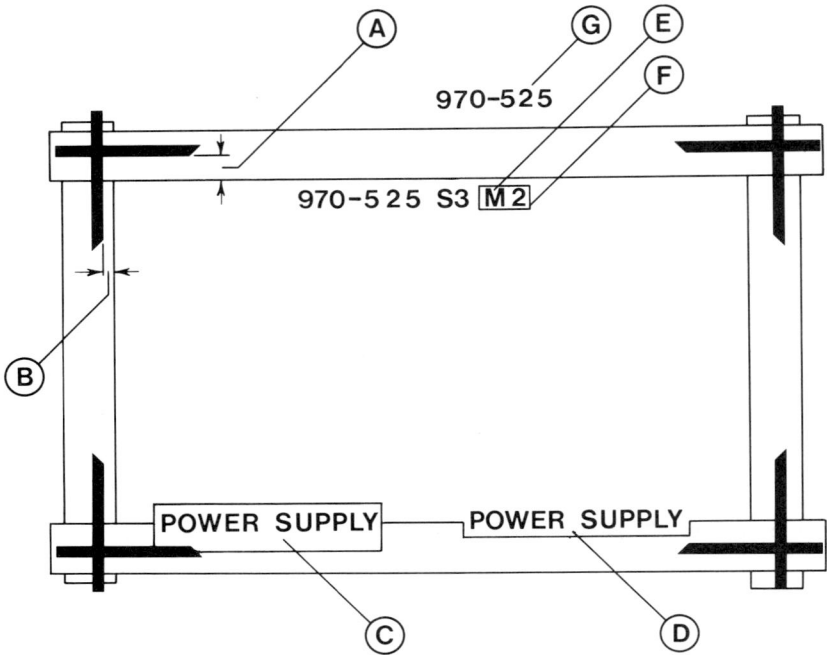

Fig. 8.40 Preparation of the artwork for the solder mask.

will be applied, or the lithotape can be withdrawn a little so that the inscriptions are not covered (D), i.e. they will be covered completely by the solder mask.

The artwork for the solder mask is identified in the same way as the artwork for the component notation. The solder mask will, therefore, have the designation:

970–525 M3

The 'M' indicates that the artwork is for the solder mask, and the figure indicates the issue.

In the same way as for the component notation, the 'M3' is to be placed on the artwork so that it appears on the finished board just after the code number of the solder side (see E). The finished board shows the following designation:

970–525 S3 M3

It should be noted that the 'M3' will be negative-reading. Owing to the rather faint colour contrast usually found between the base material and the solder mask, the legibility may be seriously impaired. An improvement can be obtained by taping on the artwork for the solder side a little rectangular area which, on the finished board, forms a metallized background (F) where the designation 'M3' appears.

The title block and the revision block are applied and filled in the same way as for the solder and component sides.

Sometimes the title block is omitted on the reduced master film, for which reason it may be difficult to identify the master film. As a precautionary measure, the code number can be written just outside the contour of the board.

8.8.4 Applying the Solder-mask Pads

The solder-mask pads are taped in the same way as the ordinary solder pads. Since they do not have a centre hole it is necessary to use transparent pads so that it can easily be checked that they are applied concentrically with the solder pads. As pointed out in Section 6.9, complex boards with very high packaging densities do not leave the manufacturer a great tolerance for screen printing the solder mask and the solder-mask pads must therefore be applied with great care. Applying the solder-mask pads directly on top of the pads of the diazo print is in this respect a very great help. It is, however, necessary that the artwork of the solder side, or preferably a diazo print thereof, is placed underneath to enable the critical places to be identified.

At the critical places the clearances around the pads is approximately 0.3 mm (0.012 in) which means that the solder-mask pad types A1, B1, and C1 as indicated in Table 6.6 (Section 6.9.3) are used. In order to facilitate the screen printing of the solder mask, it is recommended to use a larger clearance around the pads at all the uncritical places, preferably not under 0.5 mm (0.02 in) which is achieved by means of the solder-mask pad types A2, B2, and C2. It is also possible to use these pads at slightly critical places as the draughtsman, by displaying some dexterity, is able to apply the pads eccentrically or to trim them as illustrated in Fig. 6.44.

Referring to Section 6.9.4 where the adapted requirements for the solder mask have been dealt with, we find that no matter how well the draughtsman prepares the artwork it is not possible in practice to obtain a perfect solder mask if a 0.3-mm (0.012-in) clearance around the pads is demanded. It is, in fact, necessary to accept a slight overprinting of the pads corresponding to a reduction of the clearance from 0.3 mm (0.012 in) to 0.15 mm (0.006 in) in the artwork scale 1:1. The solder-mask pad types A1, B1, and C1 are applied but they are done so eccentrically or trimmed as shown in Fig. 6.44. It is not the intention to use smaller solder-mask pads which would cause overprinting along all the pad rims.

IC pads are included among the critical areas. Owing to the fact that they are applied as fixed-pad configurations (as is the case with other solder-mask pads), just one critical place (one conductor routed between two of the pads) makes the whole group critical. Such solder-mask pads are specified in Table 6.6, and it is seen that only one type complies with the requirements of the grouping 'critical places'. It should be noted that the clearance in one direction is only 0.27 mm (0.011 in) so a very careful application is necessary.

The whole of the above discussion applies to plated through boards. In the case of non-plated through boards, it is possible to improve the adhesion of the pads by letting the solder mask cover the rim of the pads and the 'ears' (see Fig. 6.14), if any. The solder-mask pads are therefore chosen in the same size, or slightly smaller, than the solder pads, which causes an overprinting of 0.1–0.2 mm (0.004–0.008 in). On the other hand, the solder mask should not cover the solder pad so that the soldering area is too small, i.e. the uncovered part of the annular ring should have a minimum width of 0.5 mm (0.02 in). By utilizing this technique, the adhesion of the solder pads can be improved considerably, very important for service and maintenance work. Practical tests show that it is possible to perform more than five soldering/desoldering operations on such pads without any sign of defective (loose) pads.

8.8.5 Dry-film Solder Masks

The above description relates to screen-printed solder masks. When dry-film solder masks become commercially accepted, the preparation of the artwork will be considerably simplified as it will be possible to use solder-mask pads of the same size as the solder pads. This implies that the finished solder mask covers up to 0.1 mm (0.004 in) of the rim of the solder pads, to which is added a possible registration error of 0.05 mm (0.002 in), in total a coverage of 0.15 mm (0.006 in) at one side of the pads. This assumes a precise registration between the pads of the artwork for the solder side and for the solder mask, and this is unlikely to be achieved except by the diazo contact-printing method.

The artwork is prepared as described in Sections 8.8.2 and 8.8.3, whereas Section 8.8.4 does not apply in this case.

8.8.6 Another Approach to the Solder Mask

Since it is extremely difficult to produce a sufficiently accurate solder mask in the case of boards with a very high packaging density, a change of sequence, in which the various artworks are made, should be considered. The normal sequence of the artwork preparation is:

1. Common pad master
2. Solder side
3. Component side
4. Solder mask
5. Component notation
6. (Master drawing)

When the artwork for the solder mask is being prepared, it is to be hoped that the conductors have to be laid down on the solder-side artwork with due consideration to the subsequent preparation of the solder mask. This means, among other things, that the conductors are not lying too close to the pads. Since this is not always carefully attended to in preparing the artwork for the solder side, the artwork for the solder mask and for the solder side may have to be modified alternately to produce an acceptable compromise.

A novel but rather efficient measure is to change the sequence as follows:

1. Common pad master
2. Solder mask
3. Solder side

4. Component side
5. Component notation
6. (Master drawing)

The diazo printing process makes it possible and attractive to prepare the artwork for the solder mask before the artwork for the solder side. The solder-mask artwork, which is now taped with all the solder-mask pads, can then be placed underneath the layout sketch and the solder-side artwork for taping the conductors of the solder side. By placing the emphasis on the solder-mask artwork, the draughtsman not only considers more effectively the clearances between the solder-mask pads and the solder pads themselves, he can also better ensure that conductors, where they pass close to solder-mask pads, are laid so that they do not touch or cover the solder-mask pads and therefore fail to be covered, at such places, by the solder mask on the finished board. There will, of course, remain places at which the solder mask will encroach on the rim of the pads or leave conductors uncovered, but in the author's experience, this novel method leads to a nearly perfect result.

8.9 Master Drawing

8.9.1 Introduction

The controlling document of the package sent to the PCB manufacturer is the master drawing. It contains the specification of the board and collates the various artwork and records the current issues to which the boards have to be manufactured. The general opinion is that printed circuit boards are subject to many changes so it is very important to keep the list of the various issue levels up to date.

The dimensional stability of the master drawing is not important and it is preferably printed on diazo paper.

Below we shall first deal with a registration system and afterwards with the content and the practical preparation of the master drawing.

8.9.2 Identification Numbering System

A register of printed circuit boards is best maintained on the basis of the code number assigned to the master drawing, which identifies the complete board. Other documents which are required for manufacture use the same code number with a suffix denoting the purpose of the document. Figure 8.41 illustrates a suitable system

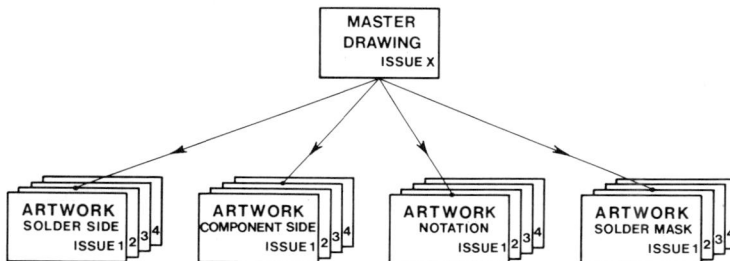

Fig. 8.41 Example of a registration system.

in which the master drawing records all changes and up-issues of all the contents of the documentation package. In this way, only the master drawing and issue number has to be quoted on the purchase order since this drawing identifies, at all times, the total up-to-date information required for manufacture. This is regarded as being much safer than quoting, on the purchase order, the issue levels of all the artwork to be used.

In Table 8.5 up-issuings caused by rectifications or amendments are given together with later up-issuings. If any of the artwork is altered, say the component side, it will be up-issued from issue 1 to issue 2, reflected in an up-issuing of the master drawing from

Table 8.5 Examples of up-issuing procedures*.

Master drawing	Issue 1	Issue 2	Issue 3	Issue 4
Artwork for solder side	issue 1	issue 1 → issue 2	issue 2
Artwork for component side	issue 1 → issue 2	issue 2	issue 2
Artwork for notation	issue 1	issue 1	issue 1	issue 1
Artwork for solder mask	issue 1	issue 1	issue 2	issue 2

➤, up-issuing caused by rectifications or amendments: - - ➤, later up-issuing.

issue 1 to issue 2. The next change shown in Table 8.5 concerns the solder-side artwork which is up-issued from issue 1 to issue 2. This change makes it necessary to amend the solder-mask artwork which is also up-issued from issue 1 to issue 2. Both changes are reflected in an up-issuing of the master drawing from issue 2 to issue 3. The last type of changes to be discussed here concerns only the master drawing itself. Here a change has been introduced without affecting the artwork, for example a change in the plating thickness. The table shows an up-issuing of the master drawing from issue 3 to issue 4, all artwork issues being unchanged.

We have previously seen how individual artwork is identified by means of a code number (group number and serial number) followed by a type designation and the issue number, for example, 970–525 C3. In this way, only the code number of the master drawing appears in the company's general drawing register which, for this purpose, covers the entire documentation package. A clear indication of the master drawing can be obtained by prefixing PCB and suffixing the issue level to its code number. The master drawing belonging to the case described will, therefore, have the designation:

PCB 970–525, issue 3

The author realizes that this simple but rather efficient system does not accord with the standards of the military and other large institutions, but he hopes that it may be helpful to electronics companies working in other fields.

It may seem troublesome to change the issue number for minor modifications but .. must be stressed that unless the system is strictly followed in all cases the risk is incurred of receiving boards in the wrong issue. It is, of course, assumed that the purchase office is informed about all up-issues so that the orders can be written correctly.

The master drawing is provided with a title block which corresponds closely to the title blocks of the artwork and with a revision block which indicates the status of the various artwork issues. A practical example of such a title block and revision block is shown in Figs. 8.42 and 8.43.

RADIOMETER A/S EMDRUPVEJ 72 DK-2400 COPENHAGEN NV DENMARK	TEL.: (01) 69 63 11 TELEX: 15411		
PCB MASTER DRAWING **PCB 970 – 5 2 5 , ISSUE** **SOLDER SIDE**	⊠	2	
NAME OF BOARD: POWER SUPPLY FIRST USED IN: MS30	SCALE: 4 : 1		

Fig. 8.42 Title block of master drawing.

		ISSUE	⊠	⊠	3						
MASTER DRAWING		DATE	6.1.75	12.3.75	9.6.75						
RELATED ISSUES OF ARTWORK	SOLDER SIDE	S	1	1	2						
	COMPONENT SIDE	C	1	2	2						
	NOTATION 970– 525	N	1	1	1						
	SOLDER MASK	M	1	1	2						
RELEASED FOR PURCHASE OF:		DRAWN	AEH	BJo	AEH						
PROTOTYPE BOARDS	PL	CHECKED	KJK	KJK	GuP						
PRODUCTION SERIES	PL	APPROVED	jph	jph	jph						

Fig. 8.43 Revision block of master drawing.

8.9.3 Mechanical Specification

Since the artwork shows only the pattern and not the machining characteristics, for example the hole sizes, the mechanical requirements form an essential feature of the master drawing which, for this reason, is sometimes referred to as the drilling drawing or the dimensional drawing.

The easiest way to prepare the master drawing is to make an extra diazo print as described in Section 8.4.2c. The outline is drawn on the basis of the corner marks which may remain, and all the dimensions of the board including any cut-outs are indicated in the conventional manner. On the other hand, the dimensional position of mounting holes and the like need not be indicated because all the drilled holes of the board are digitized when the NC drilling tape is prepared. Such dimensioning would not necessarily be identical with that obtained in digitizing from the master film and might confuse the manufacturer.

Only the positions of the drilled holes appear in the master film. Their diameters must be specified in the master drawing. In those cases where the master drawing is based on a diazo print of the common pad master, all the solder pads are marked. The hole sizes can now be indicated in two different ways: either by encircling holes of the same size in different groups such as shown in Fig. 8.44, or by marking the holes by

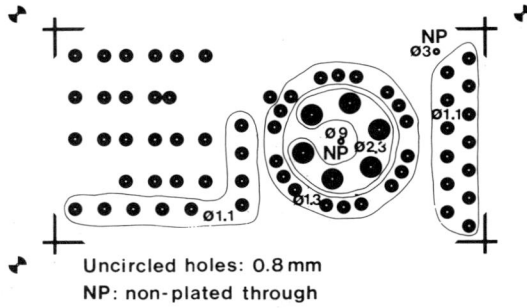

Uncircled holes: 0.8 mm
NP: non-plated through

Fig. 8.44 Encircling the holes.

letters which are tabulated (see Fig. 8.45). Usually certain hole sizes will be dominating, for example 0.8 mm (0.031 in), so by entering a note that all unmarked holes are of a certain size, quite a lot of work can be saved. Marking by letters is more usual and frequently the tabulation includes the number of holes of each size. The author prefers the first method because it is considerably faster for the draughtsman and offers no drawback when the board is being digitized.

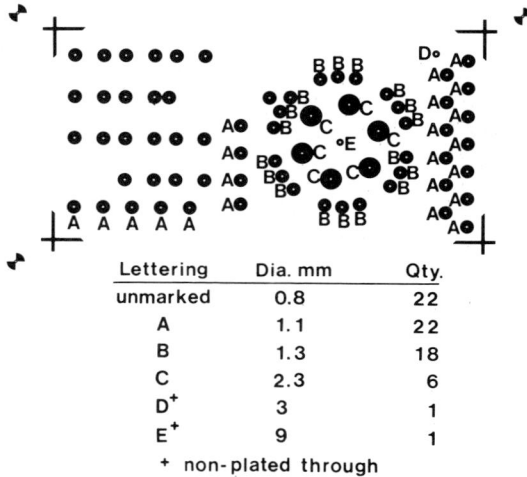

Lettering	Dia. mm	Qty.
unmarked	0.8	22
A	1.1	22
B	1.3	18
C	2.3	6
D+	3	1
E+	9	1

+ non-plated through

Fig. 8.45 Tabulating the holes.

Unnecessarily close tolerances are to be avoided. When they are found, it is usually due to extending the tolerancing practice of the mechanical drawing office to printed circuit boards. If the general tolerances have been stated in the purchase specification, it is superfluous to state them in the master drawing. Realistic tolerances, i.e. tolerances which can be fulfilled by the PCB manufacturer without forcing him to take very special measures, have been discussed in Section 3.8.

In some cases, it is necessary to link the outline to the pattern more accurately than when the outline is defined by the corner marks. In these cases a reference system is introduced in which a point of the pattern, for example, the centre of a tooling hole, is taken as datum. This has been discussed in Section 4.14.1, and here we shall refer to Fig. 8.46 which shows how the contour of the board is determined on the basis of such a reference system. The coordinate axes which are named datum lines and marked 'datum' on the drawing, are placed with the zero-point at the centre of one of the tooling holes. The second datum point which is needed to define the direction of the X-axis can conveniently be an additional pad placed either inside or just outside the outline of the board. The other tooling hole as well as the contour and the polarizing slot of an edge connector are dimensioned with respect to this reference system. In this way, a far more accurate linking of the said parts to the pattern is achieved, the pattern being represented by the first tooling hole.

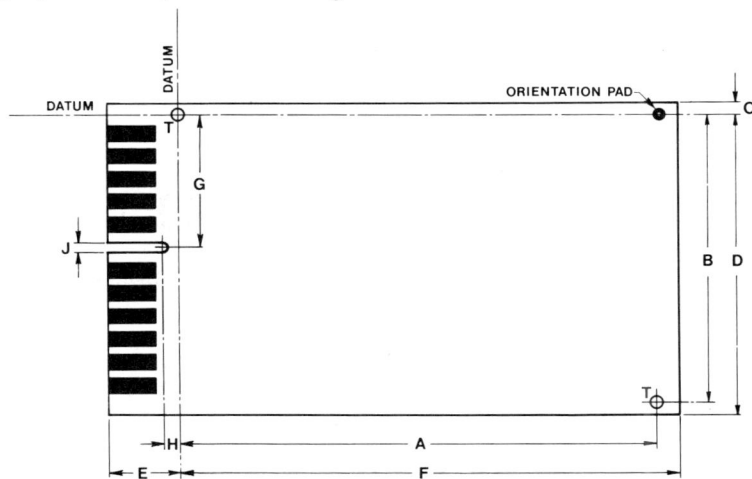

Fig. 8.46 Using a datum reference system with zero point in one of the tooling holes (T).

The zero-point of the reference system can alternatively be laid in a corner of one of the contact fingers of the edge connector, if any, but when the board (the master film) is to be digitized, it is rather difficult for the operator to perform an exact determination of the zero-point when it is not located at the centre of a pad. For easy checking the board, it is an advantage to start from a hole as the centre can be assumed not to move due to the plating. The checking is less positive if it is based on a corner which moves a little because of the plating and/or the etching.

Holes which are to be drilled in relation to the reference system can conveniently be marked on the artwork by means of pads. In order to ensure that they are introduced on the basis of their coordinates and not digitized together with the pattern, a note to this effect must be included in the master drawing.

When determining the details of the contour the draughtsman should bear the problems of manufacture in mind. We have previously dealt with inward sharp corners. In the case of an edge connector where the contact fingers are located on a tongue, the inward corner (see Fig. 8.47A), if rounded, could prevent the board from being seated fully in the receptacle. Usually it is possible, without too much loss of

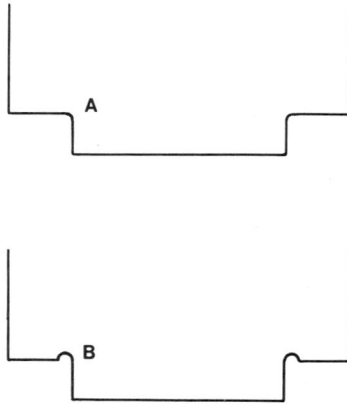

Fig. 8.47 Production of an edge connector tongue.

space, to carry the rounding upwards as shown in Fig. 8.47B so that the profile can be routed all the way round.

A polarizing slot should for the same reason not terminate in a straight edge but be rounded as shown in Fig. 8.48.

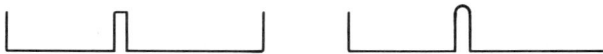

Fig. 8.48 Radiusing of a polarization slot.

Counter-sunk holes should always be avoided as they are difficult to produce owing to the large tolerance on the thickness of the base material. Normally the PCB manufacturer does not guarantee the diameter D, but rather the distance h (see Fig. 8.49).

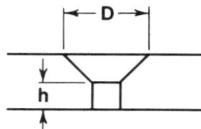

Fig. 8.49 Counter-sinking of a hole.

Small boards (for example of the size 25×40 mm ($1 \times 1\frac{1}{2}$ in) or less) can be difficult to handle during assembly and soldering. If the number of boards justifies the procurement of a blanking tool, the tool can be made with 'return to blank', i.e. the board is pressed back into the production panel which frequently comprises a great number of boards. It is a disadvantage that the panel does not remain entirely flat after blanking; this is due to internal tensions in the base material. If the PCB manufacturer uses an NC router, this disadvantage can be avoided. The contour is routed with the exception of the corners (Fig. 8.50) so that the boards after being assembled and soldered can be broken loose from the production panel.

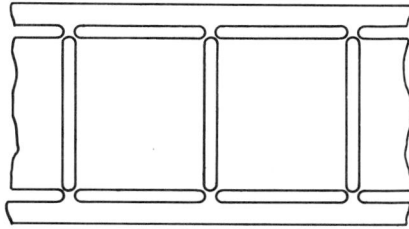

Fig. 8.50 Pin routing of boards to be held at the corners.

8.9.4 Main PCB Specification

In addition to the mechanical specification the master drawing has to provide the actual printed circuit board specification. The PCB manufacturer must have a clear idea of how the finished board will appear in order to organize the necessary manufacturing processes. There are good reasons for emphasizing that the master drawing should only prescribe the end result and not the manufacturing processes. Only where a particular process is demanded for a technical reason should a process be specified. The technical contents of the PCB specification is naturally fairly complicated. The author has therefore found it most expedient to prepare a tabular layout of a general specification in the shape of a transparent self-adhesive form which can be applied to the master drawing. The form is designed with due attention to the fact that it has to cover all the existing types of boards, and all the alternative combinations of requirements permitted by the purchasing specification. Figure 8.51 shows the form which has boxes for all relevant data, so that the combination required is easily selected by ticking the appropriate boxes. In addition to the time saved, a great advantage is gained by the standardized appearance and contents from type to type, independent of changing designers and/or draughtsmen. The standardized format can also be to the advantage of the PCB manufacturer.

Below we shall deal with the form which in Fig. 8.51 is ticked corresponding to a double-sided, plated through board.

1st Section (Board)

In this section the board is characterized by the base material, its thickness and the thickness of the copper foil. Variants are covered by ticking the empty boxes. Note that the example shows two check marks for the thickness of the copper foil which means that the PCB manufacturer may choose whether he wants to start with a 17.5 μm ($\frac{1}{2}$ oz) or a 35 μm (1 oz) copper foil. Two check marks are necessary to specify the type of board: one for the number of sides (single/double) and one for the type of holes (non-plated through/plated through), and, in rare cases, a check mark for plating of the conductors only. The dimensional tolerance is permanently ticked whereas the positional tolerance is used in special cases only.

BOARD (TOLERANCES: NEMA, CLASS 1)

→	MATERIAL	**X**	FR4, GLASS FIBRE, EPOXY RESIN			

	THICKNESS OF BOARD					1.5/1.6 mm
→						**X**

					17.5 μm	35 μm	70 μm
→	THICKNESS OF COPPER FOIL					**X**	

		SINGLE SIDED	DOUBLE SIDED	NO PLATED-THROUGH HOLES	PLATED-THROUGH HOLES	PLATED ON CONDUCTORS ONLY
→	TYPE OF BOARD			**X**	**X**	

×	TOLERANCE ON OVERALL OUTSIDE DIMENSIONS: ± 0.3 mm
	TRUE POSITION TOLERANCE INDICATED BY ⊕ ∅ A
	∅A INDICATES DIAMETER A OF TOLERANCE ZONE, THE CENTRE BEING IN THE TRUE POSITION

ELECTROPLATING Cu-PURITY > 99.5% Sn/Pb: 60/40 55% < Sn < 70%

→		HOLES		→
X	Cu-PLATING	Sn/Pb-PLATING		**X**
	AVERAGE LAYER: min 25 μm	LAYER: min. 10 μm, max. 20 μm		
	NOWHERE LESS THAN 15 μm			

→		CONDUCTORS		→
X	Cu-PLATING	Sn/Pb-PLATING		**X**
	TOTAL THICKNESS OF Cu-FOIL + Cu-PLATING min. 35 μm LAYER: min. 10 μm			

FINISH

	PROTECTIVE COATING (ROSIN BASED)
X	FUSING OF Sn/Pb-PLATING
	HOT ROLLER TINNING, min. 5 μm (CONDUCTOR WIDTH < 2 mm)

EDGE CONNECTOR

A	B	C	
			FINISH OF CONTACTS
X			Ni-PLATING, UNDER GOLD: min. 5 μm
X			Au-PLATING (HARD GOLD): min. 2.5 μm
			Sn/Pb-PLATING ⎫ TOGETHER WITH REST OF BOARD
			HOT ROLLER TINNING ⎰ FOR TEST PURPOSES ONLY
			BARE COPPER (NO COATING)
X			EDGE BEVELLED 45° ± 0.5 mm
X			POLARISING KEYWAY TOLERANCES: WIDTH: ± 0.1 mm SYMMETRY W.RESP. TO ADJACENT CONTACTS: ± 0.1 mm

HOLES (ALL DIAMETERS ARE FINISHED DIAMETERS)

X	DIAMETER OF NON-MARKED HOLES: **0.8** mm
X	SIGN ∅ AT HOLE INDICATIONS = HOLE DIAMETER
X	NP = NON-PLATED THROUGH.
	HOLE SIZES AS INDICATED IN TABLE

	TOLERANCE ON HOLE DIAMETER d (UNLESS OTHERWISE INDICATED ON DRAWING)	NON-PLATED THROUGH	PLATED THROUGH
×	d ≤ 0.8 mm	± 0.05 mm	± 0.1 mm
×	0.8 mm < d < 5 mm	± 0.1 mm	± 0.13 mm

ACCEPTABILITY OF P. C. BOARDS

×	INCOMING INSPECTION AND ACCEPTANCE CRITERIA AS DESCRIBED IN RADIOMETER'S "PURCHASE SPECIFICATION FOR RIGID P. C. BOARDS"

REMARKS

×	DATE MARKING OF BOARDS: WEEK NO. AND YEAR
X	NO EXPOSED COPPER AT TRANSITION TO EDGE CONNECTOR CONTACTS
×	ALL DIMENSIONS ARE GIVEN IN mm. 1 M = 1/10 in = 2.54 mm
×	INITIAL ORDERS: PLEASE CONTACT RADIOMETER IF MINOR MODIFICATIONS ARE ADVISABLE OR IMPLY ESSENTIAL COST SAVINGS

THIS IS A STANDARD SPECIFICATION FORM. ITEMS VALID FOR THE PARTICULAR P. C. BOARD ARE INDICATED BY A CROSS IN THE APPROPRIATE BOX

Fig. 8.51 Completed standard specification.

2nd Section (Electroplating)

The plating specification is in very close agreement with Section 4.5. Note that the draughtsman has to distinguish between holes and conductors, and, furthermore, between copper plating and tin/lead plating. This gives a higher degree of flexibility making it possible to specify normal plated through boards (four check marks) as well as boards plated on the conductors only (one check mark).

3rd Section (Finish)

Here the surface treatment is specified by ticking one of the boxes.

4th Section (Edge Connector)

It is possible to specify up to three different edge connectors which are marked on the master drawing by A, B, or C. A normal gold-plated edge connector requires two check marks for the nickel plating and the gold plating, one check mark for the chamfering, if desired, and one check mark for a polarizing slot. An edge connector which is used only for test purposes, requires a check mark for either tin/lead plating, roller tinning or just unplated copper, and possibly a check mark for a polarizing slot.

5th Section (Holes)

This section specifies the holes in detail. Very often up to 90% of the holes have the same diameter, frequently 0.8 mm (0.031 in) in digital boards, so in order to simplify the specification, such holes are left unmarked on the master drawing, and the size of unmarked holes indicated in the form. If non-plated through holes occur in a plated through board, such holes are marked NP (non-plated through), and the corresponding box is ticked.

A special box is provided for the cases where it is required to tabulate the hole sizes and the numbers of the holes, the holes being marked by lettering on the master drawing.

The hole tolerances follow closely the specification given in Sections 4.6.2 and 4.7.2.

6th Section (Acceptability of PC Boards)

This section refers to the quality as stated in the purchase specification. The box is permanently ticked.

7th Section (Remarks)

This section serves for any requirements not otherwise covered.

8.9.5 Rectifications

Rectifications of the artwork as well as the master drawing are indicated by an up-issuing as described in Section 8.9.2. The manufacturer's problem is to have the nature of the rectification defined so that the production documentation and tools can be up-dated. A practical way is to encircle the changes in red on a diazo-paper print of the master drawing supplemented, if necessary, by a diazo-paper print of the rectified artwork with the changes similarly marked. In this way the PCB manufacturer can for example easily see whether a hole has been changed in size and/or location.

Warning: A taped artwork must never be passed through a rotary printing machine. It must always be printed in a vacuum frame.

Changes are, naturally, not put into effect without incurring costs in revision of manufacturing documentation and modification of tools. If the changes include new artwork, the new master films will differ a little from the old ones, and if the pattern has close tolerances, i.e. the solder pads have very small annular rings, such deviations may make the old drilling tape unusable with the new master films. The PCB manufacturer will then find he is compelled to produce a new drill tape, thus incurring both increased cost and extended delivery time. In the case of minor changes to the artwork, it seems better to ask the manufacturer to update the master film by a simple touch-up, noting, however, the importance of raising the issue number to correspond with the artwork which is updated by the customer. The PCB manufacturer should be given the fullest possible details of all changes either by encircling in red or by hand-written notes. This small effort on the part of the customer avoids queries and uncertainties and helps to ensure trouble-free introduction of the changes.

8.10 Checking the Artwork

It is very important when the artwork is completed, to make careful checks of its exact correspondence with the schematic diagram. It can easily occur that in taping a complicated artwork, an occasional conductor is terminated on a wrong pad. Checking for such errors can be very troublesome in the case of plated through boards since the circuitry is shared between the two artwork sides.

By means of a combined contact printing technique enabling all the information given on the various artwork sheets (solder side, component side, and component notation) to be gathered on one sheet of paper in such a way that the various artwork patterns can be distinguished from each other, it becomes possible to follow a conductor from its starting point to its end point, irrespective of how many times it crosses from one side to the other. It is relatively easy to compare this single print with the schematic diagram checking each conductor from end to end and marking each off as checked to ensure that none is omitted.

The exposure times given below for the preparation of the combined print depend, of course, on the equipment used and serve only to illustrate the method. The diazo contact-printing equipment is especially suited for the method described below.

The combined contact printing of several artworks at one time is performed as illustrated in Fig. 8.52. A sheet of diazo sensitized paper and the artwork for the component notation, the component side and the solder side, in that order, are placed in the vacuum frame with the individual artwork secured in register by means of Scotch tape. The vacuum frame is then closed and the sensitized paper exposed for approximately 100 sec in order to obtain a white background. The solder side is then

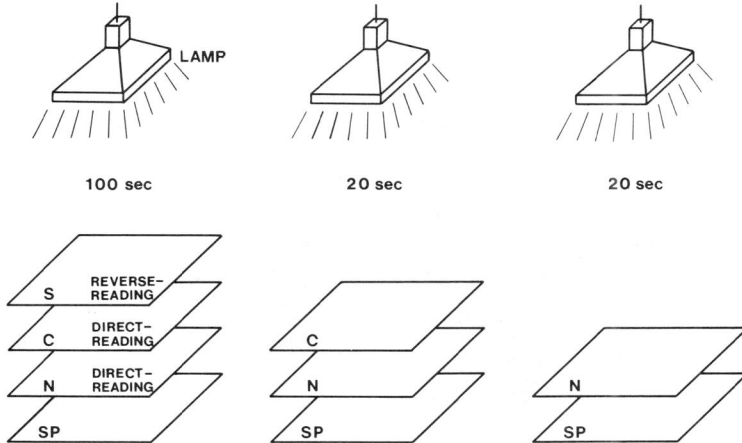

Fig. 8.52 Checking the artwork by 'multiple printing'. S, solder side; C, component side; N, notation; SP, sensitized paper.

removed, leaving the sensitized paper and the two other artworks secured in the same relative position by the Scotch tape. By exposing the sensitized paper for 20 sec, the part of the combined pattern which originates from the solder side is partially bleached. The component side is removed and the sensitized paper is exposed for a further 20 sec. Then the component notation is removed and the sensitized paper developed. Because of the bleaching effect, the three artwork patterns will appear in different densities, the component notation in heavy lines, and the component and solder sides progressively less dense. Figure 8.53 shows a combined print which, unfortunately, does not show to the best advantage owing to difficulties in the reproduction.

There is nothing to prevent the artwork from being laid in a different sequence when making combined prints, but the author has found it advantageous to get the component notation as heavy as possible for the sake of correct identification of the components.

If one of the artwork sides has a ground plane it is very important that this side is placed uppermost in order that it will be bleached as far as possible; otherwise it will be so dense that the pattern of the other sides will be obscured. Another method which is, however, more expensive, is to contact print the artwork individually on diazo films affording prints in different colours. When these copies are superimposed, a similar result is obtained with the advantage that separation between the different artwork patterns is much better. This technique is especially important when making multilayers as the difference in density is inadequate when more than three artwork items are brought together in a combined print.

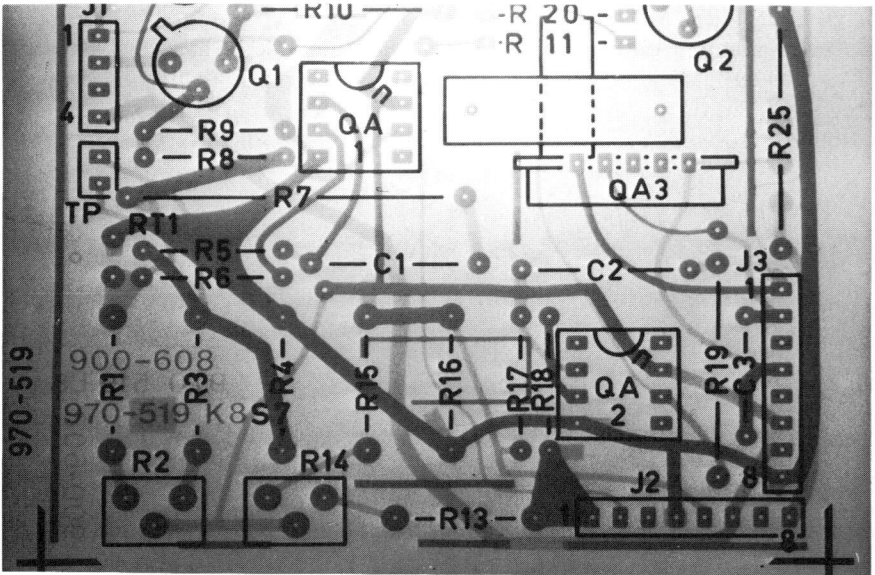

Fig. 8.53 An example of 'multiple printing'.

8.11 Filing the Artwork

It has been pointed out that the finished artwork is fragile and does not stand being rolled as this might cause some of the conductors to be displaced. Adhesive along the edges of the conductors collects dust and other particles whereby the conductors become blurred. Imprudent handling of the artwork, for example by laying the component and solder sides together with the taped sides facing each other, may cause short lengths of tape to work loose or be displaced.

For the above reasons the artwork must be stored very carefully, and the file must be arranged with due regard to their delicacy. Many drawing files are arranged for vertical suspension of their contents but these are not recommended for filing artwork even when individually placed in protective plastic bags.

It is safer to file the artwork horizontally which is achieved in a file with drawers (see Fig. 8.54). In order to keep the individual sets of artwork separated, they can be placed in large folders made of thin cardboard. If taped artwork is filed for a long time, say for 2 or 3 years, it may happen that the artwork sheets stick together somewhat. This is caused by the tape adhesive which creeps to the surface of the tape. When the sheets are to be separated, it could be that some of the tapes come loose or are displaced a little. It may therefore be wise to use thin silicone impregnated paper as intermediate layers. As the height of the drawers usually is limited to approximately 40 mm ($1\frac{1}{2}$ in) it is possible to file 10 sets of artwork per drawer. This is a very suitable number as the weight is not so large that it is impossible to remove the lowest set.

Artwork which lies about or is filed under insufficiently protected conditions will,

Fig. 8.54 Filing the artwork.

inevitably, deteriorate in quality. When the cost of one single set of artwork is considered, the procurement of a satisfactory file with drawers should be within the reach of everybody.

8.12 High-frequency Boards with Ground Planes

8.12.1 Introduction

A ground plane is often an integral part of high-frequency boards where it will normally be found on the component side. The preparation of ground-plane artwork has been described in Sections 8.3.3c and 8.3.1o, but in the case of complicated ground planes it is, however, not so simple a matter to trim the contour, cut the clearances around the solder pads, or to provide 'channels' in the ground plane for shielded conductors. Obviously it would be much easier to prepare the ground-plane artwork as a negative artwork in very much the same way as the solder-mask artwork. By means of a negative-working diazo film which has been recently brought onto the market, it is possible to reverse the ground-plane artwork. A combined contact print can then be made of the reversed ground-plane artwork and the original pad master, and used as a master for the component-side artwork by taping the conductors, which are to run in the 'channels' provided in the ground plane, and by connecting solder pads which are to be grounded, to the ground plane by means of short lengths of tape. The procedure is described in detail in Section 8.12.4.

8.12.2 Negative-working Diazo Film

A negative-working diazo film reverses the image of the original, i.e. all opaque areas become transparent, and vice versa. It is, like the well-known, positive-working diazo film described in the preceding sections, sensitive to blue–violet/ultraviolet light, for which reason the same equipment as described in Section 7.2 can be used. The resolution is also the same as for positive-working diazo films, i.e. more than 500 lines per mm (1250 lines per 0.1 inch).

The image transfer, however, is a little different as two exposures are needed: a rather short initial exposure forms the image by decomposing part of the diazo salts, and is followed by an ammonia development. The post-exposure which is about 10 times as long as the initial exposure, serves to clear the background, after which the film is developed again.

The user will soon discover that it is a little more tricky to use the negative-working diazo film than the positive-working type. It is more sensitive to daylight and should only be exposed to half-light when handled, and this only for a short time. Correct exposure time is vital for a successful printing; expsoing the film for too short or too long a time causes too 'thin' an image, so correct exposure time should always be determined before a new batch is taken into use (see next section). It is also important, after the development following the initial exposure, that the diazo film is protected against direct light during outgassing the ammonia, otherwise a further undesirable decomposition of the diazo salts takes place. After an outgassing period of 15 min the post-exposure can start. Finally, complete development is essential and can be ensured by two or three passes through the developing machine. Overdevelopment is, however, impossible.

8.12.3 Determination of Exposure Time

The exposure time depends upon the equipment as described in Section 7.2. A correct initial exposure is rather important owing to the fairly limited margin offered by negative-working diazo films. The post-exposure, however, is not particularly demanding. In principle, the optimum set of exposure times is derived in the same way as described in Section 8.4.1, but as two parameters (the initial exposure and the post-exposure) are involved, a sample with a 'two-dimensional' test pattern as shown in Fig. 8.55 is most practical. The test is carried out as described below:

Initial Exposure: The test pattern is contact printed on negative working film, rows 1 to 6 being exposed for 10, 20, 30, 40, 50, and 60 sec, respectively. Especially when the equipment is provided with a light monitor it could be tempting to expose all rows

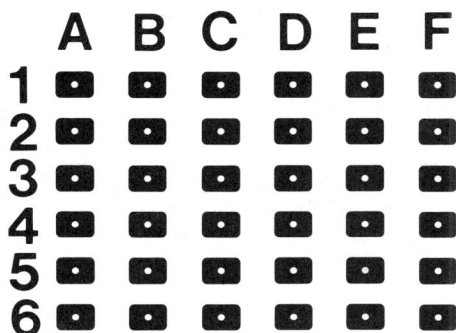

Fig. 8.55 Test pattern for determination of exposure time.

for 10 sec, then to cover row 1 and again expose for 10 sec, and so on. Due to the very narrow margin in exposure, it is recommended that each row be exposed independently of the the others, so that possible start effects of the exposure lamp will be taken into account. Care should be taken to avoid strong daylight, as this film is more sensitive to daylight than normal positive-working diazo films.

The sample is now developed in the ammonia developer, preferably with two or three passes to ensure complete development. Then the sample is left for outgassing of the ammonia for 15 min. It is important not to expose the sample to strong daylight during this period, as otherwise the diazo salts will continue to decompose.

Post-exposure: After the ammonia vapour is completely outgassed the sample is post-exposed, which means an exposure with no contact printing of the original. Columns A to F are exposed for 2, 4, 6, 8, 10, and 12 min, respectively. Due to the relatively long exposure times and the wide margin of exposure, it is now an advantage to expose the total sample for 2 min, then to cover column A, expose for 2 min again, and so on. When the last column has been exposed, the sample is developed, preferably with two or three passes through the ammonia developer.

Determining Exposure Time: A finished sample is shown in Fig. 8.56. It is now a question of determining the optimum combination of initial exposure and post-exposure, but unfortunately it is not possible to reproduce the sample to show how the

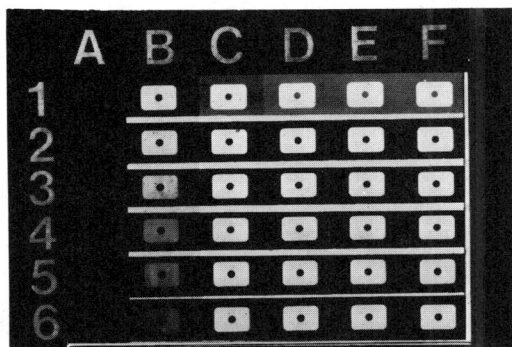

Fig. 8.56 Double exposed and developed negative sample.

282

colour varies from light yellow (a thin layer of dye) to deep red (a thick film of dye), although some variation can be seen. The transparent areas must be clear with no yellow film which would affect subsequent contact printings. On the other hand, the opaque areas should be of sufficient density to ensure correct contact printing. Until experience in assessing such samples has been gained it is recommended that contact prints are made on normal positive-working diazo film and the results studied. In order to keep the processing time as short as possible, the exposure times, especially that of the post-exposure, should be kept as short as possible. In column C all the transparent areas are clear with no yellow tint, and row 4 seems to give the surrounding opaque areas sufficient density in column C. The exposure times are therefore determined as:

Initial exposure time: 40 sec
Post-exposure time: 6 min

8.12.4 Generation of Ground-plane Artwork

Having described the negative-working diazo film in the previous sections we shall now demonstrate how such films can be utilized for preparing a ground-plane artwork. The procedure is shown in Figs. 8.57 to 8.62 below. How the films are turned during the contact printing is not described, however, as the reader presumably has gained sufficient knowledge from the previous description of contact printing positive-working diazo films.

Let us suppose that the ground plane is located on the component side of the board. In order to achieve a perfect registration between the solder and component sides, the starting point for the artwork is chosen as the normal pad master. A master for the component side including corner and registration marks and photo-reduction dimensions is now derived by contact printing the pad master (see Fig. 8.57). Before proceeding it is necessary to determine the clearance between the solder pads and the ground plane. Ordinary solder-mask pads giving a clearance of 0.3 mm (0.012 in)

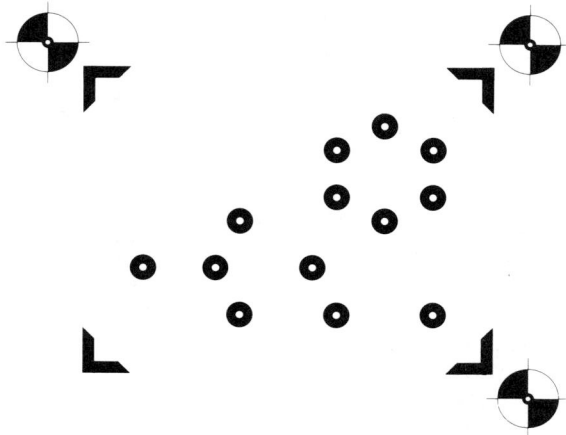

Fig. 8.57 Master for component side.

seem rather small, and experience shows also that solder bridging between a component lead and the ground plane may occur, especially if the lead is clinched 45° or more. Solder-mask pads for non-critical places ensure—as discussed in Section 6.9.3—a clearance of 0.5–0.6 mm (0.020–0.024 in) which in practice seems satisfactory. If a solder mask with such large clearances already exists, it is a very simple matter to contact print it and use it as a negative basis for the ground plane. Otherwise an artwork with the necessary larger clearances must be prepared. Another contact print of the pad master is therefore made and taped in the same way as the solder mask described in Section 8.8. The result is shown in Fig. 8.58.

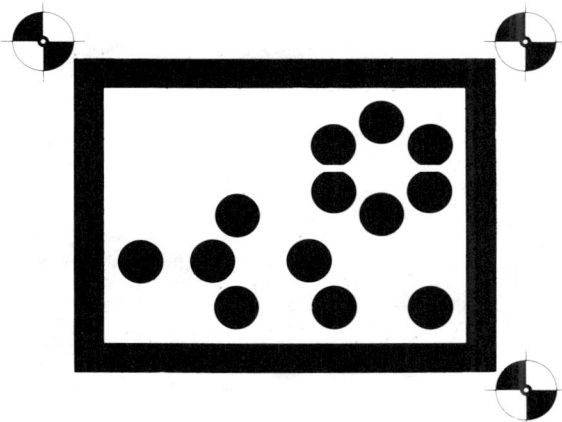

Fig. 8.58 Master for ground-plane artwork, taped with pads giving sufficient clearance between solder pads and ground plane.

It frequently happens that some conductors are to be routed on the component side, for which reason 'channels' in the ground plane must be provided. This is easily done on the artwork shown in Fig. 8.58 by taping the 'channels' with tape, or in the case of irregularly shaped 'channels', by applying red lithotape and cutting it to correct shape (see Fig. 8.59). The 'channel' width must of course be designed to give sufficient clearance on the finished board, i.e. about 0.5 mm (0.02 in) between the conductor and the sides of the 'channel'.

Large areas should be relieved as discussed in Section 6.6.4, if this can be allowed for high-frequency reasons. It is, however, much easier to relieve (break-up) the surface by applying suitable pieces of tape or rectangular pads. Oval pads should, as a matter of fact, be preferred because sharp corners should always be avoided. The finished negative ground-plane artwork is shown in Fig. 8.59.

The next step is to reverse the negative ground-plane artwork. This is done by contact printing it on to negative-working diazo film as described previously, i.e. in this case with an initial exposure of 40 sec and a post-exposure of 4 min, each exposure followed by development. The contact print, which now is a positive ground plane, is shown in Fig. 8.60.

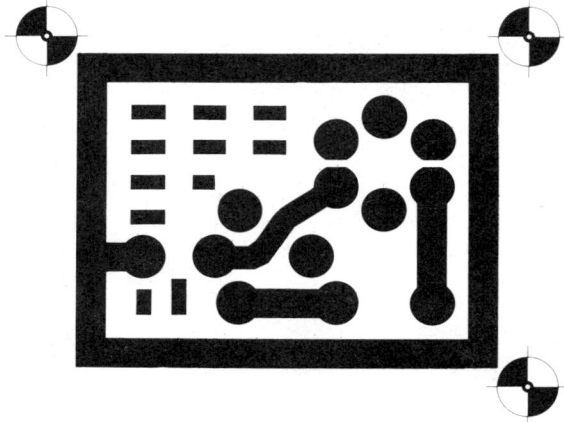

Fig. 8.59 Finished ground-plane artwork. Ground plane relieved and provided with 'channels' for shielded conductors.

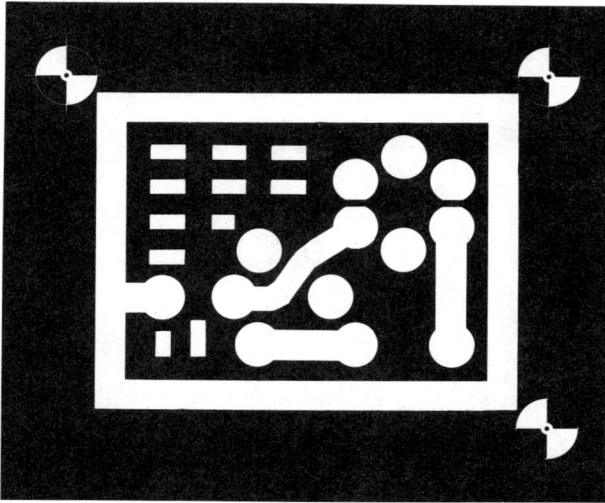

Fig. 8.60 Positive ground-plane artwork derived by contact printing artwork shown in Fig. 8.59 on negative-working diazo film.

The reversed and positive ground-plane artwork is now registered with respect to the pad master so that both can be contact printed and form the component-side artwork. Although the registration marks on the reversed ground-plane artwork are negative, it is very easy to registrate the two artwork sheets precisely. The registration marks will, however, be more or less extinguished in the contact print which would impede later registrations of the component-side artwork. It is recommended, after registration, that the two artwork sheets be taped together in such a way that it is possible to lift and cut away the corner marks of the reversed ground-plane artwork. The combined contact print is shown in Fig. 8.61. As solder pads, corner, and registration marks are derived from the original pad master, exact registration between the component and solder sides is maintained. It will not do much harm should the ground-plane artwork be slightly displaced when contact printed, as generous clearances have been chosen, but normally a very close registration is easily obtained.

Fig. 8.61 Combined contact print of master for component side shown in Fig. 8.57, and positive ground-plane artwork shown in Fig 8.60.

Fig. 8.62 Finished ground-plane/component-side artwork.

The last step is to tape the conductors which are to run in the 'channels' and to connect grounded solder pads to the ground plane by means of small pieces of tape. The finished artwork is shown in Fig. 8.62.

In Section 8.5 it was emphasized that it is very important to use an identification system. This applies, of course, also in this case, and space for a type designation or a code number can easily be obtained by providing cut-outs in the ground plane, i.e. a cut-out is taped at a suitable place on the negative ground-plane artwork. As Figs. 8.57 to 8.62 serve only to illustrate the principle, such details have not been shown.

The author's conclusion is that it is both very easy and quick to generate a ground-plane artwork by means of negative-working diazo film, the only tricky point being the contact printing itself. It is, however, just a question of adopting procedures which only differ slightly from those followed when using normal positive-working diazo films. The method will no doubt gain considerable application in the future.

9

Automated Generation of
Artwork/Filmwork

9.1 Introduction

The preceding chapters clearly show that manual preparation of precision artwork for printed circuit boards is a rather slow and demanding task, even when rational procedures are followed. The layout and taping require great care and expertise from the PCB designer and draftsman, especially when the goal is trouble-free production. Unfortunately, and strange as it may seem, the electronics industry has not had the foresight to establish training schools for this important category of personnel. It therefore is becoming more and more difficult to recruit qualified personnel.

Understandably, great efforts have been made to automate the layout and taping stages in order to shorten the turnaround time, to increase the accuracy of the layout, and above all, to reduce the demand for specially trained personnel. These efforts have resulted in systems for:

1. Automated drafting, including photo plotting of the filmwork, based on a manually prepared layout sketch and a parts list. See Paragraph 9.3.
2. CAD (computer-aided design of the layout), followed by photo plotting of the filmwork, based on a complete schematic diagram and a parts list. See Paragraph 9.5. In some cases this design work can be combined with preparation, on a CAD basis, of the corresponding schematic diagram.
3. CAE (computer-aided engineering), by which the CAD system is connected with other systems to include the total electronic design effort. In addition to the basic PCB design work, this includes:

   - Circuit development and simulation
   - Generation of parts lists
   - Development of automatic test programs

As an example, the CAD system can be connected to a materials management system so that the complete parts list for the assembled board is generated almost automatically. The automatic process is much faster and allows fewer opportunities for error than does the manual case.

9.1.1 Basis of Assessment

The following considerations are intended to give prospective users of systems for automated filmwork generation a basis for assessing the advantages of such systems.

In general, it is impossible to quantify the advantages to be obtained by changing from manual methods to CAD methods, because the conditions vary greatly from company to company. Such factors as company size, overhead factors, product mix, and development activities play a large part and should necessarily enter into the assessment. Therefore, the considerations stated below suggest ideas which the prospective user can adapt for his own conditions when proposing the procurement of a CAD system.

9.1.1a Development Phase Considerations

Development personnel have always complained about the long delay between completing the schematic diagram for a new product and having the PCB documentation ready and the prototype boards available for assembly and testing. A delay of six to eight weeks in the development project can destroy personnel motivation and possibly cause important details in the development work to be forgotten or overlooked.

When, for example, a CAD system has been introduced, the turnaround time in the PCB design department can be shortened so much that the physical manufacture of the prototype boards now becomes the main bottleneck. It should, however, be borne in mind that when the new product is microprocessor-based, the software development time is frequently another cause of delay. There are various development stages to consider when assessing the turnaround time:

(a) A functional model stage, which in the case of complex electronic circuitry freqently requires "real" printed circuit boards in order to prevent development faults. Such faults might not be discovered until a later stage where the corrections cost far more. This applies particularly to high-frequency circuits, fast microprocessor circuits, or mixed digital/analog circuits. Here a wire-wrap setup can easily give a false picture of the circuit's operation. Some possible problems are undesirable self-inductances caused by missing voltage and ground planes, uncontrollable stray capacitances, etc.

(b) A real prototype stage, where the board layout frequently has to be changed one or several times. Adding just a few ICs can mean displacing all other components so that a new layout seems to be the only rational course of action.

(c) A final correction stage before the product is released for pilot production. In many companies, the release procedure involves several departments or functions, such as:

- Sales and marketing
- Service
- Production
- Quality assurance

The release reviews frequently criticise in areas overlooked by the project group, such as components being located where they cannot be replaced with reasonable ease, or components exhibiting a doubtful reliability. The sales department may also want to add new selling features. In some cases, the necessary corrections require a totally new board layout.

(d) A final correction stage after finishing the pilot production. The experience gained during pilot production frequently reveals the need for minor corrections to the board. Since regular production is often planned to immediately follow the pilot production, little time is available to purchase the boards. The draftsman must therefore correct the boards quickly, and unfortunately, the company frequently has no time to procure and test revised prototype boards. It is therefore desirable that such corrected boards have a high probability of correct operation, correct minimum spacings, etc.

9.1.1b Current Products Considerations

In addition, the number of changes to current products can present a great problem. Boards that have been revised up to 12 or 15 times are not unheard of, or even rare. Especially in the case of high-density boards as used in modern electronics, it can be problematic to modify boards in order to accommodate product changes.

9.1.1c Change Considerations

Particularly in the case of large and complex boards, the previously described manual methods can seem quite hopeless when introducing even small changes. Not only is the necessary time quite long, but worse, there is a high risk of introducing faults into previously flawless areas which must be moved a little to make room for the changes. As a rule, such faults are not found until the prototype boards have been assembled and are ready for testing. Finally, there is an important psychological effect when a PCB designer and a draftsman have to repeat the same job but in just a slightly different way. The reader can easily imagine how depressing such a situation can be.

The manual method tempts the PCB designer to accept nonoptimum solutions in order to avoid delaying the project, or to avoid moving otherwise correct circuits because of the risk of new faults. It is a well-known fact that the layout sketch used for manual layout is not the best basis for an assessment, and even if the layout is checked before taping, small errors, such as crowded components, can easily remain undiscovered until a later stage.

With a CAD system, even large alterations of the board will not cause problems because of the existence of the data base, and a change can frequently be completed within hours, whereas the manual method would require days.

The automatic check routines of the CAD system examine all parts of the pattern for violation of the design rules, especially the minimum spacings, with an accuracy that cannot be achieved by manual checking. Along with the check routines, the auxiliary routines, which make it easy and safe to move, add, or delete components as well as conductors, are among the most important reasons for the high layout quality found in CAD boards.

9.1.1d Personnel Considerations

The reader can easily understand that a PCB department is extremely vulnerable with respect to staff turnover and easily can be paralyzed, especially in the case of manual artwork generation. All such work depends directly on the experience of the PCB designer and draftsman. Gifted and skilled PCB designers are, however, just as essential to the CAD work as to the manual PCB layout, particularly when doing the initial component layout. On the other hand, even in the case of different operators, a reasonably homogeneous layout quality is insured, partly because of the check routines.

A frequently used time factor of four between manual artwork generation and CAD work indicates an increased development capacity in the PCB department. Particularly in the case of several parallel development projects, such a capacity increase is of paramount importance to the timely completion of low-priority projects. With manual artwork generation, the general trend is to postpone low-priority projects for the benefit of the primary projects, especially when the latter encounter problems. Electronic development projects generally tend to bunch up, so many PCB layouts and taping jobs must be done within the same period. In such cases a CAD system with several work stations proves of great value.

The fact that the procurement of a CAD system considerably increases the capacity of the PCB department does not necessarily mean a staff reduction. The number of design jobs automatically increases because of the short turnaround time; furthermore, some cases will still require a manual solution. Finally, the old artwork collection will usually be maintained manually because of the expense of converting to CAD filmwork.

The increasing number of CAD jobs results, among other things, from the opportunity to enter the functional model stage with printed circuit boards instead of the wire-wrap setup normally used by circuit designers at this stage. Also, boards which are to be used in a very limited number, for example, for special test gear in the production department, can be made as CAD boards with advantages in cost and turnaround time.

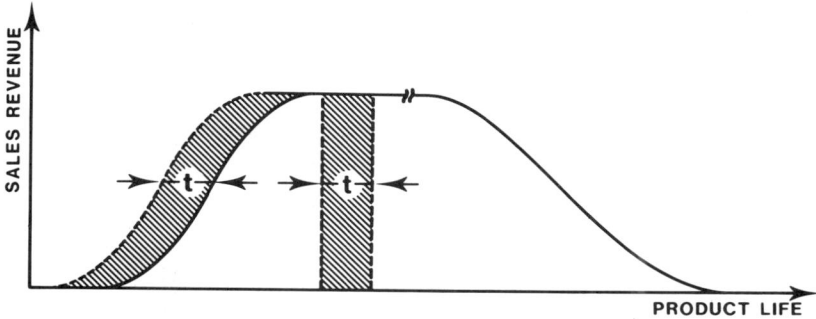

Fig. 9.1 Sales curve of a new product. The letter "t" indicates the earlier marketing time achieved by an automated generation of the artwork/filmwork.

9.1.1e Economic Considerations

When the CAD application results in a development project's being completed earlier, the saving in interest can be considered as a source of income. Since printed circuit board schedules very often determine the project deadline, the total saving in interest can be included in assessing the financial benefit.

To get a feeling for what this means, assume that a company completes development projects valued at $1,000,000 per year. Furthermore, assume that in 50 percent of the cases, the PCB design work lies on the critical path and that the time gained is 2 months. At an annual interest rate of 12 percent, the saving in interest is $10,000 per year, which can be perceived as income from the CAD system. The reader can easily convert these figures to his company's actual conditions.

Earlier marketing of a new product may have a much greater benefit than the interest savings. The sales curve of a new product usually appears as shown in Figure 9.1, which depicts the life cycle of the product. The startup period is followed by a period of stable sales. At a later time, which we cannot influence because it is determined by the market conditions, the curve declines. If we, as shown by the dotted curve, can accelerate the marketing by some months, t, then the product's lifetime will increase by the same time.

This implies that the additional sales resulting from the CAD system can be ascertained as the sales during t months in the best period. See the hatched area in Figure 9.1. We have here ignored the obvious fact that entering the market ahead of the competitors can result in a larger market share, i.e., somewhat higher sales in the stable period.

We will now convert this into money, assuming a product replacement (renewal) corresponding to a yearly sales volume of $1,000,000. If we again assume that the PCB design lies on the critical path in 50 percent of the cases, and the gain in time is 2 months, the sales increase is calculated as

$$2/12 \times 1/2 \times \$1,000,000 \cong \$80,000 \text{ per year}$$

Assuming an overhead of 50 percent, we can regard $40,000 per year as caused by the CAD system. The reader can easily convert these figures to specific company conditions.

Although the increased revenue may not be obvious to the chief accountant, it is one of the strongest arguments for procuring a CAD system. A simple calculation shows why large companies selling a large volume of new products can easily afford to buy large CAD systems.

The crux of the above economic considerations is whether the PCB design work really lies on the critical path; i.e., whether completing the PCB documentation earlier in the development stages mentioned in Paragraph 9.1.1a in fact results in an earlier marketing of the product. It is therefore recommended that the reader compile a history of various projects, so that he can produce realistic figures regarding the probable improvement in turnaround time, should the occasion arise.

9.1.2 Acceptance of CAD Systems

It is very difficult to assess the extent to which filmwork is being generated by means of CAD systems. The PCB manufacturers seem to be in the best position to make the assessment, but inquiries show a range of 50 to 95 percent of filmwork done on CAD systems. The figures depend somewhat on the PCB manufacturer's customer mix. If the customers predominantly consist of large companies, CAD filmwork will predominate with respect to manual artwork, whereas the reverse situation will apply if the customers consist of smaller companies.

The picture changes somewhat if the PCB manufacturer has established a CAD service bureau and offers to perform the design work on the basis of a schematic diagram and a parts list. This implies the advantage that the PCB manufacturer can offer to make prototype boards in a very short time because he has at his disposal all the necessary documentation in an expedient form. For regular series production, the customers tend to remain with the same PCB manufacturer, further increasing the number of CAD boards in production.

Finally, it should be mentioned that if the PCB manufacturer also produces multilayer boards, these will in nearly all cases be based on CAD filmwork, a fact which contributes to the high quote of 95 percent as mentioned above.

Although it is impossible to get an unambiguous answer, there is no doubt that the number of CAD boards is steadily growing, whereas the number of boards based on manually designed artwork is decreasing. A similar inquiry made nearly 5 years ago showed a variation from 20 to 70 percent. Relating these figures to the newer estimate of 50 to 95 percent shows that the progress is extremely high. This relative growth, which as a matter of course will continue, is conditioned by several factors:

1. Many new CAD systems have been put on the market, and older CAD systems have been updated.

2. Because of the powerful computers based on 16- or 32-bit processors, and the advanced software, CAD systems offered today can handle even very complex board designs. In many cases, the systems can also generate schematic diagrams and parts lists.
3. CAD system prices have fallen considerably in recent years.
4. More and more companies realize the necessity of introducing automated design methods, because the board complexity has made it difficult, or even impossible, to generate and update artwork via manual methods.
5. Many service bureaus have been established, and offer great expertise in PCB design.
6. There is a very strong trend within the electronics industry to minimize the development time of new products.

9.2 Automating the PCB Department

Below we will address some conditions which influence a full or gradual transition to automated PCB design.

9.2.1 Relative Time-Consumption

It is of interest to more closely consider the time consumed by the individual operations which form part of the manual artwork generation. Such an analysis can give very useful information regarding the best area in which to concentrate one's efforts toward faster and cheaper procedures. Table 9.1 shows the average distribution of time to manually prepare the artwork for a large number of boards. A typical example of such a board is shown in Figure 9.2.

Table 9.1. Time distribution for manual preparation of PCB artwork

Working stages	Percentage of total time consumed
The Layout Stage	
Rough layout of components	10
Routing of conductors	35
Checking and rectification	10
Total	55
The Taping Stage	
Common pad master	3
Solder mask	3
Solder and component sides	22
Component notation	5
Checking and rectification	9
Master drawing	3
Total	45

294

Fig. 9.2 Typical example of printed circuit boards used in analyzing the time distribution when generating manual artwork (layout and taping).

Since large variations of board complexity (density) and type (pure digital or analog, or a combination thereof) obviously occur in practice, these figures should be regarded only as guidelines.

9.2.2 Comparison of Systems

Table 9.1 shows that taping takes about half of the total time consumed. A first step towards automation could therefore be to replace the taping phase by automated drafting methods as described in Paragraph 9.3. The next step towards a fully automated artwork generation is to replace the layout phase by automated design methods, i.e., by computer-aided design (CAD) methods, of which the interactive system described in Paragraph 9.4 seems to be the most important. Both paragraphs aim to give the reader a qualitative understanding of the equipment as well as information on preparing the documentation package for a service bureau. Since purchasing an automated drafting system or a CAD system is a major step, most electronics companies would presumably prefer to gain experience in the initial stage by using a service bureau. Practical examples are therefore given in Paragraphs 9.4 and 9.6.

A rough comparison of the stages involved in manual artwork generation, in an automated drafting system, and in a CAD system is given in Figure 9.3, which on the whole is self-explanatory.

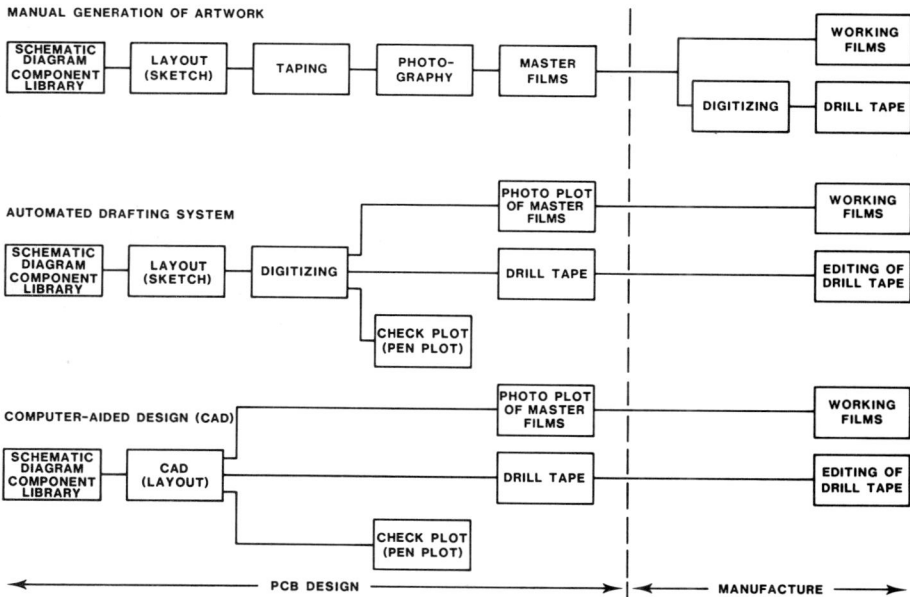

Fig. 9.3 Comparison between various methods of preparing artwork/filmwork.

Regarding the manufacturing documentation to be used by the PCB manufacturer, the primary difference between the manual method and the two latter methods is the means of producing master films and drill tapes. In the automated drafting system and the CAD system, the master films as well as the drill tape are derived from the same data base, which results in extremely high compatibility between the two. On the other hand, the PCB manufacturer frequently must edit the drill tape to make it match his equipment and general practice.

The manual method can cause relative and absolute faults. The relative fault is a misregistration between the individual layers of the board, for example the solder and component sides, and the absolute fault is a more or less incorrect reduction ratio during the photographic preparation of the master film from the artwork. The drill tape usually matches perfectly with the board side being digitized, but perhaps not particularly well with the other board side. It should be realized that raster digitizing means a risk that neither board side has a good match with the drill tape.

The main difference between the automated drafting system and the CAD system is the degree of interactivity, i.e., the operator's possibilities of working in a dialogue with the system and actively intervening in a process. Another difference is in the processors, which may be either microcomputers or minicomputers.

Although for many years automated drafting systems were pure digitizing systems with no provision for changes to the graphics data, the immense advance of modern computer technology has resulted in advanced forms of data manipulation and data editing, including routines for automatic component placement and automatic conductor routing. In this respect, the efficiencies of the more advanced drafting systems approach more nearly those of the CAD systems.

The automated drafting system no longer requires the layout sketch to be complete in every detail, since the operator can now introduce changes and rectify design faults during the digitizing process. This is caused, among other things, by the fact that it is possible to follow all transactions on a graphics display, which in many cases is a color display unit.

The automated drafting system and the CAD system both can control a photoplotter and a CNC/NC drilling machine, usually off-line. The data medium for the photoplotter can be a magnetic tape, a floppy disk, or a paper tape. In some applications, the paper tape becomes too bulky for practical use, so another data medium is preferred. The data medium for the drilling machine is usually a paper tape.

These systems can also control a pen plotter, either on-line (directly connected) or off-line (for example, by means of a tape). The pen plot generated can be used for checking the layout, for example, during the progress of the layout or just before starting the expensive photo plotting.

The photoplotter can generate step-and-repeat master films based on the data medium. The advantages are an optimum side-to-side registration, and a perfect match of the corresponding drill tape with the step-and-repeat master films. This capability, however, is used only rarely because it requires a multifold increase in plotting time, so preparation of step-and-repeat master films is usually entrusted to the PCB manufacturer, who has special photographic step-and-repeat equipment at his disposal.

9.2.3 Production Stage Considerations

From the PCB manufacturer's viewpoint, the customer's use of an automated system is a great advantage because of the considerable time saved during the preproduction tooling stage. This is first and foremost due to the better quality of the documentation. Some PCB manufacturers estimate an 80 percent reduction of the turnaround time when preparing the necessary tooling (master and working films, drill tapes, jigs, etc.). In practice this means a reduction from 2 weeks to only 2 days, which is an acceleration of $1\frac{1}{2}$ weeks. Although this corresponds to a labor reduction from perhaps 20 hours to 4 hours, it is just as important that the tooling is less dependent on the availability of a qualified staff, and fully independent of whether the digitizer is occupied or not.

An important reason for the above time saving is the higher quality of filmwork generated by automated methods compared to filmwork derived from manual artwork. There are almost no side-to-side misregistration problems, which is of utmost importance in the case of mini via holes of 0.012 to 0.016 in. (0.3 to 0.4 mm), which have now become quite popular. There is furthermore a high probability that all minimum spacings are met. Finally, it is not necessary for the PCB manufacturer to prepare the drill tape because it is usually furnished by the customer.

9.2.4 Drill Tape Considerations

A much-debated part of the customer's documentation is the drill tape provided. For economic reasons, the customer prefers the PCB manufacturer to use the provided tape.

9.2.4a Problems with a Customer's Drill Tape

In many cases the drill tape is prepared with no particular consideration for an efficient drilling sequence, and the work table of the drilling machine does not follow the most direct route, but moves back and forth at random. In extreme cases, the drill bit fouls on a locating pin and breaks.

If the drilling machine automatically exchanges drill bits, the drill tape must contain the necessary information, which of course must comply with the PCB manufacturer's practice. This will present a problem if the customer changes PCB manufacturers.

The location of the datum point of the reference system can be problematic for the drilling process, especially when a step-and-repeat process is used to prepare multiple working films for the large production panels usually employed. Sometimes every other board is reversed in order to achieve a better plating balance, which in turn requires a corresponding reversal of the drill tape. Finally, the tape may have to be converted to another code before the drilling machine can accept it. Thus, very few customers deliver satisfactory drill tapes which can be used as-is.

9.2.4b Previous Practice

A few years ago it was, for the aforementioned reasons, fairly common for the PCB manufacturer to prepare his own drill tapes by digitizing the CAD filmwork, even though this procedure could seem superfluous. The digitizing process, however, went off quite easily because most of the solder pads inherently fell exactly on the grid intersections. This implied that it was only necessary to digitize one pad of an IC configuration, and the remaining pads were digitized automatically by the machine's software.

9.2.4c Current Practice

Although most PCB manufacturers currently accept the customer's drill tapes, they usually must edit the drill tapes in order to achieve an optimum drilling sequence. The manufacturer also introduces special codes for drill bit exchange, data points, etc. The editing is normally performed by a computer, either in the digitizer itself, or in a self-contained editing equipment. The drill tape delivered by the customer rarely requires so much editing that it is more economical to digitize the board. If so, the digitizing is highly facilitated by the lock-in facility of the digitizer, which means that it is not necessary to aim at the pads very precisely.

Under all circumstances, the PCB manufacturer checks the drill tape prior to the production run by drilling a single board and inspecting the board against the master film for missing and displaced holes. Another possibility is to drill a film directly, but this requires that the film be very precisely registered in the drilling machine.

9.3 The Automated Drafting System

9.3.1 Generating the Filmwork

As stated in Table 9.1, approximately 50 percent of the total design labor, exclusive of checking and correction, is spent on the taping process. The first step in automating the PCB department could be to replace the manual drafting and taping by automated methods.

On the whole, the layout is prepared conventionally as described in Chapter 5, ending up as a sketch showing the component placement and the interconnections. Thereafter, the information shown graphically in the layout sketch has to be encoded so that it can be entered into the system. This applies to both the location and size of the pads and via holes, as well as to the width and routing of the conductors (the routing is defined by the start and end points as well as the bending points), the shape of possible ground planes, and mechanical details such as board contour, cutouts, etc. The operator encodes all this data by digitizing the locations, and supplementing them with library codes that supply details from the library in background storage, for example, solder pad diameters.

Newer automated drafting systems are organized with a large memory, a PCB data base (library), and efficient software which, in conjunction with an on-line (color) graphics display unit, allow convenient manipulation (processing) and editing of data. It is possible, by means of the pen plotter, to get an in-process pen plot so that possible layout problems can be studied and solved off-line.

A detailed discussion of the operator's possibilities for modifying the layout by means of data manipulation and data editing is given in Paragraph 9.3.2 (Menu). When all data have reached final form, they are processed in the computer and appear as output data in the form of:

1. A data medium, e.g., a magnetic tape or a paper tape, containing the PCB pattern information necessary for controlling a photoplotter, which generates the 1:1 master film. Because the plotting time for a complete set of master films can be quite long, e.g., 4 hours, it is desirable to operate the photoplotter off-line. Because the large amount of data may make it awkward to use paper tapes, magnetic tapes are used.

2. A magnetic tape containing all information about the board. Information on later PCB revisions can be entered into the system so that the revision can be started immediately.
3. A paper tape for controlling the CNC/NC drilling machine.
4. A paper tape for controlling an automatic component insertion machine or an automatic test equipment. This, however, requires special software for producing the paper tape.

All things considered, it seems that the various electronics companies and service bureaus have achieved excellent results by using automated drafting. The advantages are the improved quality of the master films, the shorter turnaround time, and the lower cost.

As mentioned earlier, the automated drafting system can be used as an intermediate step towards a full CAD system offering an automated layout feature. It should be noted, however, that it is now possible to upgrade automated drafting systems, partly with more advanced software, partly with higher computer capacity so that they approach the performance of the CAD systems.

Introducing a CAD system does not necessarily mean scrapping the automated drafting system, since it still has many applications:

1. Solving layout problems not suited for CAD, for example, high-frequency boards and single-sided boards.
2. Converting old hand-taped artwork into photoplotted filmwork. As described in Paragraph 6.6.2, taped conductors can creep a little at sharp bends after some years. They can also contract and cause interruptions. Conversion of old artwork, possibly including an updating, can ensure continued use.
3. Relieving the workload on the CAD system by performing certain preparatory jobs, for example, preparing a wiring list.
4. Drafting schematic diagrams.

9.3.2 System Configurations

Many configurations of automated drafting systems exist. Modern systems are frequently based on a so-called work station, which contains all the facilities neeeded by the PCB designer in his daily work. Figure 9.4 shows a block diagram of a typical system configuration. It comprises the following units, the functions of which are:

1. A computer with internal memory supplemented with a magnetic tape drive and a background memory, i.e., an external memory.

   Computer: Handles the data transmission between the various subunits of the system, forming the central processing unit. A typical example is a minicomputer in a 16-bit architecture, with a 500 kilobyte memory. See the note in Paragraph 9.5.10b2 regarding the definition of bytes.

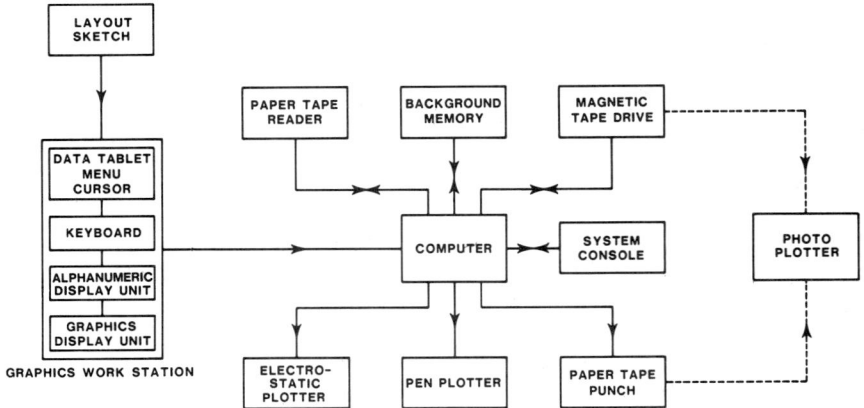

Fig. 9.4 Typical configuration of an automated drafting system.

Magnetic Tape Drive: Interfaces the external equipment and stores the design data.

Background Memory: Supports and extends the computer's internal memory and stores job files and library files applicable to the current job, as well as all fixed system programs. The capacity can, for example, be 10 megabytes.

Work Files: Contain information about the current job, frequently distributed in various levels. Some of the levels represent the individual layers of the board, for example, the pattern of the solder and component sides, solder and insulation masks, etc. Other levels represent groups which are common to the above layers, for example, the component placement.

Library Files: Represent commonly used elements common to a number of jobs. These elements, frequently called macros, are stored in one of the library files of the background memory, from which they can be retrieved during the work process.

A macro can represent such frequently repeated patterns as a component outline, or details common to a pattern, e.g., all the pads of an IC package together with the component outline, designations, etc.

2. A graphics work station comprising a digitizer tablet with a menu, a keyboard, an alphanumeric display unit, and a color graphics display unit. All these units are often built together to form a work table which offers, in a practical manner, all facilities necessary for entering, manipulating, and editing the data. Figure 9.5 shows the typical appearance of a graphics work station.

Fig. 9.5 Example of a graphics work station.

In some system configurations, the above units are physically separate. It is, however, quite common for the keyboard and the alphanumeric display unit to be combined into a console. Instead of a small data tablet (digitizer), the system can have a digitizing table with a large work area. See the example in Figure 9.6.

Below we will address the work station's units:

Digitizer: Converts the layout sketch's graphical information to digital data. The operator systematically probes the important points of the layout sketch, i.e., solder pads, via holes, conductor start and stop points as well as bends, ground and voltage plane contours, board contour, and reference points. In the following description, the digitizer is referred to as a data tablet.

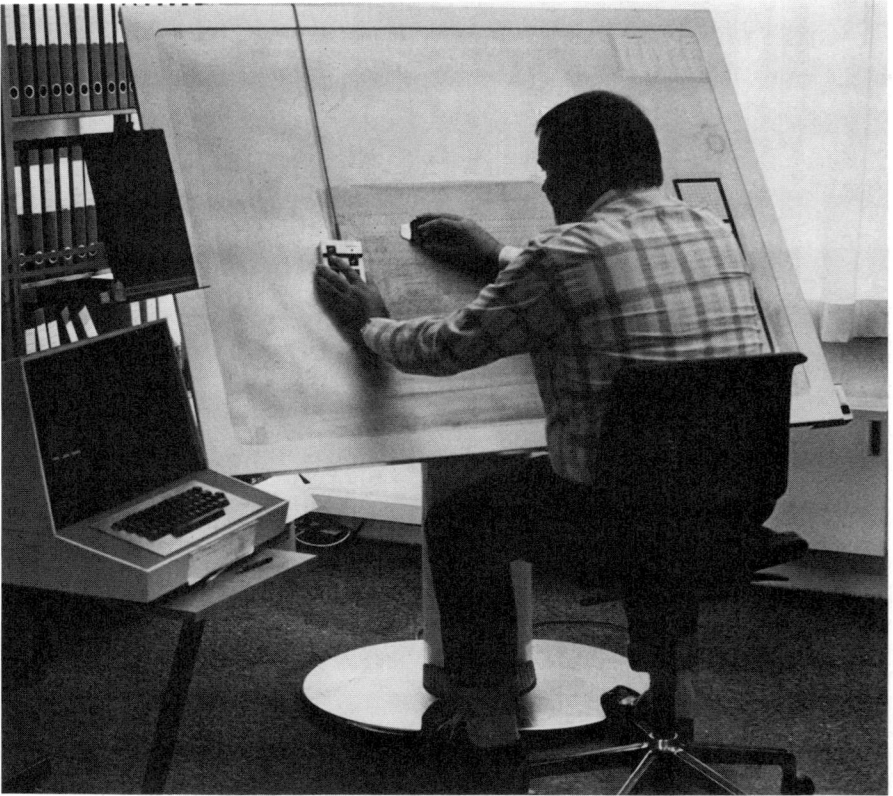

Fig. 9.6 Example of a large digitizing table.

The layout sketch is laid over the data tablet and probed by means of a reading head, i.e., a small crosshair cursor (mouse) with a bull's eye. See the example in Figure 9.7.

Under the surface of the data tablet, conductors are laid in a modular X-Y system, and by means of electric pulses transmitted over these conductors and detected by a small coil in the cursor, the electronic system can determine the position of the cursor very precisely. When the cursor is aligned with the selected point, the operator presses a push button to enter the position into the system.

In some systems the cursor is replaced by a pencil-like stylus with a detector coil in the pointed end. When the probe point is aligned with the desired point and pressed downwards, the system reads the position of the point.

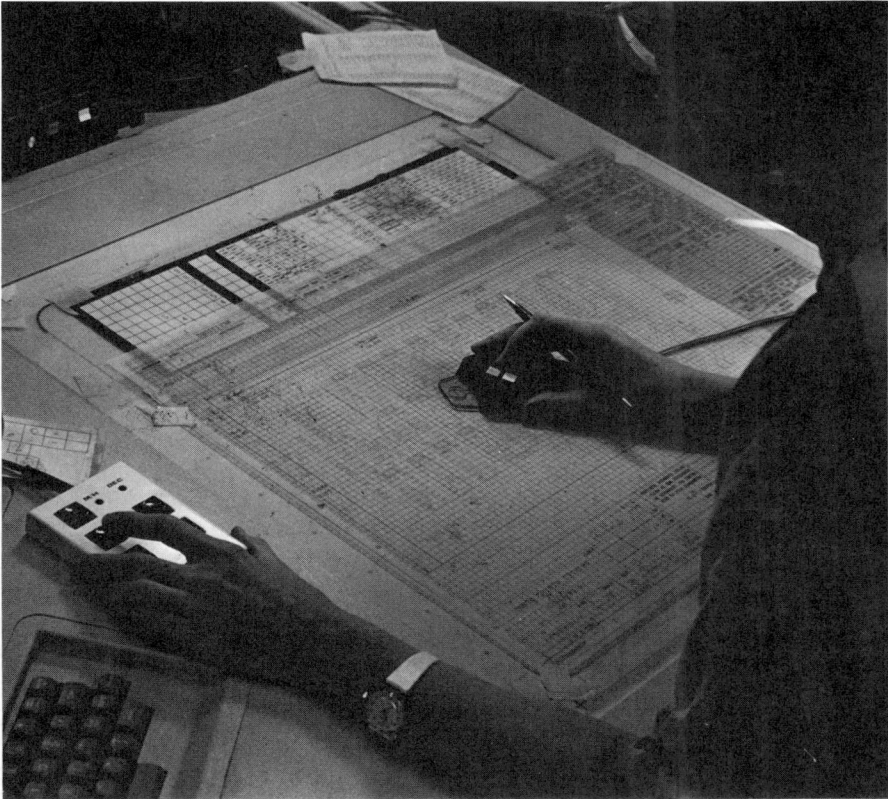

Fig. 9.7 Digitizing by means of a cursor (mouse).

Because of the grid rounding feature, the operator does not have to aim at or hit the point very precisely. Likewise, the layout sketch does not have to be drawn very precisely, although the conductor bends should be indicated fairly accurately. Points lying less than half a grid unit away from the grid lines will automatically be moved to the grid lines and will fall on the intersection of the two nearest grid lines. See Figure 9.8. Therefore, it is most practical to base the layout on a grid and to use a scale of at least 2:1.

Menu: A menu can be perceived as a paper keyboard placed on the data tablet. See Figure 9.5. The menu comprises a number of boxes, each having its own function. A function is activated by digitizing the box in question with the cursor. Below we will discuss a selection of these functions:

Placement: Every element of the layout sketch can, by means of the menu, be assigned one or more codes. In conjunction with the coordinates

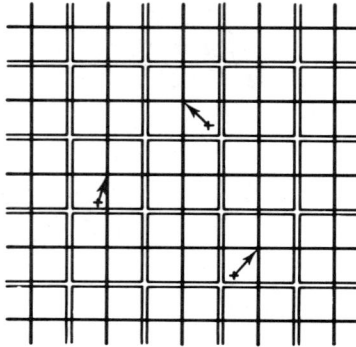

Fig. 9.8 Grid rounding moves an off-grid point to the nearest intersection of grid lines.

of the element, the codes form the input data for the system. By way of example, a code can indicate the dimensions of an element on the finished photoplotted master film, i.e., the diameter of a solder pad or the width of a conductor, etc. The code can also indicate various details, for example, how to fill out a ground or voltage plane or how much a conductor should be withdrawn in order not to cover the pad's center hole, if any. It is, however, a great inconvenience to the PCB manufacturer to receive filmwork having solder pads with center holes since these have to be filled by touch-up. Withdrawal should therefore be used as an exception only.

Data Manipulation: Other menu boxes allow an automatic fine adjustment of the mutual conductor spacings when two or more conductors are routed between two solder pads. In this way the conductors will be located with equal spacings, both mutually and with respect to the solder pads.

It is also possible to change the grid, for example, going from an inch raster to a millimeter raster when placing a metric component. In the case of difficult conditions, for example, a densely packed area, the operator can change the scale to improve the clarity.

Element groups that frequently recur in the job in question only have to be defined once. They can then be placed at the selected positions by means of suitable menu commands. Similarly, by means of multiple menu commands, a macro can be called from the library file, placed at the desired location, and turned the right way.

In some systems, input data are checked automatically to verify that they meet the design rules which have been entered via the alphanumeric keyboard. By way of example, it is possible to check the distance between two solder pads, between two conductors, or between a conductor and a solder pad.

Data Editing: The system can search a specified area or point for certain data (elements), for example, solder pads or conductors. The operator activates the search by digitizing the editing boxes of the menu, and then digitizing one of the data tablet points localized by means of the graphics display unit. The result appears on the alphanumeric display unit.

Selected elements, for example, a conductor, are deleted from the job file in a similar way. The start and end points of the conductor, however, should be digitized.

When it is desired to move a part of a complex conductor, it is unnecessary to delete the whole conductor and enter the changed part of the conductor together with the remaining unchanged part. Special menu commands make the move possible by deleting, entering, and adding conductor segments.

Finally, it is possible to move a macro by means of a simple menu command followed by digitizing the present and future location on the data tablet. All other data concerning the macro, for example, hole diameters shown in the drill film, are automatically moved to the new position. It is also possible to rotate the macro, for example, 90 degrees, by a simple menu command.

Keyboard and Alphanumeric Display Unit: These units enable the operator to communicate with the system without using the data tablet. By way of example, basic design data can be entered conveniently into the data base when starting a new job. The computer generates guiding questions and instructions which are displayed on the alphanumeric display unit. The latter can, for example, contain 8 lines each with 32 characters. The operator responds to the guidance by means of the keyboard, in some cases supplemented by menu commands.

The display unit, for example, can put the following questions to the operator:

• What is the job name?
• What is the desired grid scale?
• Is the module inch or metric?
• What symbol disc is to be used in the photoplotter?
• What minimum spacings apply:
    Pad-to-pad?
    Conductor-to-conductor?
    Conductor-to-pad?

The operator enters the answers into the system via the keyboard. The system immediately echoes the answers on the display.

Graphics Display: This usually consists of a full color display based on a raster scan system and a graphics processor. The operator can follow the job details by looking at the design, or parts of it, being displayed. This improves his view considerably, especially when he has to introduce modifications with respect to the layout sketch.

Most systems have various auxiliary functions which can be activated by menu commands. An example is defining a window to show part of the design on the display. It is also possible to zoom a specified area, for example, within the window, so that the area is shown enlarged. Furthermore, the operator can pan rapidly or slowly in a certain direction, frequently in multiples of 45 degrees, so that he can scroll design areas across the display. When the raster overlays the display, the solder pads and conductors that usually are shown as crosses and lines, respectively, can be displayed in their actual size and shape, corresponding to the symbol disc of the photoplotter.

3. A system console, often a visual display unit (VDU), i.e., an alphanumeric display with a keyboard, which is the operator's interface to the operating system. The operator can activate system commands, generate on-line programs and data processing programs, and respond to system prompts and questions.

4. A paper tape reader used for loading stored data into the system. It also loads data defining the solder pads and conductor widths required for a specific job to the photoplotter's symbol disk. Furthermore, "constructional data" for the most frequently used font types can be read for direct plotting by the photoplotter. The characters can, for example, be defined on the basis of a 7×7 point raster and composed of straight lines and arcs of a circle.

5. A paper tape punch which generates a number of different control tapes, for example, a plotter tape or a drill tape. In some drafting systems the drill tape can be generated in a version which optimizes the work table's route according to certain selectable criteria.

6. A pen plotter which can generate a check plot on paper, either during the work phase or after the job has been finished, so that details can be studied and approved before starting the expensive and relatively slow photoplotting. Pen plotters, which usually can plot in various colors, are normally operated on-line with the computer.

In certain cases, the pen plotter is replaced by or supplemented with an electrostatic plotter which within a few seconds generates a so-called hard copy of the image shown on the graphics display unit. This requires, however, that the plotter be connected directly to the display screen (storage or raster scan). Depending on the complexity of the image, a pen plotter would perhaps take 10 minutes to plot the same image. A disadvantage is that the resolution of the electrostatic plotter is much lower than that of the pen plotter, namely 0.005 in. (0.127 mm) against 0.001 in. (0.025 mm). Smaller electrostatic plotters have a fixed paper format, for example $8\frac{1}{2} \times 11$ in. (216×279 mm), for which reason the details of a complex image can become very small. Large electrostatic plotters use paper rolls in widths of, for example, 20 or 36 in. (508 or 916 mm). This implies, of course, a certain enlargement of the details.

9.3.3 Photoplotter

The last part of the automated drafting system shown in Figure 9.4 is the plotter, which plots the data originating from the graphics-to-digital conversion described. The plotter, which is a precision-made machine, can be purchased in various types, but common to all are a drawing head and a table, which move with respect to each other.

In some plotters the table remains in place while the drawing head moves in the X and Y directions. This is achieved by means of an X-axis carriage which rolls on precision steel tracks mounted along the two edges of the work surface. A Y-axis carriage, which is mounted on the X-axis carriage, carries the drawing head. In other types, the table moves along the X-axis, and for the Y-axis movement, a carriage carrying the drawing head is driven along a fixed bridge mounted across the support structure.

The aforementioned plotters are called flatbed plotters because they have a flat working area as shown in Figure 9.9. This distinguishes them from drum plotters, in which the drawing head moves along the Y-axis, whereas the X-axis

Fig. 9.9 Example of a photoplotter.

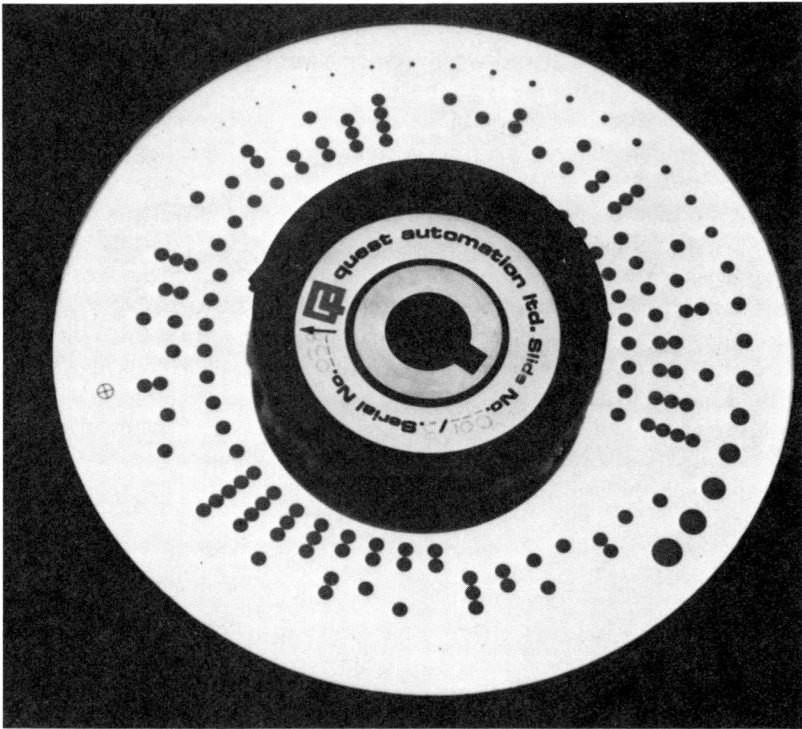

Fig. 9.10 Example of a symbol disc.

movement is achieved by rotation of the drum which carries the film. The inherent lower accuracy along the X-axis, however, makes this type of plotter less suited for close-tolerance filmwork.

The drawing head is normally a light-spot projector and the drawing medium a photographic film. For drawing conductors, the film is exposed continuously as the projector moves across it, whereas the film is flashed under stationary conditions for pads and other symbols. In this way, a filmwork master is produced directly on the photographic film in a 1:1 scale. The light-spot projector contains a symbol disc allowing a certain number of different conductor-drawing apertures and pad-flashing apertures to be selected by the input program. See Figure 9.10.

An alternative is cut-and-peel material which is cut with a knife mounted in place of the spot-light projector. See Paragraph 6.3.10. This type of material is sometimes used for the preparation of thin- and thick-film hybrid circuit masks, but could also be used for some types of artwork, especially artwork for large ground planes, etc.

Some plotters have a totally enclosed drawing table and, once the film is loaded, such plotters can be operated in daylight. Others require a darkroom since the drawing table is not enclosed.

For production of schematic diagrams, etc., it is possible to use a special matte film. The film is exposed by the light-spot projector, and the result is suitable for making printing plates directly, which can save on technical publication costs.

As printed circuit boards increase in complexity, the need for higher filmwork accuracy increases steadily. One of the most important characteristics of a plotter is, therefore, the positional accuracy of the plotted pattern. The accuracy of a precision plotter depends somewhat on its working conditions. By a static plot is understood a plot made by a stationary light-spot projector, i.e., solder pads are being plotted. The accuracy can be as high as ± 0.0004 in. (± 0.01 mm) within the central area of the table, for example, 12 × 12 in. (300 × 300 mm), and ± 0.0006 in. (± 0.015 mm) over the entire area, for example, 24 × 32 in. (600 × 800 mm).

By a dynamic plot is understood plotting with a moving light-spot projector, i.e., conductors are being plotted. The accuracy can, as an example, be ± 0.0014 in. (± 0.035 mm) over the entire plotting area.

Also, the repeatability is important when a plotter is evaluated. It can be ± 0.00024 in. (± 0.006 mm) in the case of unidirectional movements, and ± 0.0005 in. (± 0.013 mm) for multidirectional movements. Moreover, the resolution or the smallest step size is of interest, and 0.0001 in. (0.0025 mm) is frequently specified for precision plotters. The resolution and the plotting speed, however, are interrelated so that a tradeoff between resolution and speed is to be expected. Also, the exposure requirements may limit the maximum speed, a typical value being 4 in. (100 mm) per second.

9.3.4 Laser Plotter

The newest development within the plotter technology is the laser plotter, which exposes the film by means of a laser beam. Just as in the photoplotter, the film is placed on an X-axis carriage rolling on precision tracks, and the translation is performed by means of a precision lead screw and a servocontrol system. In place of the photoplotter's Y-axis carriage, a scanning mechanism is used, carrying an optical system which allows the laser beam to sweep across the film in the X-axis direction. By means of on-off modulation of the very narrow laser beam, the laser spot of only 0.001 in. (25 μm) exposes the individual pattern elements sequentially. This requires that the data generated by the computer (of the automated drafting system or of the CAD system) and recorded on the magnetic tape, be converted into a binary raster scan data stream in real time. The exact exposure position is controlled by a reference beam which scans a reference mask and thereby generates the signals necessary for modulating the laser beam.

Two parallel conductors are imaged as shown in A, Figure 9.11. Since the laser spot has a diameter of 0.001 in. (25 μm), an edge definition as shown in B, Figure 9.11, might be expected, but because of the energy distribution, the laser spot is not defined very sharply. Therefore, the intermediate areas receive a certain exposure, resulting in a highly regular edge definition as shown in C, Figure 9.11.

310

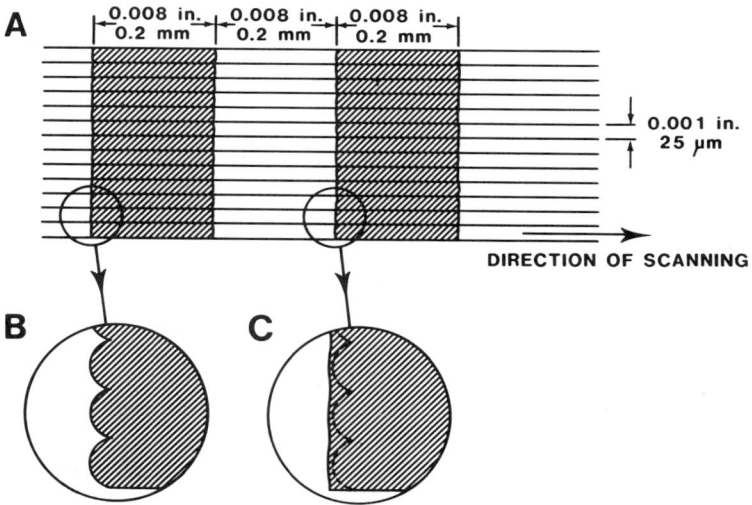

Fig. 9.11 Working principle of a laser plotter.
A: Drawing of two parallel conductors.
B: Edge definition of conductor without regard to the energy distribution of the laser spot.
C: Practical edge definition of conductor caused by the rim effect of the laser spot (exposure of the intermediate areas).

For example, exposure accuracies of ± 0.001 in. (± 25 μm) over the entire pattern area of 18 × 24 in. (450 × 600 mm), can be achieved. The medium can be a film or a glass plate with a coating of silver halide, but it can also be a UV sensitive paper which is used for check plots.

The edge definition can be very high, about 0.0005 in. (13 μm). Slanting lines and circular arcs are composed of 0.001 in. (25 μm) steps. It is possible to plot conductors having a width of 0.003 in. (76 μm), the width tolerance being ± 0.0005 in. (± 13 μm). The plotting time is very short, for example, a few minutes, where it would be hours when using a photoplotter.

A graphics terminal is connected to the computer of the laser plotter so that the operator can communicate with the system to create, among other things, a comprehensive library of solder pads in many different shapes. The PCB designer, therefore, has a free choice regarding solder pads because the laser plotter does not use a symbol disc, as does the photoplotter.

It should be noted that the laser plotter can be used for direct imaging of the pattern onto the board if a photo-resist has been applied. This removes the necessity for overlaying a working film. The result is a higher edge definition and possibly narrower conductors than are achievable when using conventional imaging methods. Provided that the exposure cycle can be shortened, PCB manufacturers are expected to take an interest in using laser plotters for direct imaging, particularly in the case of prototype boards, fine-line boards, or inner layers of multilayer boards.

9.3.5 Layout Considerations

At first glance, a layout sketch prepared for digitizing looks rather like a layout sketch for manually taped artwork. Closer inspection, however, shows major differences. When the layout sketch is to be digitized, all conductors must be straight lines and preferably parallel to the X-axis on one side of the board and to the Y-axis on the other side of the board; via holes are introduced where a conductor is transferred from one side to the other. A higher degree of flexibility is achieved by permitting the conductors to be routed on a 45-degree slope instead of always routing them perpendicularly. Several systems allow any angle for conductors, not just 45 degrees.

The layout sketch does not have to be drawn with the same degree of accuracy as a sketch for manually taped artwork. This is due to the grid rounding feature of the digitizing system, i.e., pads lying less than half a grid unit away from a grid line automatically move to the grid line and fall on the intersection of the two nearest grid lines. See Figure 9.8. This is of great importance to the precise location of solder pads on the master film. The conductors should be drawn with greater care because otherwise the digitizing process can be overwhelmingly difficult to survey.

A practical solution is to base the layout on a grid and use an enlarged scale, for example 2:1 or 4:1, depending on the area of the work table. It should, however, be noted that a modern work station provided with a (color) graphics display unit and offering the possibilities of data manipulation, data editing, etc., makes it possible for the operator to introduce modifications during the progress of work.

The design rules given in Chapter 6 are based on manual taping, and take into account the inevitable inaccuracies and instabilities. When the layout sketch is digitized, the data are very exact, and the design criteria may change. The minimum spacing, for example, can be reduced a little, but not more than the PCB manufacturer needs for a reasonable safety margin in his processes.

It stands to reason that when the PCB designer choses to sketch the layout on a grid, the dimensions of conductors and solder pads as well as the minimum spacing between these elements must match the grid size. Some additional constraints must be considered, namely the actual dimensions of the symbols on the photoplotter's symbol disc.

When preparing manually taped artwork, the PCB designer can always take the catalogue of the drafting aids supplier and hunt for suitable solder pad sizes among the copious selection. In the case of photoplotting, the selection of symbols is very limited, and the PCB designer must compromise regarding the optimum size and shape of the solder pads. A special symbol disc is rather expensive, and if the photoplotting is to be performed by a subcontractor or a service bureau, it is also problematic to persuade him to procure another special symbol disc.

The conclusion is therefore that when formulating design rules, the symbols in the existing symbol discs should be taken as the starting point. When these symbols are known, and the grid has been chosen, all details of the pattern can be determined, for example, the minimum spacing, the number of conductors passing through a narrow gap, the width of the annular rings, etc.

9.3.5a Practical Considerations

For practical reasons, the layout sketch should be drawn in a scale which permits digitizing on the work area of the data tablet or the digitizing table. Since even a rather small board becomes quite awkward to handle in a scale of 4:1, a scale of 2:1 is recommended. Therefore a module of 0.1 in. (2.54 mm) corresponds to 0.2 in. (5.08 mm) on the layout sketch.

The drafting material for the layout sketch must be stable so that changes in temperature and humidity do not cause dimensional changes affecting the digitizing. It is therefore recommended to use a stabilized polyester drafting film with a preprinted grid of 0.2 in. (5.08 mm). The grid can be printed as shown in Figure 9.12. The intersections between the slanting lines indicate 1/4, 1/2, and 3/4 grid points, which facilitate the sketching of conductors having spacings of 1/40 in. (0.635 mm), or bends of 45 degrees.

9.3.5b Digitizing Considerations

In many digitizing systems, a grid unit can be subdivided by means of a matrix located in the menu. See Figure 9.13. The matrix is divided into 10 × 10 microfields, and the operator can use a menu command to indicate whether he wants to use all 10 × 10 microfields, or just 8 × 8 microfields. In the first case, the operator works in a decimal mode, i.e., with subdivisions of 1/10 of the grid unit, and in the latter, he works in an octagonal mode, i.e., with subdivisions of 1/8 of the grid unit.

For example, if the bend of a conductor lies as shown in A, Figure 9.14, the operator will choose the octagonal mode, whereas he will choose a decimal mode if the bend lies as shown in B, Figure 9.14. The coordinates of the bend are derived as the coordinates of the zero point of the grid unit plus the coordinates of the microfield.

When the operator has to digitize a large number of points lying on 1/4-, 1/2-, or 3/4-grid units, using the aforementioned matrix becomes a little inconvenient because the cursor or stylus has to activate a menu command for each digitizing operation. This means that the cursor or stylus must be moved backwards and forwards between the digitizing area and the menu.

Some systems are therefore provided with a detached unit having push buttons for the three fixed subdivisions shown in A, Figure 9.15, i.e., for 1/2-grid unit locations. The unit is operated by the free hand during the digitizing. When

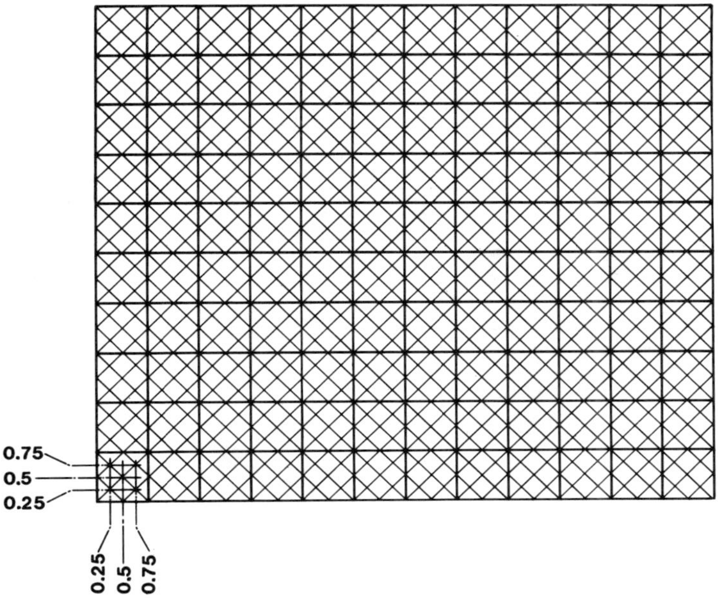

Fig. 9.12 Example of a grid particularly suited for layout work because of the slanting lines.

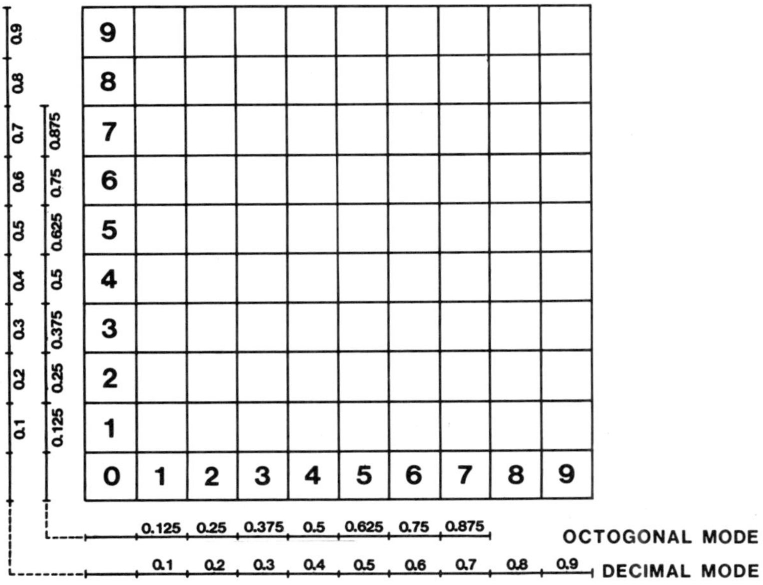

Fig. 9.13 Subdividing a grid unit by means of a matrix.

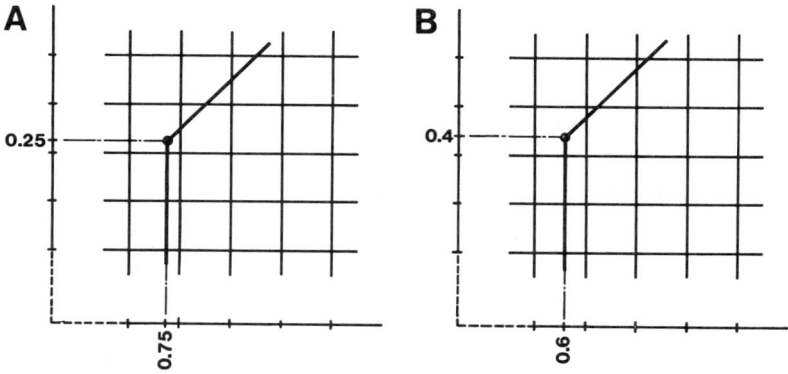

Fig. 9.14 Digitizing the bending point of a conductor.

    A: Octogonal mode. The coordinates of the bending point in the grid field are determined as

$$X = 0.75 \ (= \tfrac{6}{8}) \text{ and } Y = 0.25 \ (= \tfrac{2}{8} \ ).$$

    B: Decimal mode. The coordinates of the bending point are determined as

$$X = 0.6 \ (= \tfrac{6}{10}) \text{ and } Y = 0.4 \ (= \tfrac{4}{10}).$$

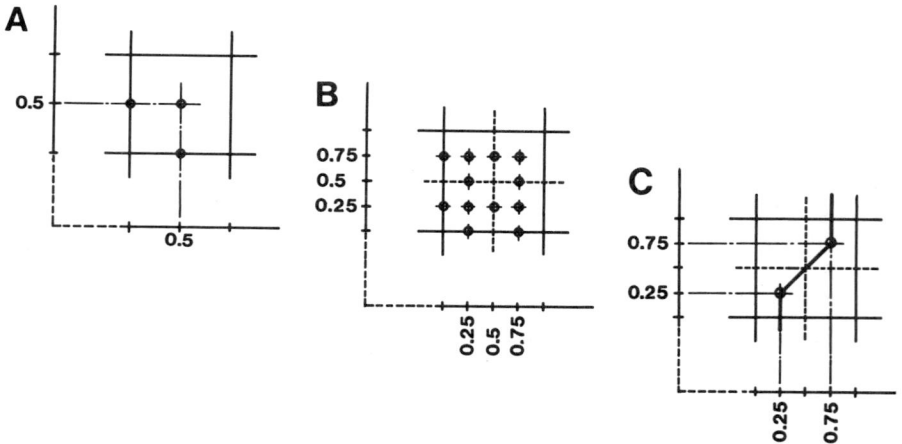

Fig. 9.15 Fixed subdivisions on detached unit (handset).

    A. Subdivisions corresponding to 1/2-grid unit locations offer three different possibilities: (0.5, 0), (0.5, 0.5) and (0. 0.5).

    B: Subdivisions corresponding to 1/4-grid unit locations are achieved by changing the scale factor by 2 and using the 1/2-grid unit subdivision stated at A. This gives 12 possibilities.

    C: Bending points of a conductor determined on the basis of the 1/2-grid unit locations shown at B.

Table 9.2. Common conductor widths and solder
pad diameters

Conductor width		Solder pad diameter	
in.	mm	in.	mm
0.006	0.15	0.050	1.27
0.010	0.25	0.055	1.40
0.013	0.33	0.063	1.60
0.016	0.40	0.080	2.03
0.025	0.64	0.100	2.54
0.040	1.02	(0.020)	(0.51)

digitizing a number of points with 1/4- or 3/4-grid unit locations, the operator can, by means of a menu command, change the scale with a factor of 2. In this way, the initial grid unit is divided into four subfields as shown in B, Figure 9.15, and by means of the detached unit it is easy to digitize points with a 1/4- or 3/4-grid unit location as shown in C, Figure 9.15.

9.3.5c Pattern Considerations

Although many systems can check minimum spacings, the PCB designer must realize beforehand what minimum spacings it is possible to achieve when considering the limitation discussed in Paragraph 9.3.5b. The problem arises first and foremost in the case of high-density boards, and is therefore limited to the narrowest conductors and the smallest solder pads. The considerations regarding minimum spacings and pattern dimensions are also valid for CAD conditions. The common conductor widths and solder pad diameters are examined in Table 9.2. In the examples given in Paragraphs 9.3.5d to 9.3.5f, the results have been slightly rounded. Therefore, adding the partial distances does not necessarily give the grid unit expressed as a whole number.

9.3.5d Minimum Spacing between Conductors

We are now going to determine the minimum spacing achievable on the finished master film when the layout is based on a drafting film with a 0.2 in. (5.08 mm) preprinted grid, a scale of 2:1, and commonly used dimensions of the symbols occurring in symbol discs, cf. Table 9.2.

The starting point is the assumption that the PCB designer has placed two conductors so that they follow two grid lines, i.e., with a center-to-center distance of one grid unit. See A, Figure 9.16A. We shall now see how many conductors we can place between the first two conductors when the producibility of the board is considered, i.e., the conductor width and the spacing must exhibit reasonable values. In agreement with the practical conditions, it is also assumed that in all cases the same conductor width is used, and furthermore, that there is the same spacing between the individual conductors.

A

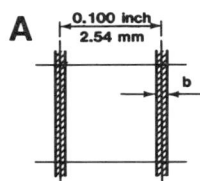

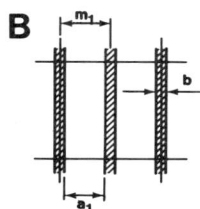

m_1		b		a_1	
inch	mm	inch	mm	inch	mm
0.05	1.27	0.013	0.33	0.037	0.94
		0.025	0.64	0.025	0.63
		0.040	1.02	0.010	0.25

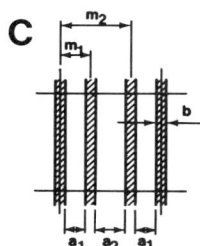

B

m_2		m_1		b		a_1		a_2	
inch	mm	inch	mm	inch	mm	inch	mm	inch	mm
0.070	1.78	0.030	0.76	0.010	0.25	0.020	0.51	0.030	0.77
				0.013	0.33	0.017	0.43	0.027	0.69
				0.025	0.64	0.005	0.12	0.015	0.38
AUTOMATIC DISTANCE DISTRIBUTION									
0.066	1.68	0.033	0.84	0.010	0.25	0.023	0.59	0.023	0.59
				0.013	0.33	0.020	0.51	0.020	0.51
				0.025	0.64	0.008	0.20	0.008	0.20

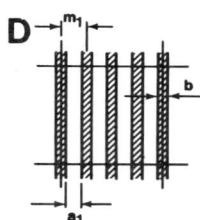

C

m_1		b		a_1	
inch	mm	inch	mm	inch	mm
0.025	0.64	0.006	0.15	0.019	0.49
		0.010	0.25	0.015	0.39
		0.013	0.33	0.012	0.30
		0.016	0.40	0.009	0.23

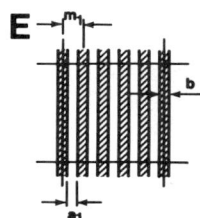

D

m_1		b		a_1	
inch	mm	inch	mm	inch	mm
0.020	0.51	0.006	0.15	0.014	0.35
		0.010	0.25	0.010	0.25
		0.013	0.33	0.007	0.18

E

Fig. 9.16A Minimum spacing between parallel conductors, shown as a function of the conductor width and the number of conductors.

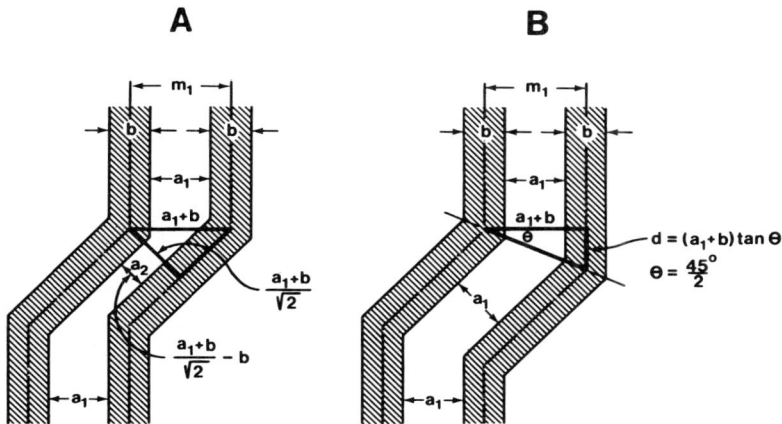

m_1		b		a_1		a_2		d	
inch	mm	inch	mm	inch	mm	inch	mm	inch	mm
0.025	0.64	0.006	0.15	0.019	0.49	0.012	0.30	0.010	0.25
		0.010	0.25	0.015	0.39	0.008	0.20	0.010	0.25
		0.013	0.33	0.012	0.30	0.005	0.12	0.010	0.25
		0.016	0.40	0.009	0.25	0.002	0.05	0.010	0.25

Fig. 9.16B Minimum spacing between parallel conductors being bent 45°.

    A: Bending points at same level severely reduce the minimum spacing between the angled conductor segments.

    B: Bending points being shifted in order to maintain the minimum spacing along the whole conductor length.

Figure 9.16A shows in B, C, D, and E the conditions when 1, 2, 3, and 4 conductors, respectively, are placed between the conductors shown in A. The mutual distances achieved by varying the conductor widths are shown to the right of the drawings. Case C shows a minor complication because the grid must be trisected to give space for the two additional conductors. Because trisectioning is impossible, both in the decimal mode and in the octagonal mode, center-to-center distances of a 0.3, 0.4, and 0.3 grid unit have been used. Some drafting systems allow equal spacing by data manipulation, which in this case results in a center-to-center distance of 1/3-grid unit. As stated in Paragraph 9.3.5b, it is easier for the PCB designer or the operator to use 1/4-, 1/2-, and 3/4-grid units, for which reason he usually prefers the execution shown in D, but with one conductor deleted.

It is always simplest to use the same conductor width as much as possible. Obviously, a conductor width of 0.010 in. (0.25 mm) can be used in all four cases because the smallest spacing in case E becomes 0.010 in. (0.25 mm), which today is fully acceptable to most PCB manufacturers. If only three conductors are placed between the first two conductors, a conductor width of 0.013 in. (0.33 mm) is applicable.

318

It is very common to bend conductors 45 degrees, but it should be noted that the way a group of parallel conductors are bent 45 degrees affects the spacing. This appears from cases A and B shown in Figure 9.16B. At A, the bending points are at the same level, which causes a reduction of the minimum spacing along the bent length. At B, the bending points have been shifted a little so that the minimum spacing is maintained along the whole conductor length.

By applying trigonometric relations to the two triangles shown in heavy lines, one can determine the reduction of the minimum spacing and the shifting of the bending points for cases A and B, respectively.

Case A: The reduced minimum spacing a' is found as:

$$a' = \frac{a_1 + b}{2} - b = \frac{m_1}{2} - b$$

Case B: The shifting c of the bending points is found as:

$$c = (a_1 + b)\tan\theta = m_1 \tan\theta, \text{ where } \theta = \frac{45°}{2}$$

Since Case D, Figure 9.16A is the most common way of routing conductors, the values of a' and c are shown in Figure 9.16B, valid for the same values of the minimum spacing and conductor width as used in Case D, Figure 9.16A. The table shows that in Case A, the best compromise is a conductor width of 0.010 in. (0.25 mm) and a minimum spacing of 0.015 in. (0.39 mm), since the reduced minimum spacing becomes 0.08 in. (0.20 mm). It is quite obvious that case B is superior, and the shifting of 0.01 in. (0.26 mm) is easily established by operating the digitizer in the decimal mode, cf. Paragraph 9.3.5b.

9.3.5e Minimum Spacing between Conductor and Solder Pad

The next point of interest to the PCB designer is the minimum spacing between a conductor and a solder pad. For practical reasons, it is assumed that the solder pads lie on one of the grid lines, and also that the conductor is displaced according to the decimal or octagonal mode, preferably 4/8- = 5/10-grid unit. In other words, the conductor lies in the middle of the grid unit, or possibly at 6/8-grid unit, so that the sketch can be drawn by eye, but supported by the intersections between the slanting lines shown in Figure 9.12. The results appear in Figure 9.17, and it is again seen that conductor widths of 0.010 and 0.013 in. (0.25 and 0.33 mm) are suitable in all three configurations.

9.3.5f Minimum Spacing when Routing Conductor(s) between Two IC Pads

We shall now see which minimum spacings can be achieved when 1, 2, or 3 conductors are to be routed between two IC pads, i.e., transversely to the IC package. The conditions are determined for the pad sizes of 0.050 in. (1.27 mm) and 0.055 in. (1.40 mm) in diameter.

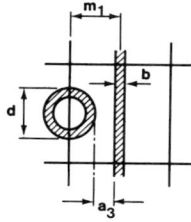

	d inch	d mm	m₁ inch	m₁ mm	b inch	b mm	a₃ inch	a₃ mm
A	0.050	1.27	0.050	1.27	0.010	0.25	0.020	0.51
					0.013	0.33	0.019	0.47
					0.020	0.51	0.015	0.38
					0.025	0.64	0.013	0.32
B	0.055	1.40	0.050	1.27	0.010	0.25	0.018	0.44
					0.013	0.33	0.016	0.41
					0.020	0.51	0.013	0.33
					0.025	0.64	0.010	0.25
C	0.063	1.60	0.050	1.27	0.010	0.25	0.014	0.34
					0.013	0.33	0.012	0.30
					0.020	0.51	0.009	0.22
			0.060	1.52	0.020	0.51	0.019	0.47
					0.025	0.64	0.016	0.41
					0.040	1.02	0.009	0.22
D	0.080	2.03	0.060	1.52	0.010	0.25	0.015	0.38
					0.013	0.33	0.014	0.35
					0.020	0.51	0.010	0.25
E	0.100	2.54	0.075	1.90	0.010	0.25	0.020	0.51
					0.013	0.33	0.019	0.47
					0.020	0.51	0.015	0.38

Fig. 9.17 Minimum spacing between conductor and solder pad. The minimum spacing is shown as a function of the conductor width, the solder pad diameter, and the distance between the conductor center line and the solder pad center.

In the case of just one conductor between the pair of IC pads (cf. A, Figure 9.18) all the conductor widths investigated are suitable because the minimum spacings are sufficiently large.

When we route two conductors between the pair of IC pads (cf. B, Figure 9.18) a conductor width of 0.010 in. (0.25 mm) is still possible, but a conductor width of 0.006 in. (0.15 mm) gives a little larger minimum spacing. It is obvious that autoplacing (equispacing) is an advantage.

Three conductors routed between the pair of IC pads (cf. C, Figure 9.18) imply a conductor width of 0.006 in. (0.15 mm), preferably in conjunction with autoplacing. The solder pad diameter should not exceed 0.050 in. (1.27 mm). It is recommended to make the conductors a little wider outside the narrow passage, as shown in D, Figure 9.18. Outside the passage, the conductors can follow the 1/4- and 3/4-grid unit line. The dimensions can be taken from Case D, Figure 9.16A. It is, for example, possible to use a width of 0.010 in. (0.25 mm) and a spacing of 0.010 in. (0.25 mm).

A

d		m₁		b		a₃	
inch	mm	inch	mm	inch	mm	inch	mm
0.050	1.27	0.050	1.27	0.010	0.25	0.020	0.51
				0.013	0.33	0.019	0.47
				0.020	0.51	0.015	0.38
				0.025	0.64	0.013	0.33
0.055	1.40	0.050	1.27	0.010	0.25	0.018	0.45
				0.013	0.33	0.016	0.41
				0.020	0.51	0.013	0.33
				0.025	0.64	0.010	0.25

B

d		m₁		m₂		b		a₃		a₂	
inch	mm	inch	mm	inch	mm	inch	mm	inch	mm	inch	mm
0.050	1.27	0.040	1.02	0.060	1.52	0.006	0.15	0.012	0.30	0.014	0.35
						0.010	0.25	0.010	0.25	0.010	0.25
AUTOMATIC DISTANCE DISTRIBUTION											
		0.040	1.02	0.060	1.52	0.006	0.15	0.013	0.33	0.013	0.33
						0.010	0.25	0.010	0.25	0.010	0.25
0.055	1.40	0.040	1.02	0.060	1.52	0.006	0.15	0.009	0.23	0.014	0.35
						0.010	0.25	0.008	0.20	0.010	0.25
AUTOMATIC DISTANCE DISTRIBUTION											
		0.042	1.07	0.059	1.50	0.006	0.15	0.011	0.28	0.011	0.28
		0.041	1.04	0.059	1.50	0.010	0.25	0.008	0.20	0.008	0.20

C

d		m₁		m₂		b		a₃		a₂	
inch	mm	inch	mm	inch	mm	inch	mm	inch	mm	inch	mm
0.050	1.27	0.038	0.96	0.050	1.27	0.006	0.15	0.009	0.24	0.007	0.17
AUTOMATIC DISTANCE DISTRIBUTION											
		0.036	0.91	0.050	1.27	0.006	0.15	0.008	0.20	0.008	0.20
0.055	1.40	0.038	0.96	0.050	1.27	0.006	0.15	0.007	0.18	0.007	0.17

D

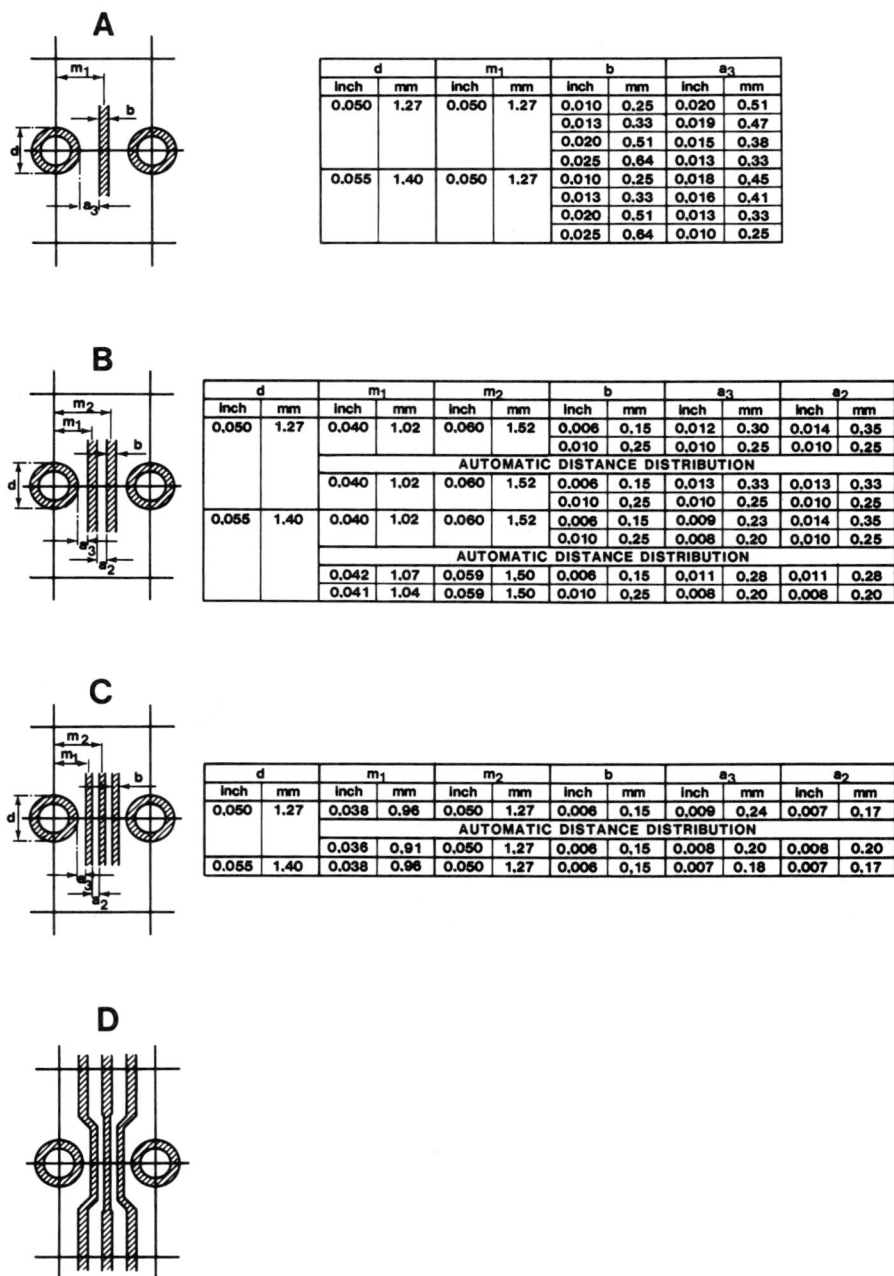

Fig. 9.18 Minimum spacing between conductors and two IC solder pads. The minimum spacing is shown as a function of the conductor width, the solder pad diameter, and the number of conductors. At C, m₁ corresponds to 3/8 grid unit, inclusive of rounding.

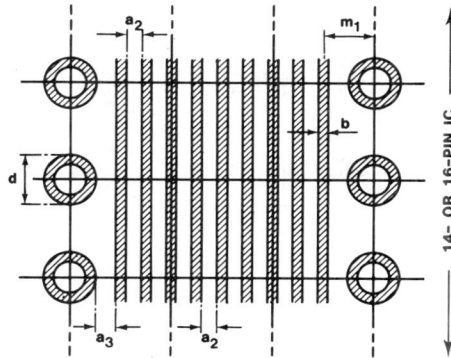

d		m_1		b		a_3		a_2	
inch	mm	inch	mm	inch	mm	inch	mm	inch	mm
0.050	1.27	0.050	1.27	0.010	0.25	0.020	0.51	0.015	0.38
				0.013	0.33	0.019	0.47	0.012	0.30
0.055	1.40	0.050	1.27	0.010	0.25	0.018	0.45	0.015	0.38
				0.013	0.33	0.016	0.40	0.012	0.30

Fig. 9.19 Minimum spacing between conductors as well as between conductors and IC solder pads when routing the conductors longitudinally through an IC package. The minimum spacing is shown as a function of the conductor width and the solder pad diameter, valid for 1/4-grid (1/40th in.) unit locations of the conductors.

Finally, we shall see how many conductors we can route longitudinally through an ordinary 14- or 16-pin IC package, when using conductor widths of 0.010 in. (0.25 mm) and 0.013 in. (0.33 mm), and demanding reasonably large minimum distances. The conditions appear in Figure 9.19.

For ease of layout sketching, it is expedient to place the conductors along the 1/4-, 1/2-, and 3/4-grid unit lines, and the minimum spacings are therefore determined by means of Figures 9.16A and 9.17. The outermost conductors are placed as shown in A, Figure 9.19, in a distance of 1/2-grid unit from the grid lines through the pads, and the minimum spacings are taken from Case A or B, Figure 9.17. By filling up with conductors in a 1/4-grid unit pitch, it is possible to accommodate nine conductors in total, cf. Figure 9.19. The minimum spacings can be taken from D, Figure 9.16A.

Routing a number of conductors longitudinally through an IC package normally requires that one or more conductors be brought out or branched to the solder pads. This implies the use of via holes so that the conductors can connect to the other side of the board.

Recently, mini via holes (sometimes called micro via holes) have become very common, and in a general execution, the pad diameter is 0.020 in. (0.51 mm) and the drilled hole diameter is 0.016 in. (0.40 mm). Usually no tolerance is specified for the diameter of the plated and reflowed hole, which possibly closes when the board is reflowed. To improve reliability, a minimum copper plating of 0.0008 in. (20 μm) should be specified.

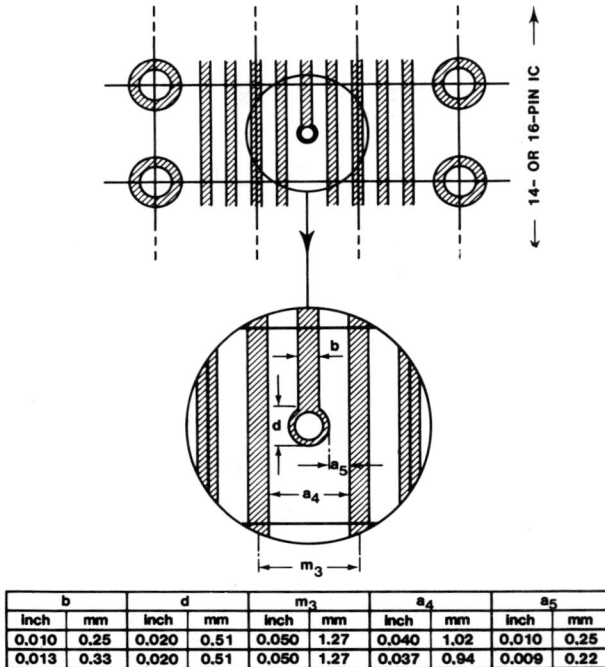

b		d		m₃		a₄		a₅	
inch	mm	inch	mm	inch	mm	inch	mm	inch	mm
0.010	0.25	0.020	0.51	0.050	1.27	0.040	1.02	0.010	0.25
0.013	0.33	0.020	0.51	0.050	1.27	0.037	0.94	0.009	0.22

Fig. 9.20 Minimum spacing around a mini via hole located in a group of conductors running longitudinally through an IC package.

Figure 9.20 shows how a mini via hole can be placed without having to reduce the number of conductors longitudinally through the IC package. The conductor widths are 0.010 in. (0.25 mm) and 0.013 in. (0.33 mm). The table shows a nominal spacing between the pad of the mini via hole and the adjacent conductors of 0.010 in. (0.25 mm) and 0.009 in (0.22 mm), respectively.

9.3.5g General Design Considerations

Going through the design considerations in Paragraph 9.3.5a to 9.3.5f has shown that it is possible to establish some general design rules for the layout of densely packed boards, for example, for digital/microprocessor applications. These general rules are:

1. The layout is performed on a drafting film with a preprinted grid. The scale should be 2:1, requiring a grid unit of 0.2 in. (5.08 mm).
2. The IC pad diameter is chosen as 0.050 in. (1.27 mm), or alternatively, 0.055 in. (1.40 mm), on a 1:1 scale.
3. The width of signal-carrying conductors is chosen as 0.010 in. (0.25 mm), or alternatively, 0.013 in. (0.33 mm), on a 1:1 scale.

4. The conductors are placed with a pitch of 1/4-grid unit, i.e., with a center-to-center distance of 0.025 in. (0.635 mm), on a 1:1 scale.
5. Via holes beneath an IC package can be executed as mini via holes, i.e., with a pad diameter of 0.020 in. (0.51 mm) and a drilled hole diameter of 0.016 or 0.014 in. (0.40 or 0.35 mm), on a 1:1 scale.

9.3.5h Solder Mask Considerations

As mentioned in Paragraph 6.9.2, it is necessary to have a certain nominal clearance between the solder pad and the corresponding aperture in the solder mask. For a screen-printed solder mask, a clearance of 0.008 in. (0.20 mm) between the pad and the mask aperture is chosen. To ensure that the edge of a conductor which passes by the solder pad can be covered by the solder mask, an additional distance of 0.008 in. (0.20 mm) is required. The distance from the conductor to the solder pad should therefore be minimum 0.016 in. (0.40 mm). In the case of a dry-film mask, the above values should be halved, i.e., the distance from the conductor to the solder pad should be minimum 0.008 in. (0.20 mm).

Based on the calculations shown in Figures 9.17 through 9.20 regarding the conductor-to-solder pad distance, the PCB designer can easily determine whether the design requires a dry film solder mask, or it can do with a screen-printed solder mask. It is obvious that the decisive criterion is whether more than one conductor passes between two IC pads.

1. One conductor between two IC pads: Here one uses conductor widths of 0.010 in. (0.25 mm) and 0.013 in. (0.33 mm) in conjunction with a screen-printed solder mask. It is, however, a prerequisite that when branching off from conductors passing longitudinally through the IC package, due regard be given to the spacing conditions around the via holes. Even when using mini via holes it is not possible to route nine conductors longitudinally through the IC package and achieve full edge coverage, unless the mini via holes are overprinted.

   It is furthermore seen that a dry film solder mask does not impose any restrictions with respect to routing nine conductors longitudinally through the IC package, provided mini via holes are used.

2. More conductors between two IC pads: When routing two conductors between two IC pads, it is necessary to use a dry film solder mask. Even with 0.006 in. (0.15 mm) wide conductors, the spacing becomes too small to make it possible to achieve full edge coverage when using a screen-printed solder mask.

   When routing three conductors between two IC pads, full edge coverage implies that the conductor width be reduced to 0.006 in. (0.15 mm) in the narrow passage, and also that the pad diameter be less than 0.050 in. (1.27 mm).

   The conditions when routing conductors longitudinally through the IC package are the same as described in Item 1 above.

9.4 Using a Service Bureau for Automated Drafting

A number of service bureaus have started up in recent years. The design services offered by these bureaus include complete design, i.e., layout based on the customer's schematic diagram (or sketch), or automated drafting based on the customer's layout sketch. Only rarely does the customer rent the equipment and do his own digitizing, although it is possible to get assistance from the service bureau.

9.4.1 Complete Design at Bureau

Here, the design is based on the customer's schematic diagram and parts list. The documentation package must contain the following information:

1. Mechanical requirements

   - Length and width of board
   - Possible cutouts
   - Definition of an edge connector slot, if any
   - Components needing a specific location; for example, a display which has to fit into a window, or pushbuttons which are brought out through the front panel

2. Type of board

   - Single-sided
   - Double-sided and plated-through
   - Multilayer board, with possible ground and voltage planes indicated

3. Electrical requirements

   - Complete schematic diagram, possibly sketched
   - Complete parts list
   - Component library with component dimensions specified, e.g., their physical size, including the terminal spacing, and the diameter of the component leads or of the solder holes
   - Possible external recognition, e.g., a UL recognition which sets certain demands on the design execution. See "How to Design Printed Circuit Boards for UL Recognition," published by Bishop Graphics, Inc. (Order No. 10004).

4. Pattern details

   - Size and shape of the solder pads
   - Conductor widths
   - Minimum spacings
     Pad-to-pad
     Pad-to-conductor
     Conductor-to-conductor
     Conductor-to-board edge

- Definition of any edge connectors
  Shape of contact tabs
  Contact tab pitch

Note: The above data must comply with the practice followed by the service bureau, i.e., the symbol disk(s) of the photoplotter. Furthermore, data must be coherent as described in Paragraph 9.3.5 regarding design requirements, so that they can be implemented on a board.

5. Special conditions

The more detailed the information provided to the service bureau, the more accurate the finished filmwork. Many important layout requirements cannot always be ascertained from the schematic diagram and the parts list, so the documentation package should contain information on such matters.

Although checklists 1 and 2 in Paragraphs 5.18 and 5.21 cannot be used directly, notes about conditions to be considered by the service bureau can be based on many of the checklist items.

9.4.2 Partial Design at Bureau

In some cases the customer does his own layout, but leaves the digitizing and photoplotting to the service bureau. Such cases require less extensive documentation. The service bureau actually needs only information on the pattern detail; see Item 4, Paragraph 9.4.1. If the service bureau must also prepare the component notation, the customer must supply component outlines and therefore he should include the component library in the documentation package. The component designations and all required text should be indicated, preferably on the layout sketch.

It is important that the layout sketch be accurate enough for the digitizing operator to accurately determine on which grid lines the pattern details lie, cf. Paragraph 9.3.5. Conductors on the component side should be drawn in a different color from those on the solder side if the entire layout is drawn on one sheet. The documentation package for a multilayer board should include a separate layout sketch for each layer.

9.4.3 Additional Service Bureau Offerings

In addition to the primary delivery of a photoplotted positive filmwork on a 1:1 scale of the component and solder sides, and also of the inner layers in the case of multilayer boards, most service bureaus can offer:

- Solder mask filmwork
- Insulation mask filmwork
- Component notation filmwork
- Drill tape for CNC/NC drilling

326

- Drill drawing indicating hole sizes by symbols
- Mechanical drawing for the master drawing
- Finished schematic diagram
- Parts list
- Tape for automatic component insertion
- Tape for drilling a bed of nails
- Tape for electrical tests

It is important that the customer specify which of these items is to be delivered. Even when the common practice is for the service bureau to forward a check plot for approval before doing the photoplotting, the customer should always require a check plot, usually a pen plot or a hard copy.

9.4.4 Price Calculations

Most service bureaus can quote the cost of a job, so the customer does not risk an unpleasant surprise upon delivery. Although the turnaround time is much shorter than in the case of manual taping, the customer should not assume that the job will cost less. Not only does the service bureau need a reasonable profit, but the large investment in the digitizer, pen plotter, photoplotter, etc., results in a much higher hourly charge than in the case of manual taping.

Two decisive factors form part of the price calculation system:

1. The number of holes, defined as all IC pins, component leads, contact pins, straps, etc., that is, all solder holes. In the case of surface mounted components, the terminal lands are included. Add to this all edge connector tabs and all test points. Via holes are normally disregarded, at least when the service bureau is responsible for the design, since the customer has no influence on the quantity.
2. The component density, which is calculated as the total number of component holes according to the definition in Item 1, divided by the available board area.

The following examples of different cost calculation systems comprise layout, digitizing, a check plot, and photoplotting of the filmwork for the solder and component sides for a double-sided, plated-through board. They do not include the drill tape.

In some cost calculation systems, the price is primarily based on the number of holes, but corrected by means of a complexity factor which varies more or less with the component density. See the curves A and B in Figure 9.21. When the component density is low, i.e., up to eight or ten component holes per square inch (6.45 cm^2), the complexity factor is 1, or possibly a little lower at very low component densities. This means that the price, on the whole, is proportional to the number of component holes. Increasing the component density makes the layout process more difficult, and the complexity factor increases approximately with the component density, varying somewhat from one service bureau to an-

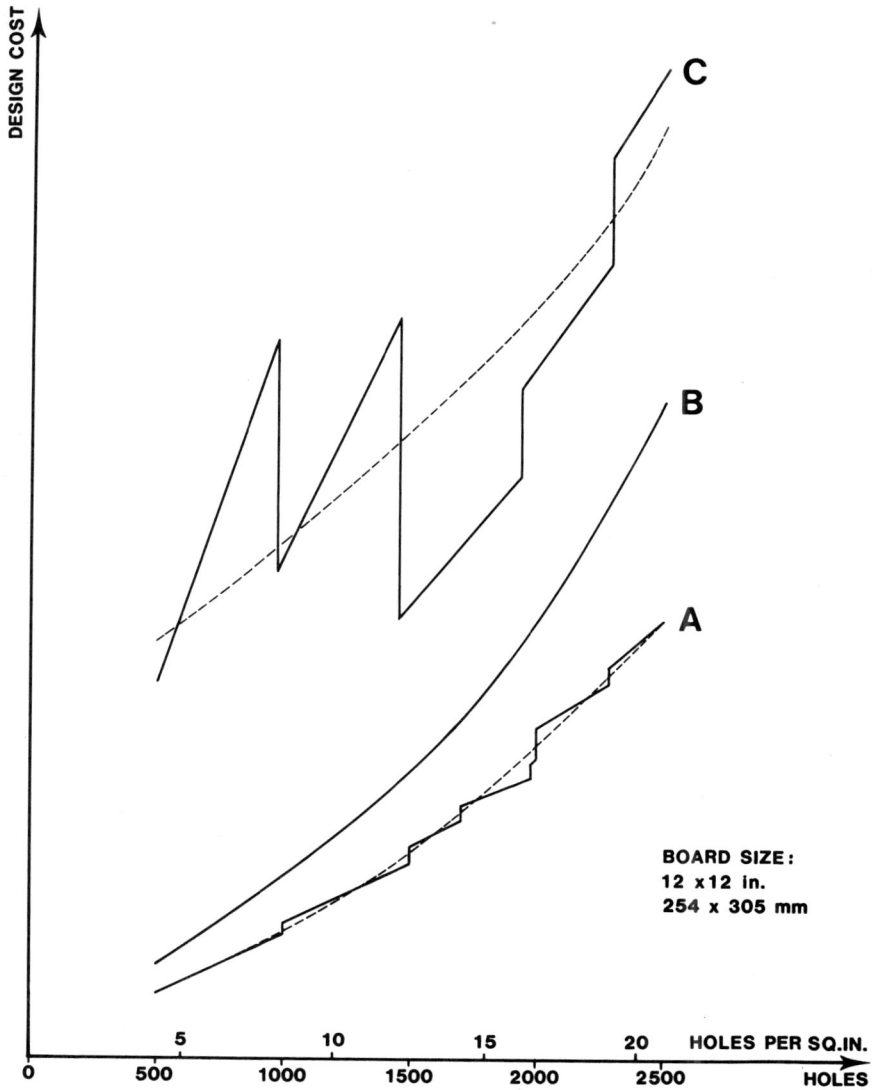

Fig. 9.21 Examples of different cost calculation systems.
  A: The complexity factor varies in steps, but moderately, with the component density. The dotted curve is smoothed.
  B: The complexity factor varies gradually with the component density.
  C: The steps between the price groups of different component densities are very large and result in an illogical curve. The dotted curve would presumably be more acceptable to the customer.

other. The complexity factor valid for curve A varies in steps, whereas it varies gradually for curve B; the shapes of the curves reflect this distinction.

The peculiar appearance of curve C is caused by the fact that the cost calculation system is based primarily on the component density, which is calculated as described above. This determines the price group, and the total design cost is determined by multiplying the price group by the number of component holes. This looks quite captivating, but the steps between the individual price groups have a direct effect on the appearance of the curve. This effect influences the optimum number of component holes. It is quite obvious that a board with 1400 holes, for example, should be provided with 100 additional holes, which causes the price to drop 40 percent.

The cost calculation system of the service bureau selected can easily be assessed on the basis of a few representative board sizes, the design costs being calculated as a function of the number of holes. It is most expedient to start with the number of holes and/or the component density corresponding to the transition from one price group to the next. This method by itself can show the customer any possible inconsistency in the cost calculation system.

9.4.5 A Practical Example

A practical example of a layout sketch is shown in Figure 9.22. Figure 9.23 shows the corresponding photoplotted solder and component sides.

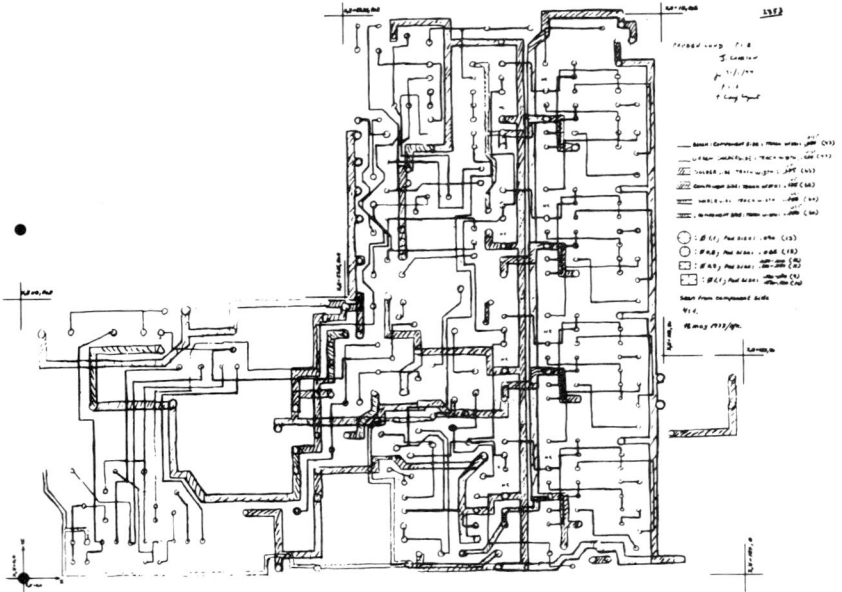

Fig. 9.22 Example of a layout sketch.

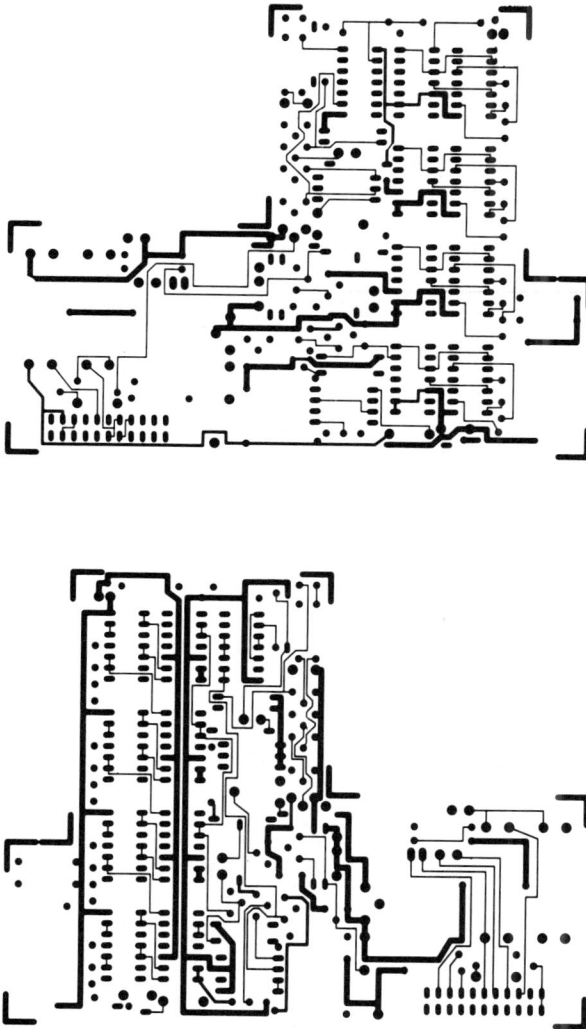

Fig. 9.23 Photo-plotted filmwork of the solder and component sides, corresponding to the layout sketch shown in Figure 9.22.

9.5 Computer-Aided Design (CAD)

9.5.1 Introduction

The next step towards fully automated filmwork generation is to replace the manual layout with automated design techniques. The analysis of the time distribution for preparing the artwork, cf. Paragraph 9.2.1, shows that the layout consumes more than 50 percent of the total time. The advantage of replacing the manual layout with automated methods is, however, greater than the 50 percent figure indicates.

A major problem within the electronics industry is the vast number of corrections and modifications of boards in current production, and issue levels of ten or higher are not unknown. When boards are changed manually (manual layout and taping), especially in the case of high-density boards, the entire board often has to be redesigned, since any apparently small change may require adjacent areas to be rearranged to accommodate the change.

On the other hand, when the PCB designer has an automatic system at his disposal, subsequent changes are introduced easily and quickly since he has only to modify the data which was stored during the original design. The operator can add, delete, and move components and conductors, and let the computer finish the new design. He saves considerable development time and dramatically reduces the turnaround time.

In practice, there are two different methods to choose from: a fully automated system based on batch input, cf. Paragraph 9.5.2, and an interactive system in which the operator plays an important part, cf. Paragraph 9.5.3. The fully automatic system will receive relatively little attention here, whereas the interactive system will be discussed in detail because it is the most widely used today within the electronics industry. The back end of both systems—the optional plotter—is the same as previously described in Pragraphs 9.3.3 and 9.3.4.

As in the case of automated drafting, we shall also discuss the preparation of a documentation package intended for a CAD service bureau.

9.5.2 Batch Input System

A batch input system is characterized by the fact that all data are loaded into the system at the beginning of the process. With no provision for operator intervention, the system processes the data as much as possible and submits the result as output data.

The operator places the components and routes the conductors, using special routines in order to optimize the component placement and thereby shorten the length of the interconnections. These problems are solved iteratively, i.e., by repeated calculations. Since each iteration is carried out at the end of a run, the processing time is fairly long.

On the whole, the computer software is based on the same routines as used in the interactive system, the major difference being that the batch input program uses built-in criteria to determine when to shift between various routines. The use of built-in criteria reduces the flexibility of the system, but on the other hand, the operator need not have any experience in layout techniques.

In the long run, it must be unsatisfactory to prevent a PCB designer from using his creativity. The computer has a limited range of action, whereas the designer can survey the whole situation, use his intuition, and go beyond the limits of the computer algorithms.

Current printed circuit boards are frequently so complex that a batch input system requires an extremely long processing time, or worse, aborts the layout.

This occurs because the time used for the interative process increases exponentially with the number of conductors that the system cannot route immediately.

9.5.3 Interactive Systems

We shall now address interactive systems, which allow the operator to intervene in the process. Most systems in use are based on the same principles. The purpose of this chapter is not to enumerate the advantages and disadvantages of a particular system, but rather to give the reader, who perhaps is unfamiliar with CAD systems, a general understanding of the working principles as well as some insight into the possibilities offered by such systems.

The most conspicuous difference between the interactive systems and the batch input system is that the operator engages in a dialog with the computer, forming part of a feedback loop and thereby becoming the controlling element. The computer takes over the tedious routine tasks, one of which is to ensure that the design complies with the established design rules. This implies that the operator can concentrate on more challenging problems and use his ingenuity to solve those that exceed the capacity of the computer.

Complexity and packaging density affect the computer's likelihood of completing a design; in many cases it is only possible for the computer to solve, for example, 95 percent of the design problems. The operator can now intervene and help the computer by taking the initiative, for example, by relocating components which obstruct further conductor routing.

9.5.4 Algorithms for Wiring the Board

We shall now elaborate on the principles which make it possible for a computer to solve so complex a problem as laying out a printed circuit board, and especially routing the conductors in the most expedient way. Although innumerable solutions exist, the computer must find an acceptable solution within a reasonably short time.

The computer program follows a predetermined procedure by which the computer starts from some given conditions (board size, wiring list, component library, etc.) and reaches the desired result (the finished layout solution) in a finite number of steps.

Such a procedure, or set of rules, is called an algorithm, and every CAD system is based on one or more types of algorithm. The following paragraphs briefly describe three principal forms of board-wiring algorithms in general use. Paragraph 9.5.5 elaborates on algorithms for component placement.

9.5.4a Channel Router

The channel router algorithm divides the board into a number of channels and tries to route an entire channel at a time. In the process, it manipulates both the conductors that belong to the channel and the conductors which have to pass through the channel from one place to another. The channel router, which is very complex in its mode of operation, interchanges and moves the conductors, possibly down to the next channel, until it finally shows the entire channel solution on the graphics display unit.

When the channel router can find a solution, it is very fast, but there is a certain risk that it will make some decisions that will block the progress at a later stage and stop the channel router. Since the operator cannot intervene in the routing process, the channel router is not particularly suited to interactive work.

The channel router is mostly used for batch input operations when very large layouts must be completed. Conductors frequently occur in a high concentration at certain places on the board, whereas other places can remain quite open. A weakness of the channel router is that some conductors may be routed through dense areas, not because they must, but just because the global router has difficulty choosing the most expedient channel.

As a small digression, we shall now consider using a channel router for the design of integrated circuits. In the case of very complex integrated circuits, the channel router is the best wiring algorithm because the task is so huge that the designer cannot evaluate the possible solutions. In return, the chip area becomes a great deal larger than necessary because the channel router is not particularly efficient at minimizing the chip area. Among other things, this is because it is difficult for the channel router to choose the sequence for placing the individual conductors in the channel in order to make it as narrow as possible. In practice, a printed circuit board cannot have variable-width channels. The channel router is therefore best suited for the design of integrated circuits.

9.5.4b The Lee Algorithm

The Lee Algorithm, also called a maze router, uses a penalty system to find the shortest (least penalizing) and most expedient way through the maze formed by the board. Since the Lee algorithm is frequently used, Paragraph 9.5.5 elaborates on this approach.

The Lee algorithm is much simpler and slower than the channel router, but if a valid solution exists, the Lee algorithm can find it. This is not completely true of all other algorithms. As previously mentioned, the main problem of the channel router is deciding which channel it allocates a conductor to, whereas the problem of the Lee algorithm is deciding which conductor to route first.

9.5.4c The Line Search Algorithm

The line search algorithm is something between a channel router and a Lee algorithm, actually a highly simplified and adapted form of the Lee algorithm. Instead of trying to find the shortest way by searching the entire maze in small steps, it takes large steps in plausible-seeming directions. This approach makes the algorithm very fast, but there is a risk that it will go in the wrong direction. There is no guarantee that it will find an acceptable route, even if one exists. A contributing factor is that the number of segments (or bending points) is normally limited because the algorithm would otherwise be too slow.

In some CAD systems using a line search algorithm, it is possible to leave the remaining conductors to a Lee algorithm because this algorithm can always find a way, provided that one exists.

9.5.5 Algorithms for Component Placement

9.5.5a Introduction

Following the above description of algorithms for wiring the board, we shall now elaborate on algorithms for placing components. Here we also face difficult problems in determining where the algorithm starts to place the components and where it ends.

Presumably, a perfect auto-interactive placement algorithm does not exist, supposedly because the operator cannot fully evaluate the component placement in the limited time available. The basic problem is to place the components so that their interconnections become as short and simple as possible.

When n components are to be placed in n fields, the first component can be placed in any of the fields. For the next component there are n-1 fields available, and so on. The total number of combinations therefore becomes

$$n \times (n-1) \times (n-2) \times \ldots 3 \times 2 \times 1 = n!$$

In order to find the best solution, all these combinations must be evaluated one by one. Five components in five fields yield 120, because $5! = 120$.

On the other hand, when n components are to be placed in a fields (a > n), the first component can be placed in any of the a fields. The next component can be placed in any of the remaining (a-1) fields, and the very last component can be placed in any of (a-(n-1)) fields. This gives the following number of combinations:

$$a \times (a-1) \times (a-2) \times \ldots \times (a-(n-2)) \times (a-(n-1)) = \frac{a!}{(a-n)!}$$

Here also all the combinations must be evaluated one by one so that the best solution can be found, but the number of solutions is insurmountably high.

When the number of fields in the above example is increased from five to ten, the five components result in 30,240 combinations, and it is easily seen that the number will become astronomically high if the number of fields and components reaches any practical size. Just 70 components in 70 fields yield $70! \simeq 10^{100}$ combinations. Establishing and evaluating all these combinations at a speed of one combination per second would require a time exceeding the age of the universe.

It is therefore absolutely necessary to find an algorithm which does not test all possible combinations, but limits the number in order to find an acceptable solution. The components can, for example, be placed at random, whereupon a successive two-by-two swapping is performed until all components have been tentatively swapped with all other components. If there are n components, the number of swaps in each run is

$$\frac{n}{2} \times (n\text{-}1)$$

In practice, a limited number of runs can be made within a reasonable time as described in Paragraph 9.5.5e.

Nevertheless, the algorithm for automatic component placement can be somewhat doubtful in practical use. In some CAD systems, the initial component placement is performed manually according to the operator's assessment of which component groups belong together. It is then possible to use some auxiliary algorithms for optimal placement. See also Paragraphs 9.5.5d, 9.5.5e, and 9.5.5f.

9.5.5b Placement Algorithm Based on Cell Division

Some CAD systems use an automatic component placement algorithm, preferably one which applies to ICs. The algorithm implies that the board has been divided into cells with room for one IC per cell. Because of the division into cells, it is necessary to disregard all discrete components such as resistors, capacitors, etc., since the space utilization would otherwise be too low. Since the cells are usually made somewhat larger than the ICs, the operator can always find space for the discrete components later.

The next step is to identify certain fixed points, frequently in the form of input and output connectors. The placement routine first finds the IC having the most connections to one of the connectors and places it in the nearest cell. The algorithm next finds the IC with the most connections to the one just placed, and possibly also to the connector, whereupon that IC is placed in the cell nearest the first IC. The algorithm proceeds in this way until all ICs have been placed. Optimum placement can then be ascertained by some auxiliary algorithms as mentioned in Paragraphs 9.5.5d, 9.5.5e, and 9.5.5f. Finally, all discrete components are located manually.

A disadvantage of this algorithm is the assumption that the ICs are of equal size. It works best in the case of 14- and 16-pin ICs, and not nearly as well when ICs with 28 or more pins also occur.

9.5.5c Placement Algorithm Based on "Rubber Bands"

Automatic component placement can also be performed with the rubber-band method, in which the individual connections are regarded as rubber bands. Because of the rubber bands, the individual components are exposed to a fictitious force depending on the number and length of the rubber bands. The result is a contraction of the components into groups belonging together. With this method also, the algorithm can be supported by the auxiliary algorithms mentioned in Paragraphs 9.5.5d, 9.5.5e, and 9.5.5f.

9.5.5d Auxiliary Algorithm Based on Minimization Cuts

Whether the first component placement is performed automatically or manually, it can be optimized by using an auxiliary algorithm which minimizes the number of interconnections. After successively placing imaginary cuts across the board, the computer can check whether the number of interconnections crossing the individual cuts can be reduced by swapping some of the components. The operator can extend the area of investigation to a certain zone at each side of the cutting lines. The algorithm then tries to achieve a more even distribution of the conductors by successively swapping the components.

9.5.5e Auxiliary Algorithms for Successive Component Swapping

Another way of optimizing the initial component placement is to use an auxiliary algorithm for successive two-by-two component swapping. Different criteria can be used to ascertain whether better results are achieved by the individual swaps. A possible criterion may be whether the total connection length becomes as short as possible, or whether the area of a component group becomes as small as possible. Here it is advantageous to divide the board into sections beforehand and perform the successive swaps within the individual sections.

A board with 200 ICs, for example, has

$$\frac{1}{2} \times 200 \times (200 - 1) = 19,900$$

possibilities of successive swaps. Experience shows that after four runs, additional runs produce no improvement worth mentioning. At one swap per second, four runs of the swapping program take about 22 hours.

Instead of trying to swap each component with every other one, it is possible to introduce a limit so that the program swaps only within a certain field around the individual components. If the field dimensions are 1/4 the board length and width, the field will contain 1/16 of the components, assuming a uniform component density. The total number of swaps during a single run therefore becomes

$$16 \times \frac{1}{2} \times \frac{n}{16} \times (\frac{n}{16} - 1) = \frac{n}{2} \times (\frac{n}{16} - 1)$$

With 200 ICs as in the above case, the number of exchanges becomes

$$\frac{200}{2} \times (\frac{200}{16} - 1) = 1150$$

This is approximately 1/17 the time required in the first case. The algorithm is approximately 17 times faster, and performs the four runs in $1\frac{1}{4}$ hours, with a result nearly as good as in the first case.

9.5.5f Density Curves

In one of the CAD systems on the market, the operator's assessment of possible swaps is supported by density curves for the X and Y directions. The density curves are calculated by the computer, based on the actual component placement. The displayed curves show how many conductors must be routed through any transverse and longitudinal cross section.

Large peaks on the density curves indicate areas where layout presumably will be difficult. By suitable trial-and-error swapping of the components, the operator can usually level the density curves and reduce the number of conductors through the areas of concern to a reasonable number. See the example of density curves shown in Figure 9.61C. Additional leveling of the density curves, corresponding to a few percent reduction of the number of conductors passing through the places in question, make layout substantially easier.

Incidentally, experience shows that it is much easier to lay out boards that have even density curves than boards that have low density curves with a few high peaks.

9.5.6 Penalty System of the Lee Algorithm

The Lee algorithm is a quite general algorithm which can be employed for many different purposes. Various variants of the algorithm used in CAD systems are adapted to the individual companies' philosophy regarding the procedures. In principle, the different activities of the layout procedure are subject to penalty points having a predetermined weight. The optimum layout is characterized by the lowest possible number of penalty points. This is illustrated by the following example, where the penalty P is calculated according to the expression

$$P = P_1 + P_2 + P_3 + P_4 + P_5 + P_6$$

The partial penalties P_1 through P_6 represent the weighted penalties for each individual activity. The importance of the penalties can vary according to the CAD system, and in the same way, the number of different penalties and their mutual weighting can vary. Therefore, the following discussion should be accepted only as a practical example, taken from an actual CAD system based on the Lee algorithm.

9.5.6a Penalty for Movement (P₁)

This penalty is assessed each time the length of a conductor is increased by a grid unit (a data structure unit, DSU), e.g., 0.050 in. (1.27 mm) or 0.025 in. (0.635 mm), in one direction or another, i.e., when the conductor moves up, down, left, or right, or shifts from one board side to the other through a via hole.

> *Note:* In usual PCB terminology, a module is always 0.100 in. (2.54 mm). It should not be confused with a grid unit, also called a system module, or by some CAD companies, a data structure unit (DSU).

If the penalty is zero, any conductor is just as good as any other conductor. This implies that it can become infinitely long and/or thread its way through the entire board. In order to make the conductor as short as possible, the penalty must have a positive value.

In some systems, the operator can vary the penalty point from job to job. In this way he can, for example, achieve different minimum values of the conductor segments (conductor lengths between bending points) before shifting sides. Thus it is possible to adapt the layout to the conditions valid for digital and analog boards, of which the latter usually have the largest solder pads because the component leads generally are thicker. If a conductor starts from a solder pad and shifts to the other side through a via hole, the center distance between the solder pad and the via hole pad should be a little larger on the analog board than on a digital board, to allow routing a conductor between these pads on the other side of the board. This can be regulated by changing the penalty weight of P_1.

9.5.6b Penalty for Making Via Holes (P_2)

It is undesirable to introduce more via holes than absolutely necessary; they take up room, and although the drilling cost of a via hole is relatively low, it should not be disregarded. Therefore, a special penalty for each via hole can be introduced. If the penalty is fixed too low, the system will tend to generate too many via holes. On the other hand, the penalty should not be too large, because the system will introduce so few via holes that, at a later layout stage, it becomes difficult for the algorithm to find reasonable paths. The penalty must therefore be suitably balanced.

9.5.6c Penalty for Wrong Side (P_3)

The most expedient layout of dense boards is achieved by keeping the horizontal conductors (in the X-axis direction) on one side of the board, and the vertical conductors (in the Y-axis direction) on the other side. This can be achieved to a reasonable extent by introducing a penalty for routing a conductor on the wrong board side, i.e., at right angles to the normal direction on the board side in question.

In the extreme, the penalty implies that a conductor is not allowed to make a small bend immediately before landing on a solder pad. See A, Figure 9.24. It must instead shift to the other side through a via hole. See B, Figure 9.24. This is not very expedient, and the best way to counteract it is to let the penalty P_2 for via holes be a little higher than the penalty P_3 for wrong side.

Fig. 9.24 Bending of conductors.
     A: The conductor is bent before landing on the solder pad.
     B: Strict distinction between vertical and horizontal conductors on the two board sides requires the introduction of a via hole.

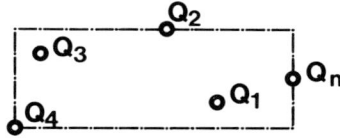

Fig. 9.25 Example of a homing algorithm. By drawing a rectangle around the group of points Q_1 to Q_n which are to be interconnected, wrong search directions are defined as directions leading out of the rectangle. Introducing a penalty on such directions makes it a homing algorithm.

9.5.6d Penalty for Wrong Direction (P_4)

In order to make the algorithm as fast as possible, it must be prevented from searching in the wrong direction, i.e., it may not search in directions leading away from the target(s). The wrong directions appear in Figure 9.25, where the starting point is Q_1 and the targets are Q_2 through Q_n, which are to be connected mutually as well as with Q_1. The rectangle shown is drawn so that it contains all the points Q_2 through Q_n. Search directions leading out of the rectangle and/or away from the target are wrong directions, and cause a penalty to be inflicted because the distance to the target increases. (This approach is called a homing algorithm.) Nevertheless, the penalty should not be so great as to prohibit the algorithm from searching in the wrong direction when there is no legal path within the rectangle.

9.5.6e Penalty for Obstructing Via Holes (P_5)

The penalty P_5 opposes a conductor going where it may later be necessary to place a via hole. This penalty is particularly important when only a few via holes are possible; for example, when a standard hole pattern for solder holes and via holes has been mandated. A conductor is therefore not allowed to pass through a predetermined via hole location without using that via hole.

A more realistic example is a multilayer board with many signal layers where buried via holes are undesirable for economic reasons. Buried via holes are via holes present in some of the inner layers, but not in the outer layers. When the last layers are to be wired, the only possibility of placing via holes is where there are no underlying conductors in the layers previously routed. Penalty P_5 prevents the algorithm from placing conductors at the few remaining locations which can be used for via holes.

Because a minimum distance is required between a solder pad and a via hole, for example, 1 module = 0.100 in. (2.54 mm), the penalty causes the conductor to run close to the solder pad, specifically between the solder pad and the position of the via hole, whether the hole exists or the position is just reserved.

When the board design does not restrict placement of via holes, the penalty should be fixed at zero.

9.5.6f Penalty for Crowding (P_6)

This penalty opposes the placement of conductors too close to solder pads and via holes. Contrary to penalty P_5, this penalty is imposed on conductors placed where via holes are not allowed. Because via holes normally are not allowed close to solder pads or preexisting via holes, penalty P_6 tries to pull the conductors farther from the solder pads and via holes.

During the first layout run, the penalty is fixed at zero, but when the finishing touch is given to the board, it is desirable to move the conductors away from the solder pads. Penalty P_6 therefore serves to prevent conductors from being placed too close, for example, 0.005 in. (1.27 mm), to the solder pads.

9.5.7 Interactivity

Paragraph 9.5.3 briefly described the fundamental operation of an interactive system. Since the meaning of the interactivity concept is not always completely clear in the sales brochures and manuals from the various CAD companies, there is every reason to explore this concept in more detail.

Interactivity is a procedure wherein the operator conducts a dialog with the computer, which responds to each command with a new or changed display on the graphics display. In some systems, the operator can actively intervene in a process. He can, for example, stop a process about to go wrong, modify some of the preconditions, and let the computer resume its work. In other systems, interactivity means only that the operator can work manually.

A process of the greatest importance to the layout speed is the wiring of the board, i.e., the routing of all the conductors. Particularly within this process, some uncertainty regarding the degree of the interactivity prevails, i.e., the extent to which the operator can intervene in the process and modify it.

In CAD systems using a line search algorithm, the automatic processes operate all the time. The operator can indeed stop an automatic process and possibly also restart it, but it is impossible to introduce even small modifications. As an example, the operator cannot route just one conductor, have it displayed on the graphics display, and perhaps modify it a little if the position is not optimum. Fundamentally, the computer is operating in a batch input mode, trying to solve the problems within a certain area of the board. The result is not displayed until a solution has been reached or the computer has given up.

Some CAD systems employ various algorithms in turn. As an example, the first algorithm can be a so-called bus router capable of routing a number of conductors in a bus, i.e., performing a fairly simple layout. The next algorithm can be a line search algorithm which is used as the global router, but in principle, this algorithm too can cope only with relatively simple connections.

Because of the high processing speed it is impossible to follow a layout accomplished by a line search algorithm. Frequently, a number of conductors are missing at the end of the process because the operator could not intervene. The missing conductors can be very difficult and time-consuming to route because it is often necessary to relocate a large number of conductors already laid out.

A disadvantage of a batch program like the line search algorithm is that it routes the conductors in such a convoluted way that the operator has little opportunity to lay out the remaining conductors manually. Experience shows that the best way of solving a complex board is to use only the bus router, after which the operator places the large number of remaining conductors manually, one by one. Although a third algorithm could be a Lee algorithm, it could probably route relatively few of the remaining conductors, e.g., only 5 percent, so it is seldom used.

Conditions differ somewhat in CAD systems based on the Lee algorithm. This algorithm can, by itself, always find a solution if one exists; it therefore eliminates the aforementioned problems. The Lee algorithm is somewhat slower because of the systematic search for possible paths through the maze.

To follow up this overview, we shall now study the most decisive requirements for a CAD system to be called an interactive system. The interactive system must provide for:

1. Dealing with a single group at a time, i.e., partial solutions comprising one or more conductors chosen by the operator.
2. Keeping the operator informed of progress, i.e., the system must keep the display up-to-date with the partial results.
3. The operator's stopping and modifying the layout of a selected group, cf. Item 1 above, and resuming the layout procedure with the next conductor, with a jump backwards, or a jump forwards in the sequence.

These demands indirectly require the use of a Lee algorithm, although in principle they can be met with a line search algorithm. A particular advantage of the Lee algorithm is that it can show the search wave's propagation on the graphics display unit, cf. Paragraph 9.5.9 and Figure 9.61F.

If the algorithm does not succeed in wiring a certain connection, the operator can call up a display of the search wave and see how far it propagates, first from one side and then from the other side. This is a very convenient way of identifying the area in which the obstruction occurs.

Similar conditions as mentioned above apply to interactive component placement. The operator can always intervene, and use his experience and intuition to optimize the placement. Even if the CAD system can automatically place components with an optimizing procedure, for example, by successive component swaps or by gate swaps as shown in Figure 9.26, it can sometimes be faster and perhaps also better if the operator makes a few manual swaps at especially difficult locations.

9.5.8 Wiring Principles

A CAD system incorporates one of two principles, depending on the algorithm used by the system. The line search algorithm allows only pad-to-pad (point-to-point) wiring, whereas the Lee algorithm also allows conductor-to-conductor wiring, i.e., from an accumulation of points to another accumulation of points.

9.5.8a Pad-to-Pad Wiring (The Line Search Algorithm)

When three pads, points a, b, and c, are to be interconnected, it is natural for the operator, when preparing the wiring list, to connect a to b and then b to c;

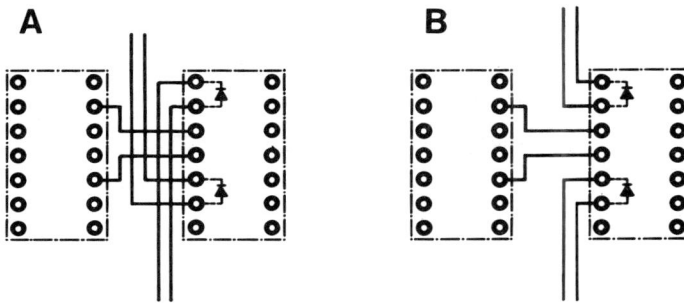

Fig. 9.26 Gate swapping. Swapping the gates shown at A gives a much simpler pattern as shown.

that is, from point to point. See A, Figure 9.27. In practice, it may be inexpedient, or even impossible, to go from b to c, but possible to go from a to c. Because of the fixed wiring sequence of a to b to c, the algorithm will not accept the wiring of a to b and a to c as an alternative solution. See B, Figure 9.27.

One solution to this problem is to change the wiring sequence stated in the wiring list to be b to a to c. It is, however, troublesome to go back to the wiring list, and an immediate solution is for the operator to manually (interactively) supplement the connection a to b with a connection a to c. If so, the computer will, at a later stage, report a short circuit from a to c because this connection is not stated in the wiring list. The computer will also report that the connection from b to c is missing.

Another problem is shown at A, Figure 9.28, where a T-branch is to be established at point d. In theory, this cannot be done unless there is a solder pad (a point) at d, which naturally is inexpedient. The solution is for the computer to perceive the configuration as two different conductors, specifically, a conductor from a to b, and another conductor from a to c with a bend at d. See B, Figure 9.28. Because the two conductors are superimposed along the segment from a to d, one of the conductors disappears from a to d during the photoplotting.

9.5.8b Conductor-to-Conductor Wiring (The Lee Algorithm)

The Lee algorithm makes it possible to route from pad to pad, from conductor to pad, or from conductor to conductor. In A, Figure 9.29, the operator chooses,

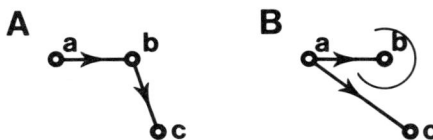

Fig. 9.27 Pad-to-pad wiring.
  A: The natural wiring sequence is from a to b to c.
  B: When the wiring cannot be established as shown at A because of blocking around B, it can possibly be executed from a to b and from a to c. This is, however, not accepted by all CAD systems.

342

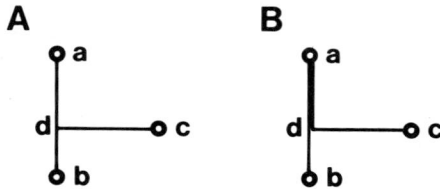

Fig. 9.28 T-branching.
    A: The desired execution of a T-branch
    B: In some CAD systems it is only possible to establish a T-branch as two
       conductors merging along the distance from a to d.

for example, c as the starting point, whereupon a search wave propagates at the same speed in all directions away from c. Assuming that all the penalties P_3, P_4, and P_5 are zero, the front of the search wave will everywhere exhibit the same penalty.

By chance, the front of the search wave hits pad a first, and by backtracking from a, the computer can define the conductor from a to c. This conductor is the shortest conductor possible because all points in the front have the same penalty. In the next turn, the search wave starts from the conductor a to c just established, i.e., from all points at the same time. The point closest to pad b determines the conductor which is branched off to b. See B, Figure 9.29.

Finally, we shall see how the Lee algorithm manages to interconnect two conductors S and R and a pad P. See A, Figure 9.30. In the first pass, the search wave propagates from conductor S (source). The front of the search wave appears as shown in B. Where the front hits the other conductor R, the desired connection is established as shown in C. If the conductor R lies on the other side of the board, a via hole V is introduced. In the next pass, both conductors S and R are perceived as the source, and the search wave propagates with a front as shown in D. The connection to P is thus established as shown in E.

9.5.8c The Steiner-Tree Algorithm

A Lee algorithm can only find the cheapest way, in terms of penalties, between two points. When there are more than two points, the computer can come up

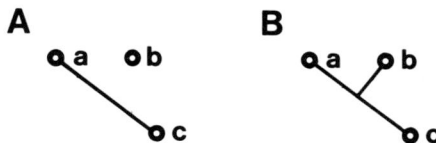

Fig. 9.29 Wiring methods.
    A: Wiring from pad to pad.
    B: Wiring from conductor to pad.

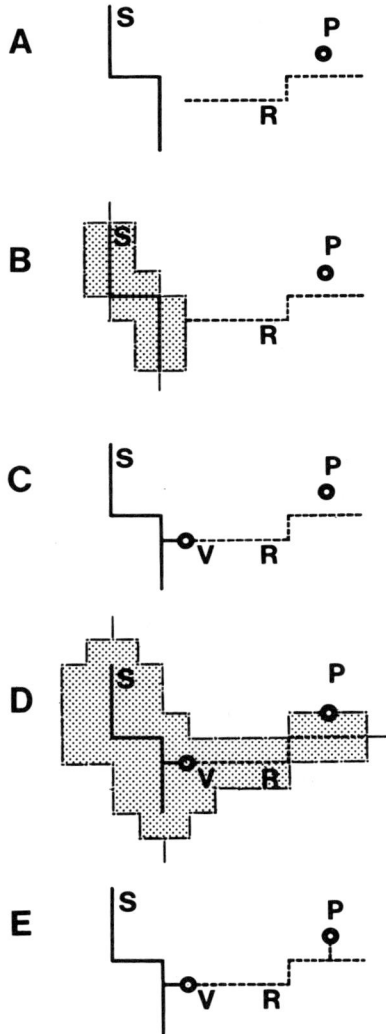

Fig. 9.30 Wiring from conductor to conductor and solder pad.
A: S and R indicate conductors on each board side, and P is a plated-through hole.
B: The search wave propagates from conductor S within the shaded area.
C: The connection between conductors S and R is established by means of a via hole V.
D: The search wave propagates from conductors S and R within the shaded area to the solder pad P.
E: The connection between the solder pad P and the conductors R and S is established.
Note: The symmetrical shape of the search wave preassumes that the penalties P_2, P_3, P_4, and P_5 are zero.

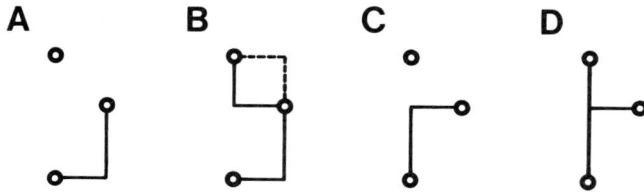

Fig. 9.31 Different solutions when interconnecting three pads.

with several different solutions, some of which are inferior to the others. This is shown in Figure 9.31, where three pads are to be interconnected. If the first part of the solution becomes a laterally reversed L, as shown in A, it is unimportant how the second part of the solution looks. The solution shown in B is poor because the total conductor length is excessive.

If the first part of the solution becomes an inverted L as shown in C, and the second part as shown in D, this solution is better because the total conductor length is shorter than in case B.

The two solutions remain within the rectangle mentioned under penalty P_4 in Paragraph 9.5.6d, and the target-homing tendency of the Lee algorithm is independent of whether the solution becomes as shown in B or in D. This is caused by the fact that the Lee algorithm cannot distinguish between the two solutions.

In order to force the best solution (D) it is possible to introduce a penalty for moving away from the center of the cluster of points. Therefore, the various partial solutions will move towards the center. In this way we have obtained a simplified Steiner-Tree algorithm and the example shown in D, Figure 9.31, is the simplest form of a Steiner-Tree.

If there are many points in the cluster, one or more center lines can be placed, on a trial basis, through the cluster in its longitudinal and/or transverse directions. Subsequently, the connections (conductors) are as nearly perpendicular as possible to the various center lines. The computer calculates the total conductor length of each configuration in order to find the optimal execution. In this way the Steiner-Tree appears as shown in Figure 9.32.

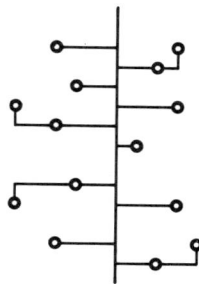

Fig. 9.32 Example of a Steiner-Tree.

Several CAD companies have tried to utilize the Steiner-Tree algorithm but have had to give it up even though the wiring has a pleasing and rational appearance. The reason is that in practice it takes too long to determine the center point or the center line, and then to check whether the layout advances in the desired direction. Another factor is the algorithm's relatively low percentage of success.

9.5.9 Algorithmic Modes of Operation

We shall now concentrate on how the algorithms function in practice. In order to do this in a reasonably organized way, we shall choose a Lee algorithm in approximately the same form as used in a commmercially available CAD system but in a much simpler application.

9.5.9a Layout with Only Two Penalties

This treatment is based on a very simple case, namely layout in one place with only two types of penalty: a penalty of 1 for taking a horizontal step and a penalty of 2 for taking a vertical step. The penalties are indicated in the bottom left corner of the diagrams shown in Figures 9.33A to 9.33H.

The principle of a Lee algorithm is best illustrated by always letting the search wave propagate from the lowest-penalty field to all unoccupied neighboring fields, whereupon the starting field is marked occupied. The wave propagation continues until a way of reaching the target field with the lowest penalty has been found.

Phase A, Figure 9.33A The diagram shows the initial condition of the problem to connect the starting field 0 with the empty target field to the right. These fields

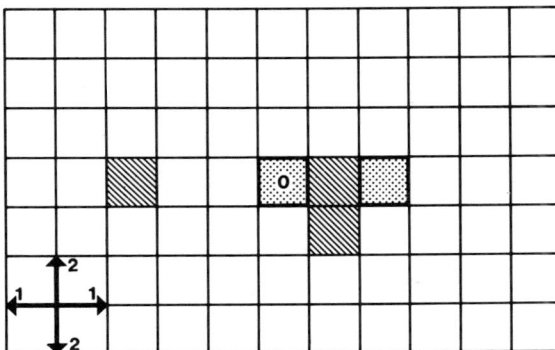

Fig. 9.33A Phase A: Illustration of the Lee algorithm's mode of operation when establishing a connection between the starting field 0 and the target field located to the right of the starting field. Both fields are shaded. The obstructions are cross-hatched. The penalties are stated in the lower left corner.

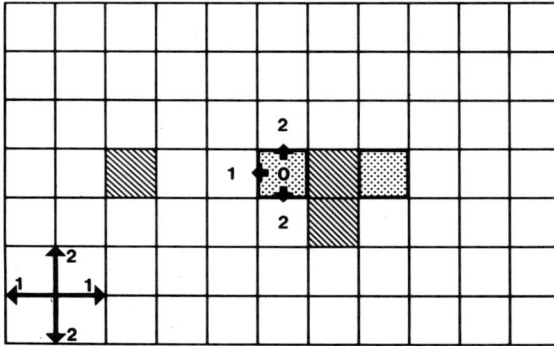

Fig. 9.33B Phase B: The search wave propagates from the starting field 0.

are shown with a grey shade. The hatched fields represent obstructions that prevent the fields from being used during the layout.

Phase B, Figure 9.33B The search wave has three possible propagation paths. It can reach the fields marked 2 by going up and down with a penalty of 2, or it can reach the field marked 1 by going to the left with a penalty of 1. The starting field is marked occupied.

Phase C, Figure 9.33C We are again starting from the field exhibiting the lowest penalty, i.e., from the field 1, excluding the occupied starting field 0 and the fields 2. There are three possible directions: up and down to the fields 3 with penalties of 2, and to the left to the field 2 with a penalty of 1. The penalties marked in these fields are the sum of the old and the new penalties.

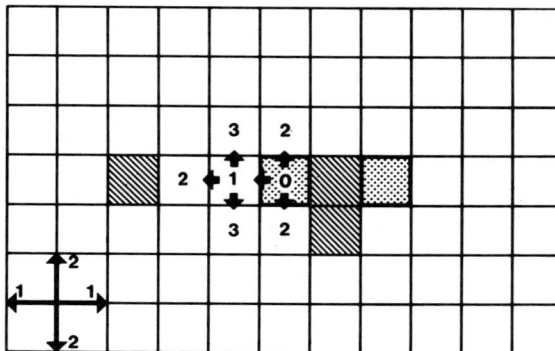

Fig. 9.33C Phase C: The search wave propagates from the field marked 1.

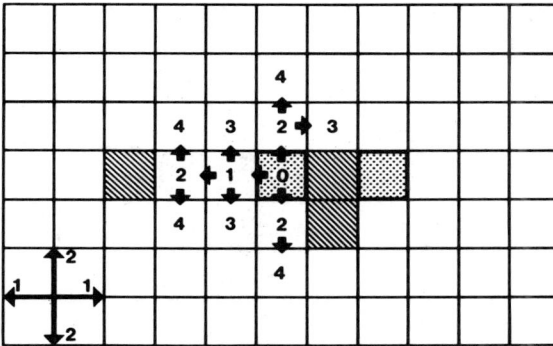

Fig. 9.33D Phase D: The search wave propagates from the fields marked 2.

Phase D, Figure 9.33D In this phase, the search wave uses the fields marked 2 as starting fields, and in the example it is immaterial in which field we start. We can, for example, take the field 2 just above the starting field 0 and go up with a penalty of 2 to the field marked 4 and to the right with a penalty of 1 to the field marked 3.

We could also have gone to the left with a penalty of 1 and reached a field with a total penalty of 3, i.e., the same penalty obtained by moving up with a penalty of 2 from the field marked 1. The Lee algorithm, however, is not so consistent (homogeneous and noncontradictory) as the example supposes. The penalty for going other ways can vary a little.

For the algorithm to generate short and attractive conductors, a field should not be marked occupied before the alternative possibilities have been investigated. Nevertheless, to keep the example as simple and clear as possible, we shall deviate

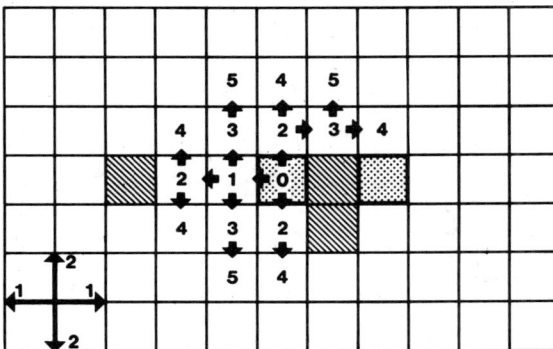

Fig. 9.33E Phase E: The search wave propagates from the fields marked 3.

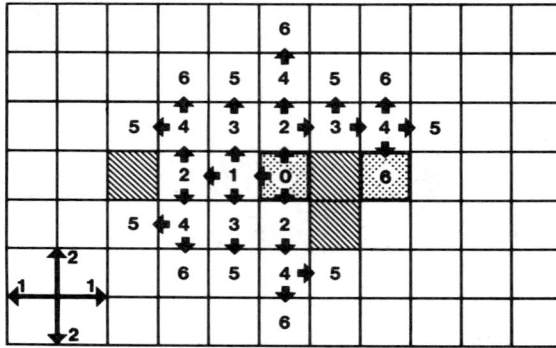

Fig. 9.33F Phase F: The search wave propagates from the fields marked 4, and reaches the target field with a penalty of 6.

from actual practice and mark such fields occupied. Incidentally, this makes the algorithm run faster, but in return, it is not certain that the route will be as short as possible.

The lower field marked 2 just below the starting field 0 is developed downwards with a penalty of 2 to the field marked 4, and the left field marked 2 is developed upwards and downwards with a penalty of 2 to the fields marked 4. All the fields marked 2 have now been developed, and only fields with a penalty of 3 or 4 are in the front of the search wave.

Phase E, Figure 9.33E The search wave has now been developed from all the fields marked 3 with a penalty of 2 for vertical movements and a penalty of 1 for horizontal movements. The result is fields having penalties of 4 or 5. The "old"

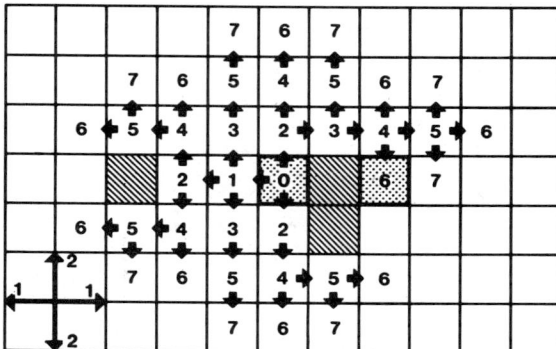

Fig. 9.33G Phase G: The search wave propagates from the fields marked 5 to ensure that there are no "cheaper" ways to the target field.

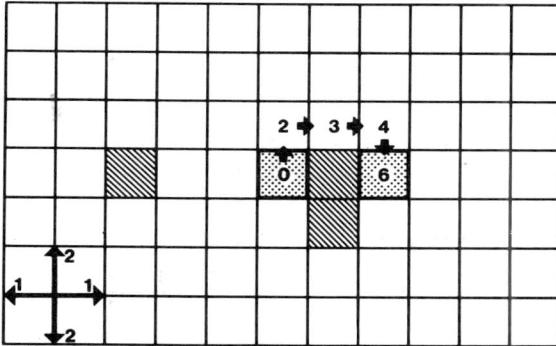

Fig. 9.33H Result H: The resulting conductor is found by back-tracking from the target field to the starting field.

fields marked 4 remain untouched and the front of the search wave contains only fields with penalties of 4 or 5.

Phase F, Figure 9.33F In the same way as before, the search wave has developed from all fields marked 4 with a penalty of 2 for vertical movements and a penalty of 1 for horizontal movements. The result is fields having penalties of 5 or 6. The "old" fields marked 5 remain untouched, and the front of the search wave contains only fields with penalties of 5 or 6. The target field has been reached with a penalty of 6.

Phase G, Figure 9.33G Although the target field has been reached, it is not yet certain that there are no other and less penalizing ways to the target field. The algorithm could quite conceivably reach the target field with a penalty of 1 by moving from a field lying to the right of the target field. If so, two routes to the target field, with the same penalty, have been found. If the difference in penalties had been greater, and the obstructions had been placed otherwise, it could very well happen that the latter route would exhibit a lower penalty.

The search wave continues to propagate, and all fields marked 5 are developed to fields with penalties of 6 or 7.

Result H, Figure 9.33H Not until the search wave has been propagated from the fields with a penalty of 6 is it absolutely sure that the target field cannot be reached with a lower penalty than 6. This is explained a little more clearly in Paragraph 9.5.9b. For the sake of simplicity, the further development of the search wave is not shown in the sequence of Figures 9.33A through 9.33H.

Because all movements have been stored in the computer, it is easy to reverse the procedure and determine the conductor route. The result of traversing the maze is shown in diagram H, and it is apparent that in order to find a very simple connection between two points, a large search has been performed.

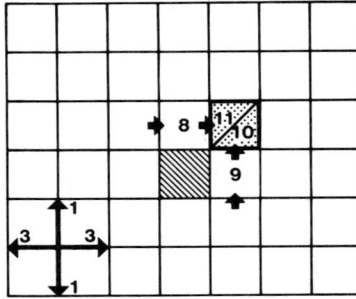

Fig. 9.34 Consistency of the Lee algorithm. Different ways can lead to the target field with different penalties. Since the first way is not necessarily the cheapest way, the algorithm is not consistent.

9.5.9b Consistency of the Lee Algorithm

In phase D above we briefly mentioned the inconsistency of the Lee algorithm, which means that it can possibly find one or more alternative ways, either with the same penalty or with different penalties. For example, if it is impossible to place via holes at certain board locations, or if certain fields are obstructed, the Lee algorithm will not be consistent.

The problem of the consistency of the Lee algorithm is illustrated in detail in Figure 9.34, where the penalty for moving vertically and horizontally is 1 and 3, respectively. When the search wave is propagated to the right from the field marked 8, the target field is reached with a total penalty of 11, and the position of the field, expressed by its coordinates, is stored in a stack together with the total penalty obtained. Somewhat later, the search wave is propagated from the field marked 9, and the target field is reached with a total penalty of 10. Just as before, the position and penalty of the target field are stored in a stack.

The continued search from the field marked 11/10, possibly an intermediate target, takes the lowest penalty stored in the stack as its starting point, i.e., the penalty 10. The search wave now propagates from the fields marked with a penalty of 10.

It is not correct to say that the target field has been hit when the search wave has propagated from the field marked 8 and reached the target field with a penalty of 11. Nor is it correct to say that the target field has been hit when the search wave has propagated from the field marked 9 and reached the target field with a penalty of 10. Not until the target field is removed from the stack when propagating the search wave from all fields marked 10, can one be sure of having found the least penalizing route. An additional propagation of the search wave from the fields marked 11 is therefore unnecessary.

9.5.9c Layout with Three Penalties

As shown in Phase G above, the Lee algorithm is forced to search large areas, and the time consumption is proportional to the area. On an empty board, the search wave of the Lee algorithm propagates over a large number of fields, but as the layout progresses, and more and more fields are marked occupied, the search process speeds up. Consequently, on a dense board, the Lee algorithm can determine at lightning speed whether or not a passable way exists.

To make the Lee algorithm faster, a penalty for moving in the wrong direction can be introduced as a sort of detour penalty. This was detailed in Paragraph 9.5.6 under penalty P_4. By means of the two examples in Figures 9.35A and 9.35B, we shall now see the effect of such a detour penalty on the layout.

In Figure 9.35A, the search wave has propagated with penalties of 2 and 1 for vertical and horizontal movements, respectively, but with a detour penalty of zero. When propagating the search wave from fields with a penalty of 5, the target field has been reached with a penalty of 7. However, as mentioned in Paragraph 9.5.9b, it is necessary to continue the propagation of the search wave until the target field (with the penalty 7) is taken out of the stack. It should be noted that the target field can easily be removed from the stack as the last field marked 7. The propagation of the search wave therefore continues from all fields marked 6 and 7.

The front of the search wave, which now comprises fields with penalties of 8 ($= 6 + 2$) or 9 ($= 7 + 2$), forms a parallelogram. The area searched by the search wave corresponds to 87 fields, not including the starting field and the field marked (9) developed from the target field.

Figure 9.35B shows how the conditions are changed by introducing a detour penalty of 4. This additional penalty implies that the penalties for moving away from the target field in vertical and horizontal directions become 6 ($= 2 + 4$) and 5 ($= 1 + 4$), respectively, but unchanged from 2 and 1 for moving towards the target field in vertical and horizontal directions, respectively. It should be noted that when the front of the search wave moves farther to the right or upwards than corresponding to the target field, the movement leads away from the target. Consequently, the detour penalty should be included. The revised penalties are indicated at the bottom left corner of the diagram.

As in the previous case, the target field has been reached with a penalty of 7 by propagating the search wave from fields having a penalty of 5. The propagation of the search wave is continued as before from fields having penalties of 6 and 7. The front of the search wave, which comprises fields with penalties from 7 to 13, achieves the appearance of a polygon. The area searched by the search wave corresponds to 28 fields which is approximately 3 times less than in Figure 9.35A. The examples show clearly how it is possible to affect the algorithm's mode of operation and speed by changing the penalty conditions.

The examples described are based on a limited number of fields, and it is possible to show that when the number of fields becomes very high, the ratio will no longer be 3 but will approach 5, assuming that the penalty conditions remain the same.

The fact that the Lee algorithm has become faster can also be demonstrated on the basis of the example given in Paragraph 9.5.9a, if the same layout is performed with an additional detour penalty of 4. Again we have the extended definition of the wrong direction. Because the starting field and the target field

352

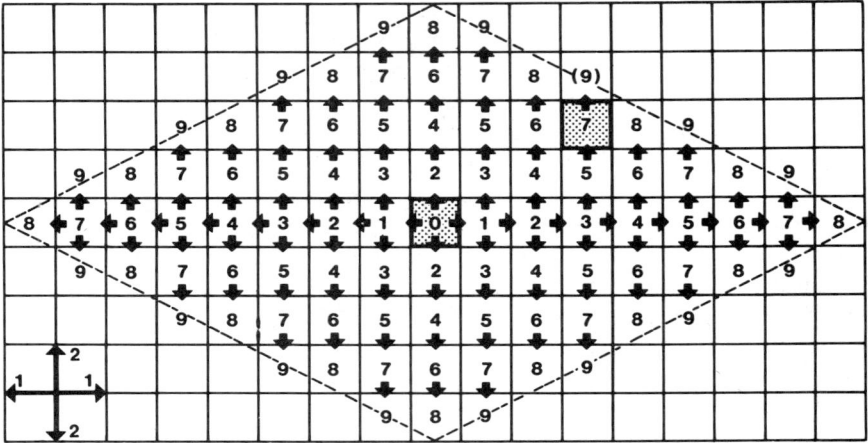

Fig. 9.35A Propagating the search wave without a detour penalty. The target field, located above and to the right of the starting field, is reached with a penalty of 7, and 87 fields have been searched.

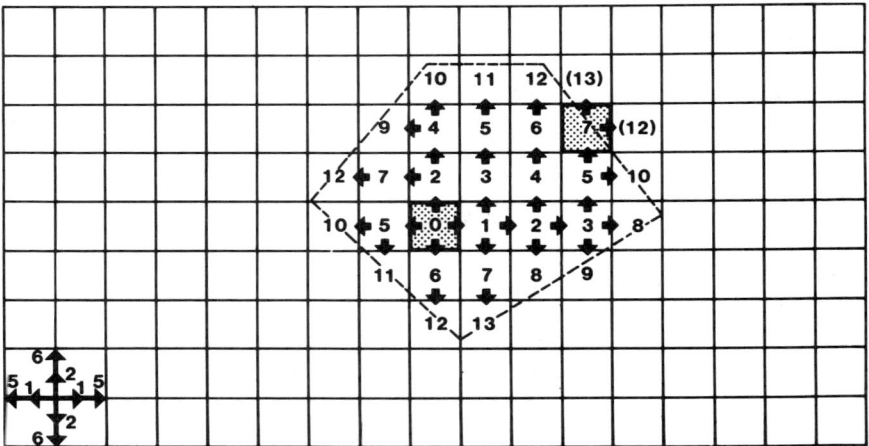

Fig. 9.35B Propagating the search wave with a detour penalty. The target field is reached with a penalty of 7, but only 28 fields have been searched.

are located in the same row (equally high in the diagram), the penalty for vertical movements becomes $6 (= 2 + 4)$. The penalty for a horizontal movement towards the column in which the target is located is 1, whereas the penalty for all other horizontal movements is $5 (= 1 + 4)$.

When propagating the search wave, a diagram as shown in Figure 9.36A is achieved. The target field has been reached after the searching of 14 to 16 fields,

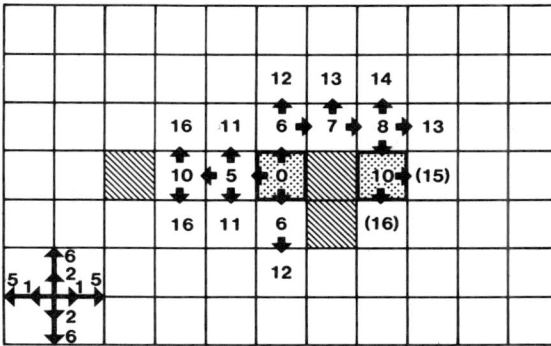

Fig. 9.36A Increasing the speed of the search wave by introducing a detour penalty. Only 14 to 16 fields have been searched. This result should be compared with the result without a detour penalty as shown in Figure 9.36B.

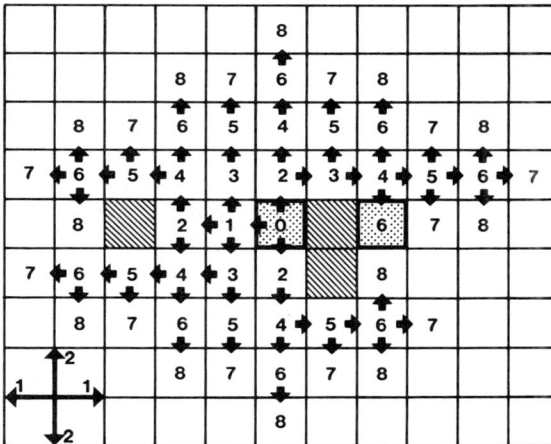

Fig. 9.36B Propagating the search wave without a detour penalty requires searching of 37 to 53 fields. The fields (15) and (16) are disregarded. The diagram shows a further development of the fields marked 6 in Figure 9.33G.

depending on when the target field was taken out of the stack. The fields marked (15) and (16), which could have been developed from the target field, are disregarded because all further searching is stopped when the target field is taken out of the stack.

In Figure 9.33G (phase G), the algorithm has performed no further development from the fields marked 6. Figure 9.36B shows this, as well as the fact that the search wave has to search 37 to 53 fields, depending on when the target field was taken out of the stack.

Table 9.3. Detour and movement penalties

	On the top side	On the underside
Penalties:		
Vertically toward the target field	4	1
Vertically away from the target field	8	5
Horizontally toward the target field	1	4
Horizontally away from the target field	5	8
Using a via hole	4	4

The examples show that the area to be searched is more than halved, with a nearly equal reduction of time. The area and time reductions are unequal because it takes the computer some time to do the necessary calculations of the detour penalty, although that time is of little overall importance to the reduction.

9.5.9d Layout with Four Penalties

We shall now see how a double-sided board is laid out. We shall use the same example as before, but with somewhat different obstructions on the top side (Figure 9.37A) as well as on the underside (Figure 9.37B). The penalties are the same as used in the example with three penalties (cf. Paragraph 9.5.8c), but supplemented with a penalty of 4 for making via holes.

In order to keep as many conductors as possible routed horizontally on the top side, the penalties for movements in vertical and horizontal directions are 4 and 1, respectively. On the underside, we prefer to have vertical conductors, so here the penalties for movements in vertical and horizontal directions are 1 and 4, respectively.

The detour penalty and the penalties for movements in vertical and horizontal directions have the combined effects shown in Table 9.3 which are indicated at the lower left corner of the diagrams. The penalty for a via hole is shown in a circle in the field where the via hole is made.

The search wave propagates simultaneously on both sides of the board. We shall now follow its propagation from the starting field to the target field. Neither the starting field nor the target field is on the underside, so to ease the survey, they are shown with a dotted line but without the grey shade.

It should be remembered that the search wave always propagates from the field having the lowest penalty. Since the conditions are far more complicated than in the previous cases because of the two sides, we have to be very systematic.

The search wave propagates from the starting field 0 on the top side. The cheapest way to reach the underside with a via hole is obviously in the starting field. The computer searches its file of penalties obtained to find the fields with the lowest penalty. In this way it reaches the field with a penalty of 4, and the search wave is propagated on the underside in the usual way.

355

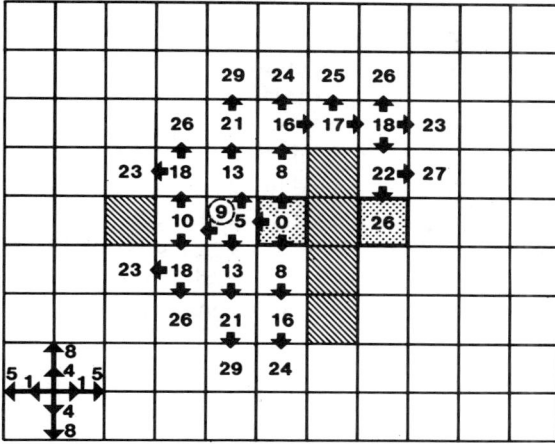

Fig. 9.37A Layout of the top side of a double-sided board.

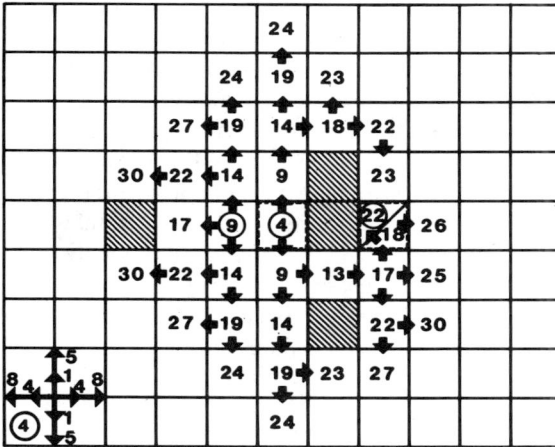

Fig. 9.37B Layout of the underside of a double-sided board.

The search wave then propagates from fields with a penalty of 5. By making a via hole in the field immediately to the left of the starting field, a penalty of 9 is achieved in the corresponding field on the underside. This is less than the penalty of 12 (4 + 8) obtained when the search wave propagates on the underside from the field marked 4. Only the penalty of 9 is marked in the field.

The computer continues to search among the penalties obtained and finds 8 as the lowest penalty, since the penalty 5 has been dealt with, and the penalties 6 and 7 do not occur. The search wave now propagates from fields with penalties of 8 and always continues to propagate from fields which have the lowest penalties not dealt with so far.

On the underside we reach the dotted target field with a penalty of 18, but in order to reach the proper target field it is necessary to make a via hole so that the total penalty becomes 22.

Actually, all additional searching could be stopped, but the searching continues until fields with a penalty of 22, including the target field, have been removed from the stack. This shows that the target field is reached with a penalty of 26 when the layout is performed on the top side of the board. It should be noticed that when two different penalties occur in a field, that is, by moving horizontally and vertically into the field, only the lowest penalty is imposed.

9.5.9e Minimizing the Layout

When the layout is complete, the CAD system can give the final touches to the layout. This process quite suitably is characterized as a minimization because the main purpose is to remove some of the via holes.

The minimization is performed automatically by the computer, which systematically removes a complete group of conductors from the board and rewires the group by using other penalty points. In order to achieve the desired effect, the penalty for making via holes is increased, and the penalty for going in the wrong direction is decreased. It is of no practical consequence that some conductors become a little longer if only the number of via holes is reduced.

In this conjunction, it should also be noted that via holes take up space. Removing a number of via holes increases the possibility of moving the conductors a little, which in turn implies a further reduction of the total number of via holes.

While reducing the number of via holes, one can possibly introduce a penalty P_ϵ which seeks to move the conductors away from nearby pads. The higher this penalty is made, the more the conductors are moved and, as a consequence, the number of via holes is reduced by a smaller amount.

9.5.9f The Lee Algorithm for Bending a Conductor

The penalty system previously discussed did not include a specific penalty for bending a conductor where it lands on a pad. As shown below, a bending algorithm introduces some practical problems which render it inapplicable. See the examples in Figures 9.38 and 9.39 showing an IC package where a conductor is routed transversely or longitudinally through the package before it lands on one of the pads.

First we shall study Figure 9.38. It is immaterial algorithmically whether the conductor has a course as shown at A or B, but from practical PCB technical viewpoints as well as wiring considerations, the course shown at A is most expedient. One reason is that the largest insulation distance is achieved, and another reason is that it is possible to route a conductor between the two neighboring solder pads. See the example shown at C. In return, the conductor shown at A has one bend more than the conductor at B.

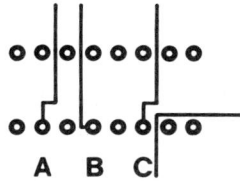

Fig. 9.38 The Lee algorithm for bending a conductor.
A: This execution is most expedient although it comprises two bends.
B: Although this execution comprises only one bend, it is less expedient than the execution shown at A.
C: Execution A of the vertical conductor permits routing a conductor between two neighbouring pads. This would be impossible if the vertical conductor were routed as shown at B. (Only one conductor between two IC pads is assumed.)

In Figure 9.39 we find the opposite conditions, and for the same reasons as stated above, the course shown at A is most expedient. It should be noted that here the conductor shown at A has one bend less than the conductor at B. Minimizing the number of bends by assigning a penalty for each bend can therefore lead to a less desirable route of a conductor where it lands on a pad. Thus the conditions become so complicated that in practice it becomes rather inexpedient to use a bending algorithm.

9.5.10 The CAD System's Mode of Operation

9.5.10a The Internal Organization of the Computer

Having examined the application of algorithms for PCB layout, we shall now study the CAD system's computer, including the interplay between the individual units. See the greatly simplified block diagram in Figure 9.40. The data, e.g., a wiring list, are communicated via an input unit to the central processing unit (CPU) as input data, including the necessary commands.

The CPU processes the data as specified by the commands (instructions) of the CAD programs. The processed data occur as output data, including necessary commands to the peripheral units, and are made accessible to the operator by means of the output unit.

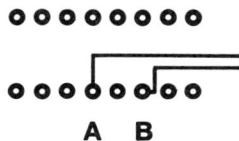

Fig. 9.39 The Lee algorithm for bending a conductor. Cases A and B show the opposite conditions of Figure 9.38.
A: The most expedient execution comprises one bend.
B: Although the execution comprises two bends, it is undesirable.

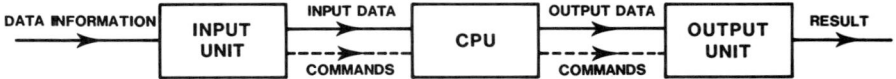

Fig. 9.40 Simplified block diagram of a CAD system.

The data information can be stored on a magnetic tape, a floppy disk, a hard disk, or a paper tape. Consequently, the input unit must be a magnetic tape station, a floppy disk drive, a Winchester disk drive, or a paper tape reader, respectively. These devices convert the data to electrical signals which the CPU can process.

The data information, however, can also come directly from the CAD system's graphics work station in the form of electrical signals, making an input unit unnecessary. These signals are generated by a keyboard, a menu, and/or a data tablet with a probe (rather like an electric pen), a cursor, or more commonly, a joystick (like the pilot's stick on an aircraft, but much smaller). These devices are used to move the crosshair on the graphics display unit so that certain data can be indicated.

Output data presented by the computer can appear on various data media, e.g., a magnetic tape, a floppy disk, or a paper tape, or the data can be printed or plotted on paper.

Of particular importance to the dialog between the operator and the interactive CAD system is the fact that the output data is presented on the graphics display unit, either as the finished result or as an intermediate result. In the latter case, the operator can assess the result and possibly help the CAD system by changing the working preconditions, e.g., by moving a component which obstructs a few missing connections.

9.5.10b The CPU's Mode of Operation

We shall now study the CPU in detail. As shown in Figure 9.41, the CPU comprises an I/O (input/output) controller, a memory, and an arithmetic unit. We shall next examine the interaction between these units.

9.5.10b1 I/O Controller

The I/O controller controls all the CPU's internal operations by executing either the commands (instructions) it receives from the input unit, or the commands contained in the stored CAD program and the operating system, which controls the execution of the CAD program. The traffic between the arithmetic unit and the memory is therefore controlled by the I/O controller. Another function of the I/O controller is controlling the external units, e.g., starting a pen plotter.

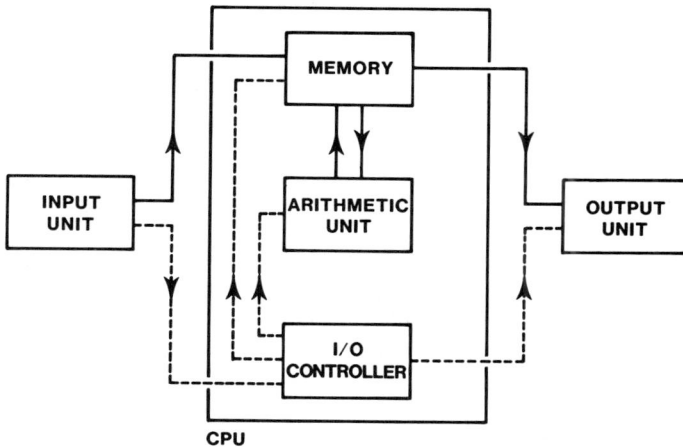

Fig. 9.41 Simplified block diagram of the CAD system's central processing unit (CPU).

9.5.10b2 Memory

The memory, also called the RAM (random-access memory), contains the CAD program, the working storage, the operating system, and the data base, i.e., all the different data required for the CAD system to function:

a. The CAD program, i.e., the instructions which implement the algorithms on which the solutions are based.

b. The operating system, which organizes and controls the execution of the CAD program.

c. The library file, containing all the data describing the components used. The data comprise the external dimensions, the location of the terminals, the size of the solder pads, etc. The library file usually includes all the components used by the operator when solving the board layout problems, but it can easily be modified as required, cf. Paragraph 9.3.2, Item 1.

d. A work file containing information regarding the current job, i.e., all input data and derived intermediate results which are to be used while the program is executed. This is also where the output data are stored.

Today the memory is always based on semiconductor memories (MOS), but earlier systems typically used a matrix of magnetic cores. The storage capacity varies from system to system, e.g., from 64 kilobytes up to 1 megabyte or more. In some systems the memory is extended by an external memory in the form of a magnetic disk unit, e.g., a Winchester disk having a capacity of 10 megabytes or more.

Note: A byte usually represents one character and normally consists of eight bits (binary digits), each of which is either 0 or 1. Since eight bits have

$2^8 = 256$ different permutations, a byte can be assigned a special value out of 256 different numerical values and thereby represent a letter, a figure, or a special sign. Several bytes can represent words, sentences, or numerals, all of which the computer can process.

In computer jargon, 1 kilobyte is 1024 bytes, and 1 megabyte is 1,048,576 bytes, as opposed to 1000 bytes and 1,000,000 bytes, respectively, in the normal electronics definition.

A characteristic decisive to the speed of the system is the access time, which is the time between sending out a data transfer command and concluding the operation. The access time of internal storage (RAM) is frequently 1 microsecond or less, whereas for a Winchester disk it can be on the order of 25 microseconds to 25 milliseconds, depending on whether the controller must move the read/write head.

9.5.10b3 Arithmetic Unit

The arithmetic unit performs all the calculations required by the external commands in conjunction with the algorithms of the CAD system, the operating system, and the stored library data. The I/O controller handles considerable traffic flowing between the arithmetic unit and the memory, which stores the intermediate results until they are needed in later calculations.

The arithmetic unit is usually based on a 16-bit processor, but many newer systems use 32-bit processors.

9.5.10c System Configurations

CAD systems for PCB designs occur in many different configurations, from multifunction systems with many possible applications beyond PCB design to dedicated single-function systems. Even if desirable, it would be impossible to discuss details of all CAD systems on the market. Since the following treatment is intended to give the reader an overview of applications and operational modes, it is based on a general system as shown in the block diagram of Figure 9.42. The configuration comprises a number of units which are described below. A typical CAD system is shown in Figure 9.43.

9.5.10c1 Computer

The computer, which constitutes the central processing unit (CPU) of every CAD system, occurs in many different versions, from a microprocessor, possibly in a 16-bit architecture, through a minicomputer to a so-called supermini in a 16-bit or 32-bit architecture. In some cases the computer can be connected to a large central computer, usually in cases where autorouting requires considerable CPU activity.

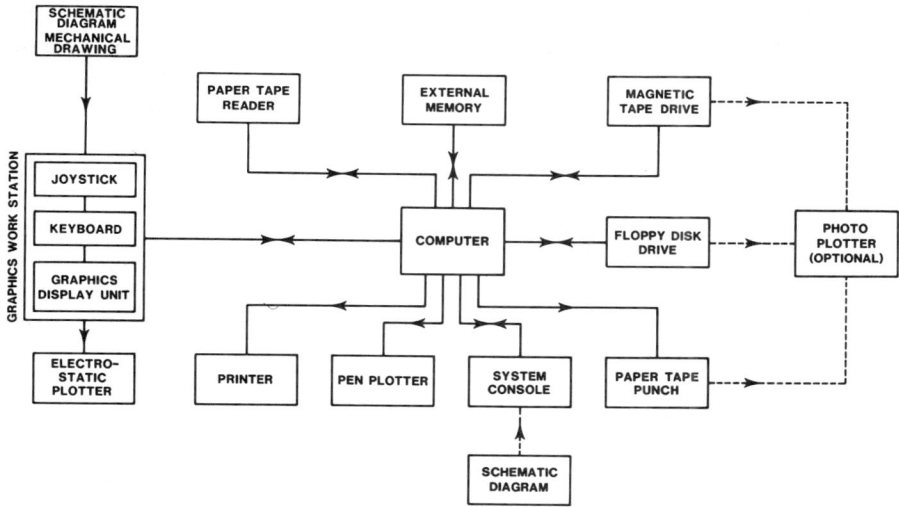

Fig. 9.42 Block diagram of the general configuration of a CAD system.

Fig. 9.43 Example of a CAD system.

On the other hand, the total system might be designed for distributed processing, and in such cases the computer is called a host computer. It contains the central data base, cf. Paragraph 9.5.10b. With the support of an I/O controller, it serves a number of microprocessor-based graphics work stations connected online, i.e., directly to the computer.

As mentioned above, the computer keeps the CAD system's program in the memory. The program is frequently delivered on a magnetic medium, such as a floppy disk, and loaded into the memory when the CAD system is installed or when the programs are updated. If a power breakdown lasts for a longer period than the internal backup battery can maintain the volatile memory contents, the operator must reload the programs and other data into the memory. Fortunately, this process requires only a few minutes.

In addition to the memory, the computer contains an arithmetic unit and an I/O controller, cf. Paragraph 9.5.10b. The I/O controller is usually an integrated part of the CPU, but it is an external unit in some systems.

9.5.10c2 Graphics Work Station

The graphics work station, mentioned briefly in Paragraph 9.5.10a, is the designer's real work location. The primary part is a graphics display unit for showing the status of the work. The display is a cathode ray tube (CRT), which may be either a storage tube or a refresh tube.

The storage tube retains (stores) the image for longer periods. To update the display, the operator must erase and rewrite the image. Depending on the number of details to be displayed, an update requires from 1 to 10 seconds; extremely complex images may require as much as 15 seconds. During interactive work, the operator usually prefers to keep the image until it becomes too difficult to distinguish between rejected solutions (which are retained on the screen) and valid solutions. The image is then erased, and only the valid solutions are displayed after a few seconds. In return, the operator has the great advantage that the display will not flicker regardless of the amount of information displayed. The image quality is very high, with a resolution corresponding to 1000×1000 pixels (picture elements).

This resolution should not be confused with the resolution of the storage tube itself, which can be about four times as high. To achieve visible lines, the lines must be made wider than the possible resolution, and to prevent the lines from melting into each other, their spacing must be made wider than the possible resolution. The result is therefore a lower effective resolution.

The refresh tube allows immediate updating so that the operator can immediately see the effects of his changes. The two main types of refresh tubes are the vector refresh tube and the raster scan tube.

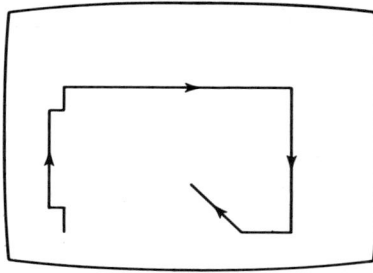

Fig. 9.44A Operating principle of a vector refresh tube.

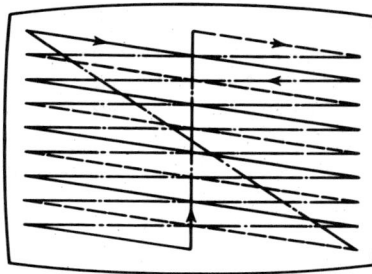

Fig. 9.44B Operating principle of a raster scan tube.

The vector refresh tube allows the beam to move both horizontally and vertically at the same time so that it is possible to draw curves and slanting lines. See Figure 9.44A. The electron beam has to rewrite the image at least 30 times per second in order to keep the flicker from being noticed. With a full screen, for example, 5000 in. (127 meters) per image (frame), this refresh rate is impossible, so the screen will flicker and cause eyestrain. The image quality is high, with a resolution corresponding to 1000 × 1000 pixels.

The raster scan tube writes the image in essentially the same way as in a TV set, frequently with two interlaced images. See Figure 9.44B. The total image is repeated 30 times per second, but the interlacing causes the image to be repeated 60 and not 30 times per second. The flickering, which is independent of the amount of data displayed, is reduced but cannot be eliminated completely. The image quality is somewhat lower than in the previous cases; for example, 500 × 500 pixels.

The display designer can eliminate entirely the flickering effect by avoiding the image interlacing technique and repeating the image 60 times per second. The image quality is higher than with interlacing; it has a resolution corresponding to 1000 × 1000 pixels.

Some CAD systems use graphics color displays so that the different layers of the board, e.g., the solder and component sides, are clearly distinguishable from each other. This does not apply to storage tubes, which can display only one color, typically green. Incidentally, some refresh tubes cannot produce blue.

The graphics work station always has one or more input devices so that the operator can communicate with the computer. Devices in general use include:

Keyboard: An alphanumeric keyboard is usually used to enter data or to submit commands to the computer. The keyboard is much like the keyboard of an electric typewriter, but it is frequently expanded with a number of special keys and/or switches which are used to activate certain functions of the system.

Joystick: A joystick is a device with a small control lever used for moving a cursor, in the form of a crosshair, on the display screen. The lever is mechanically connected to two potentiometers which deliver X- and Y-voltages corresponding to the position of the lever. These voltages are used to control the crosshair's position. The operator can therefore use the joystick to indicate a point on the display quickly and easily.

Instead of the joystick, it is possible to use two separate thumbwheels for controlling the potentiometers, but this method is not nearly as practical and therefore little used today.

Data tablet: A data tablet is designed as described in Paragraph 9.3.2, Subparagraph 2. By moving the probe across the surface of the data tablet, the operator generates control signals used to move the cursor (crosshair) on the display screen so that the desired point can be indicated. The data tablet is a little more convenient to use than the joystick, since the former does not require as steady a hand as the joystick.

Light pen: A light pen is a small probe about the size of an ordinary ballpoint pen. The tip contains a light-sensitive element, e.g., a phototransistor. When the light pen is pointed at the screen, it detects changes in the screen brightness. The change in light occurs when the electron beam passes across the points sensed by the pen. The position of the point under the pen is now defined since the time at which the light pen detects the change in light intensity can be correlated to the controlled movement of the electron beam. It is therefore obvious that a light pen can be used only in conjunction with a refresh tube, since no time reference exists in the case of a storage tube.

Along with a special program, the light pen can be used to select a point on the display, e.g., a solder pad or component, and move it to another position, or possibly erase it.

Formerly, the light pen saw much use in interactive CAD systems, and as a matter of fact, it was more or less synonymous with interactive CAD systems. In spite of this, the light pen was not very desirable ergonomically, and the current trend is to use a joystick or a data tablet with probe.

Menu: When the graphics work station is provided with a keyboard, the operator must remember all the commands which form part of the system. The number of commands can be quite large. In order to facilitate the operator's job, the graphics work station in certain systems is provided with a menu, located either within a certain area of the display screen or the data tablet. In its boxes, the menu indicates the possible commands, sometimes supplemented with explanatory text. By indicating the desired menu command by means of the display crosshair or the data tablet probe, the operator can very conveniently choose the command.

9.5.10c3 Peripheral Equipment

Figure 9.42 shows a number of peripheral devices connected to the computer. The following units have been described in Paragraph 9.3.2;

Paper tape reader (Subparagraph 4)

Paper tape punch (Subparagraph 5)

Pen plotter (Subparagraph 6)

Electrostatic plotter (Subparagraph 6)

We shall elaborate on the following additional units:

External Memory: This is used partly for storing the CAD system's programs and data without risk of loss during a power breakdown, and partly for extending the internal memory of the CPU, since some programs can be too large to fit into an internal memory of, for example, 0.5 megabyte. In a CAD system, the external memory capacity can be 10 to 20 megabytes, or even more. With this arrangement, portions of the program are moved from the external memory to the internal memory, depending on the actual requirements and the progress of the layout.

Winchester Disk Drive: The external memory is often a Winchester disk drive. This data medium consists of one or more rigid disks having a very thin and uniform coating of magnetic material on each side. The disks rotate at high speed, and the data is read and written by read/write heads mounted on movable bifurcated arms. Figure 9.45 shows a typical Winchester disk drive with two disks and two read/write heads for each disk. A servo system moves the read/

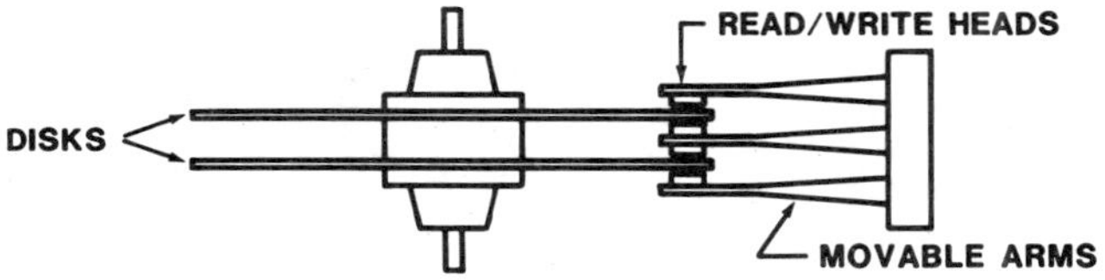

Fig. 9.45 Principle of a Winchester disk drive.

write heads across the disk until they are positioned directly over and under the desired circular data track. Each track is divided into a number of arcs, called sectors. The first part of each sector contains the sector address, and the rest the data actually used by the computer.

Floppy disk drive: A floppy disk, also called a diskette, is a flexible plastic disk thinly coated with magnetic material. The disk diameter is usually 8 in. (7.88 in. or 200 mm, to be precise), but minidisks with a diameter of $5\frac{1}{4}$ in. (133 mm) or even less are also used.

The floppy disk comes from the vendor in a square protective envelope, and both are inserted into a disk drive (two of which may be included in the system). The protective envelope has a central opening on each side so that the drive mechanism can center, engage, and turn the floppy disk. One or two read/write heads make contact with the disk through radial slots on each side of the envelope.

On the whole, the operation is much like that of the Winchester disk drive, but the floppy disk has fewer tracks. Each track is divided into a number of sectors, each containing 128, 256, 512, or 1024 bytes. The computer reads the data from the floppy disk by moving the read/write head to the proper track, and reading the data when the correct sectors pass under the head. The format, i.e., the division into tracks and sectors, varies from one CAD company to the next.

Another difference between the floppy disk and the Winchester disk is that the head actually presses against the floppy disk while reading and writing, whereas the read/write head of a Winchester disk drive never actually touches the disk except when the drive is turned off. Also, the openings in the floppy disk envelope allow foreign particles to enter, so the disk must be handled carefully and protected from contaminants in the environment, e.g., smoke, beverages, dust, etc. In contrast, the Winchester disk drive is sealed to protect it from the environment.

The floppy disks can be obtained in the following capacities:

Single-sided/single-density: 0.25 megabyte
Single-sided/double-density: 0.5 megabyte
Double-sided/single-density: 0.5 megabyte
Double-sided/double-density: 1 megabyte

A floppy disk can, for example, contain the CAD systems's algorithms, which are loaded into the internal memory from the disk drive. In the case of program changes such as updates, the CAD manufacturer delivers a new floppy disk, and the CPU reads the contents into memory.

Another floppy disk can contain the data for the finished board layout. When the board is to be revised, the data can easily be read into the CAD system.

Magnetic tape drive: The storage medium is a flexible 1/2 in. plastic tape with a thin, uniform coating of magnetic material on one side. The tape, which is wound onto a supply reel with a diameter from 6 to $10\frac{1}{2}$ in. (152 to 260 mm), closely resembles the tape used in ordinary tape recorders. The function is also quite similar except for the fact that the magnetic tape drive handles digital data.

During reading and writing, which involves nine tracks at 800 or 1600 bpi (bits per inch) per track, the supply reel is mounted on the magnetic tape drive, and the tape passes by a precision read/write head and on to a takeup reel.

The magnetic tape is used not only as an input medium but also as a storage medium for data which are to be used later, possibly off-line in a photoplotter. It should be noted that photoplotters generally read 800 bpi tapes only, and not 1600 bpi tapes.

It should also be noted that it can be more advantageous to use a magnetic tape than a paper tape for controlling the photoplotter. The reason is that the amount of data is so great that only a magnetic tape can hold all the information regarding the various layers of the board, i.e., the solder and component sides, the solder and insulation masks, the component notation, etc. Provided that the board is not too large, all the filmwork can be plotted in a single run on the same sheet of film. This implies a slightly lower price since the filmwork can be plotted unattended at night.

Cassette Cartridge: Decentralized (distributed) CAD systems comprise a central computer and a number of satellite CAD work stations, usually provided with a little less processing power than the graphics work station previously described. Post-processing the board layout on the central computer has a number of advantages, among them faster processing and certain special routines, e.g., the necking of conductors to be routed through narrow passages.

When the satellite station is working off-line, i.e., with no direct connection to the central computer, possibly because of excessive distance, the satellite station records the output data on a small cassette cartridge.

The cassette cartridge strongly resembles the tape cassette used in an ordinary cassette tape recorder although it is a little smaller. Furthermore, it is used by the satellite station as a storage device, which means that fundamentally it has the same functions as a floppy disk, but the cassette drive is cheaper than the disk drive.

System Console: Usually an alphanumeric display unit with a keyboard, the system console is the operator's interface to the operating system. When the system console is not used for this purpose, it can be used for preliminary work, such as preparing wiring lists, in order to reduce the workload on the graphics work station.

9.5.11 Setting Up a Library

9.5.11a Introduction

After purchasing and receiving a CAD system, the operator's first important task is to set up a library of all components used in daily work. In some cases, the system vendor supplies a library of the most commonly used standard components, but since most electronics companies also use a number of special components, it is necessary to supplement the library with these. The library must be updated, of course, when new components are introduced.

A complete library of all components used in the company greatly eases the operator's task. While placing the components within the available board area, he does not have to draw the individual components on the display, because they can be called up from the library in the form of library shapes. A related advantage is that the library shapes on the screen are independent entities which the operator can move or rotate as needed by means of simple commands.

A specific component shape is called up on the display screen according to its library number, for example, L7. It is traditional to indicate library numbers with an L for library. The displayed shape shows the most important component dimensions, i.e., the effective outline and the terminal locations. A single library shape can be used for multiple component types, provided that they have the same dimensions.

In addition to the information of immediate importance to the layout, the library can contain information to be used at other design stages, e.g., the component outline used in the component notation, or information regarding the sizes of the solder holes, which are indicated on the drill film by means of special symbols.

9.5.11b Library Shapes

We shall now see how the library is set up in practice. The first example is a component with axial leads, as illustrated in A, Figure 9.46. The starting point is the dimensions of the component body, determined by its length L and largest diameter D. The terminal distance T depends on how near the component body it is possible and allowable to bend the leads, cf. Paragraph 5.10. If the distance to the bend is a, the terminal distance T becomes $T = L + 2a$, which is rounded up to the nearest grid unit, frequently 1/20 in. = 1.27 mm, or 1/40 in. = 0.635 mm. This is because the center-to-center distance between the solder holes should always be a multiple of the grid unit. See B, Figure 9.46.

The library shape, which shows the component as seen from above, is defined as the rectangle which circumscribes the component, including the solder pads, and rounded up to the nearest grid unit. See C, Figure 9.46. The rectangle indicates the effective space requirement on the board, and its size is calculated as D × (L + 2a + d), where d is the solder pad diameter. The size is rounded up to the nearest grid unit.

All components are made with certain dimensional tolerances. To prevent jamming of the assembled components, the designer must take these tolerances into consideration, and should therefore always use the highest tolerance in calculations.

Entering the contour and terminal distance of a library shape into the library requires that the configuration be defined. A convenient method is to use coordinates with respect to a fixed reference point. There are several possible placements of the reference point:

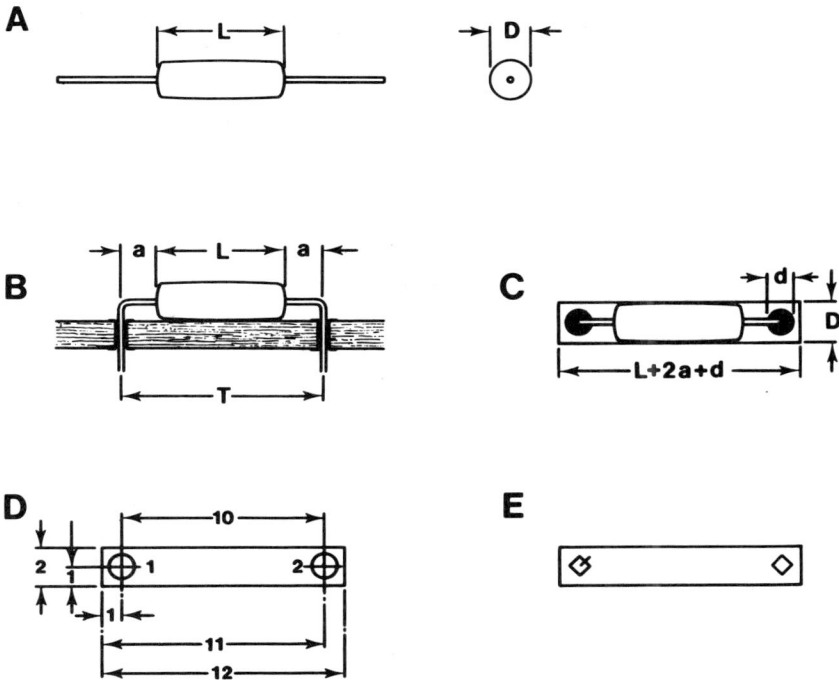

Fig. 9.46 Example of a library shape for a component with axial leads.
A: Overall dimensions of the component.
B: Determining the terminal distance.
C: Determining the area requirement.
D: Indication of coordinates.
E: Finished library shape.

- At pin number 1
- At the lower left corner of the library shape
- At the center of the library shape

The reference point location is a matter of faith or tradition, and each has its supporters. The reader should note the following effects of reference point locations:

The first two locations presuppose that a 180-degree rotation of a library shape, e.g., an IC, is followed by a parallel displacement in order to maintain merging contours. The last choice implies that the rotation can be performed without any parallel displacement because of the symmetrical characteristics. The only visible change is that the pin numbers are different and the small notch indicating pin number 1 changes side.

Another advantage of the latter method is obtained in conjunction with components having off-grid terminals, i.e., terminals not placed at the inter-

370

LIBRARY NUMBER	LIBRARY SHAPE		SOLDER PAD				COMPONENT FRAME	
	LENGTH	WIDTH	No.	X	Y	CODE	LENGTH	WIDTH
L7	12	2	1	1	1	O	12	2
			2	11	1	O		

Fig. 9.47 Example of a library list stating the coordinates of the library shape shown in Figure 9.46. The coordinates are expressed as the number of grid units @ 1/20th in. = 1.27 mm. The reference point is located in the bottom left corner of the library shape.

sections of the grid lines. By putting the reference point at the center of the component, the designer facilitates placement of the component.

The coordinates are defined as the number of grid units in the X and Y directions from the reference point. A grid unit can, for example, be 1/20 in. = 1.27 mm. In the example shown at D, Figure 9.46, the lower left corner of the contour has been chosen as the reference point in order to avoid negative coordinates.

The coordinates can be stated in a library list as shown in Figure 9.47. The library list unambiguously defines the contour of the library shape and the location of the solder pads by their coordinates. It also states the solder pad size by a code.

By means of the cursor, the operator can now draw the complete library shape on the display screen, after which the library shape can be entered directly into the library file of the CAD system. The library shape is shown at E, Figure 9.46. Pin number 1 can be indicated in different ways, e.g., by a small stroke through its pad, or by a small stroke at the corresponding corner of the library shape.

Another example of a library shape is an eight-pin IC as shown in Figure 9.48. The terminal distances are standardized as shown at B, Figure 9.48, and the library shape has coordinates as indicated at C, Figure 9.48. The corresponding library list is shown in Figure 9.49, and the finished library shape at D, Figure 9.48. As in the previous case, the operator enters the library shape into the library file by drawing the complete library shape on the display screen, and then transfers it into the library file with a simple command.

In order not to have rows of ICs located too close to each other, the contours of the library shape in the example chosen have been placed 1/10 in. (2.54 mm) = 2 grid units at 1/20 in. (1.27 mm) outside the solder pad centers. The smallest distance between rows of ICs therefore becomes 2 × 1/10 in. = 4 grid units when the library shapes are adjacent. Cf. the comments in paragraph 5.11 regarding the necessary space around an IC.

The experienced operator may find it unnecessary to set up a library list, but instead prefer to draw the library shape by eye, based on the dimensions given in the data sheet.

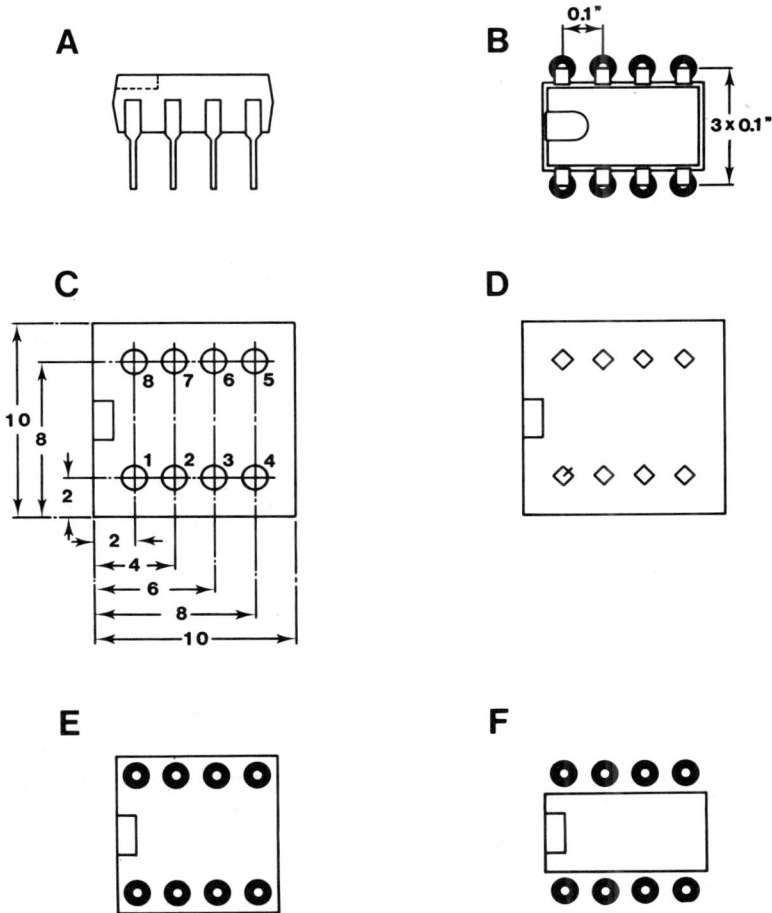

Fig. 9.48 Example of a library shape for an 8-pin IC package.
A: Side view of the IC.
B: Standardized terminal distances.
C: Indication of coordinates.
D: The finished library shape.
E: Component frame (outline) placed outside the solder pads.
F: Component frame (outline) placed within the solder pads.

At the same time he defines the details of the library shape, the operator must indicate where to place the component designation, e.g., QD17. The location can be indicated with a generic label, e.g., NAME and the computer will prompt the operator for the component designation later, during the practical work. The operator enters this information via the keyboard, after which the computer automatically places it at the selected location.

LIBRARY	LIBRARY SHAPE		SOLDER PAD				COMPONENT FRAME	
NUMBER	LENGTH	WIDTH	No.	X	Y	CODE	LENGTH	WIDTH
L201	10	10	1	2	2	0	8	8
			2	4	2	0		
			3	6	2	0		
			4	8	2	0		
			5	8	8	0		
			6	6	8	0		
			7	4	8	0		
			8	2	8	0		

Fig. 9.49 Example of a library list stating the coordinates of the library shape shown in Figure 9.48. The coordinates are expressed as the number of grid units @ 1/20th in. = 1.27 mm. The reference point is located in the bottom left corner of the library shape.

9.5.11c Pattern Detail Dimensions

In addition to defining the library shapes, it is necessary to determine the dimensions of all the pattern details which are loaded into the CAD system. Pattern details include:

• Solder pad sizes
• Conductor widths
• Insulation distances
• Text sizes

There are usually some constraints depending on the symbols (apertures) in the photoplotter's symbol discs. The symbol discs can be in a standard version, or in a custom version. Custom versions are extremely expensive and have a long lead time. When a laser plotter is used, the operator has a free hand because all configurations can be drawn, cf. Paragraph 9.3.4.

9.5.11c1 Solder Pad Sizes

In addition to the coordinates, the library lists shown in Figures 9.47 and 9.49 have a column containing a code for the solder pad size. This code is a little more complicated than is immediately apparent. It reflects a certain association between the diameter and shape of the solder pad, the finished diameter of the solder hole, and the size of the corresponding aperture of the solder mask. In the case of multilayer boards, there is also a question of the diameter of the internal pads, which normally should be a little larger than the diameter of the external solder pads. Finally, the table can include a symbol which on the drill film indicates the finished diameter of the solder hole.

When setting up the library, and also when entering updates, the operator should consider the above relationships and prepare or update a pad assignment list, as shown in Figure 9.50. On the whole, the relationships correspond to those previously stated in the design rules, but because of the higher accuracy of the CAD filmwork, the operator can choose slightly narrower annular rings and a slightly smaller clearance between the solder pad and the solder mask aperture.

CODE	SOLDER PAD				SOLDER HOLE		SOLDER MASK APERTURE			
	DIAMETER		SYMBOL DISC		DIAMETER		DIAMETER		SYMBOL DISC	
	in.	mm	No.	POS.	in.	mm	in.	mm	No.	POS.
0	0.055	1.4	001/3	15	0.031	0.8	0.078	2.0	001/3	20
1	0.078	2.0	001/3	20	0.051	1.3	0.102	2.6	001/3	23

Fig. 9.50 Pad assignment list.

The advantage of a pad assignment list is that the operator once and for all commits himself to all relationships regarding the solder pads, the finished diameter of the solder holes, the drill film's symbols for hole sizes, and the size of the solder mask apertures. There may be some differences between the apertures used in the solder mask and the insulation mask.

Once the pad assignment list is completed, the computer generates the final documentation, e.g., tapes for the photoplotter, automatically with no intervention from the operator, because the computer itself picks the necessary data from the pad assignment list. Stating just one code identifies the necessary data to be used in the actual documentation; these data include the right size and shape of the solder pads when generating the solder and component sides, and the corresponding size and shape of the solder mask apertures when generating the solder mask.

A carefully prepared pad assignment list is thus of the utmost importance, both for an effective layout and for a major reduction of time in the future.

9.5.11c2 Conductor Widths

Most CAD systems use a relatively limited number of conductor widths. Here also, we have the same relationships between the conductor width and the insulation distance as discussed in Paragraph 9.3.5d and shown in Figure 9.16. A CAD system should preferably not work in an off-grid mode, i.e., routing the conductors off the grid lines, since this can interfere with meeting and verifying the minimum spacing. The narrowest conductor and the minimum spacing must be adapted to each other.

A conductor assignment list can appear as shown in Figure 9.51. The conductor widths come from the wiring list, for example, as shown in Figure 9.56, and are retrieved automatically during the routing.

9.5.11c3 Insulation Distances

A generally accepted minimum insulation distance, also called the minimum spacing, is 0.012 in. (0.30 mm). Here also it is necessary to have the correct relationships between the size of the solder pads, the conductor widths, and the

CODE	WIDTH		SYMBOL DISC		APPLICATION
	in.	mm	No.	POS.	
0	0.012	0.30	001/3	4	SIGNAL CONDUCTOR
1	0.008	0.20	001/3	2	SIGNAL CONDUCTOR (FINE LINE)
2	0.030	0.76	001/3	9	POWER SUPPLY (HIGH DENSITY)
3	0.055	1.40	001/3	12	POWER SUPPLY (LOW DENSITY)

Fig. 9.51 Conductor assignment list.

insulation distances in order to prevent the CAD system from operating in an off-grid mode. The conditions are the same as described in Paragraph 9.3.5d. Figures 9.16 through 9.19 show that a minimum spacing of 0.012 in. (0.30 mm) and a conductor width of 0.012 in. (0.30 mm) form the basis of an acceptable procedure. An example of a spacing assignment is shown in Figure 9.52.

9.5.11c4 Text Sizes

The size of the text on the board, i.e., the type size, should also be determined. The font is usually predetermined by the CAD manufacturer and delivered as part of the CAD program. The text is drawn by the photoplotter, and the line width consequently depends on the diameter of the selected symbols in the symbol disc of the photoplotter. A suitable ratio between the text height and the line width should be chosen, cf. Paragraph 6.8.1. An example of a text assignment list is shown in Figure 9.53.

9.5.11d Component Frames

A CAD-generated component notation film usually has frames around the components, whether these be ICs or discrete components such as resistors, capacitors, etc. In rare cases, leader lines for two-terminal components are used.

In practice, there are two basic executions, since the component frame can be placed either outside the component solder pads as shown at E, Figure 9.48, or within the component solder pads as shown at F, Figure 9.48. The first execution implies the risk that the frame is screen printed across via holes, and the latter that the frame is covered by the component.

CODE	PATTERN DETAILS	MINIMUM SPACING	
		in.	mm
0	CONDUCTOR-TO-CONDUCTOR	0.012	0.30
1	CONDUCTOR-TO-SOLDER PAD	0.012	0.30
2	SOLDER PAD-TO-SOLDER PAD	0.012	0.30

Fig. 9.52 Spacing assignment list

CODE	TEXT HEIGHT		LINE WIDTH		SYMBOL	DISC
	in.	mm	in.	mm	No.	POS.
0	0.078	2.0	0.008	0.20	001/3	2
1	0.138	3.5	0.016	0.40	001/3	5

Fig. 9.53 Text assignment list.

It is expedient to link the frame with the library shape so that the computer, when generating the component notation film, can pick the component frame corresponding to the library shape from the library list. Therefore, the library list could be extended with a code defining the corresponding component frame for each library shape. The operator enters the frame data into the library file in the same way as he did the library shape, most easily by drawing it on the display screen and then transferring it to the library file. The location of the component designation is determined in the same way as with the library shape. The component designation itself is derived automatically from the component designation used in conjunction with the library shape.

In both the above cases, the frame should be placed as close to the solder pads as possible, for example, at a distance of 1/20 in. (1.27 mm). In this way it is easier to avoid screen-printing the frame across via holes, which risks ink flowing down into the holes.

9.5.11e Degree of Board Filling

The library shapes are normally designed in such a way that they can be located with their edges coinciding without risk of component jamming. Such a layout naturally offers the highest degree of board filling. Extending the library list with information on the area of the individual library shapes allows the operator or system to calculate the degree of board filling.

Since the usable board area normally is known before the layout starts, and the area of the individual library shapes can be read from the library list, the operator can calculate the actual degree of board filling as the ratio between the total area of the components and the usable board area. This allows the operator to evaluate the layout feasibility in advance, and he can consider dividing the circuitry before he starts placing the components.

The calculations can, of course, always be performed manually, but if the CAD system has a program for calculating areas, the area requirement can be automatically added up concurrently with entering the components into the system, either by means of the keyboard or by drawing on the display screen. During the entire entry procedure, the calculated degree of board filling can be followed by the operator so that he knows how close he is to the critical degree of board filling.

The CAD system can also be connected to a schematic diagram and parts list system. After the schematic diagram has been laid out, a wiring list and a parts list are generated automatically. By augmenting the system with a program for calculating the degree of board filling, the operator can obtain the actual degree of board filling, based on the parts list, the library list, and usable board area. The operator can now assess the feasibility of the board layout even before the job is started.

It is difficult to state fixed criteria for the critical degree of board filling, i.e., when the layout becomes unfeasible or at least extremely difficult, because many parameters enter into the assessment, among other things:

Type of board: Single-sided, double-sided, or multilayer
Board dimensions: Shape of board (length × width) and location of I/O terminals
Type of circuitry: Analog, digital, mixed analog/digital, high-frequency, power supply, etc.
Library shapes: Area requirement depends on company policy regarding component density.

It is obvious that a degree of board filling of 1 or greater renders the layout impossible. In the case of double-sided boards, it is generally possible to accomplish the layout at a degree of board filling up to 0.75 or 0.85. It is recommended that the operator compile empirical information from his own practice.

Not least for a CAD service bureau, it is most important to be able to assess the complexity/feasibility of the layout in advance. If so, the bureau can contact the customer about possible changes before starting a layout that proves impracticable.

9.5.12 Basic Data

The fundamental data for the layout comprises the following categories:

• Schematic diagram
• Parts list
• Mechanical specification
• Board type specification
• Special conditions

9.5.12a Schematic Diagram

The schematic diagram must define every single component by its designation (component number), e.g., C27, and show all connections, including all terminals and terminal numbers.

> *Note:* The operator should realize that in the case of ICs, some design engineers fail to indicate pull-up resistors, and possibly ground and voltage connections as well. Another problem can be inadequate (or no) indication of IC decoupling capacitors.

9.5.12b Parts List

Based on the component designations shown in the schematic diagram, the individual components are described in the parts list by their commercial type number, electrical characteristics, user company's part number, and possibly a cross-reference to the library number.

9.5.12c Mechanical Specification

The mechanical specification states the size and contour of the board, including the position of mounting holes and mechanical limitations; for example, areas with component height limits. Special placements of components, e.g., connectors and switches, must also be stated.

9.5.12d Board Type Specification

This specification indicates whether the board is to be a single- or double-sided board, or a multilayer board. The pattern details, i.e., the solder pad sizes, the conductor widths, the minimum spacings, etc., must conform to the design rules incorporated in the CAD system's programs and check routines.

9.5.12e Special Conditions

Information about special conditions regarding the component placement is always necessary in order to obtain a good design, and the circuit designers should always be questioned about such information. As an example, they could tell whether any components are especially heat-sensitive or require physical separation to avoid capacitive or inductive interaction.

Other important conditions can be shielding problems, which are solved by introducing ground planes or by making conductors as short and/or wide as possible.

9.5.13 Data Preparation

Before the operator can start the layout, certain data must be prepared and entered into the CAD system. The data capture process enters data into the following lists:

- Board list
- Component list
- Wiring list

9.5.13a Board List

The board list is prepared on the basis of the mechanical specification, cf. Paragraph 9.5.12c, which states the coordinates of all corners of the contour, including

378

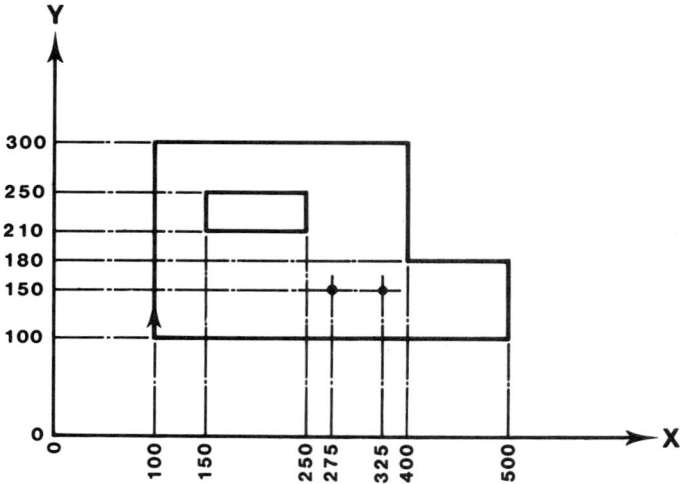

Fig. 9.54 Board list.
Top: Example of a mechanical specification based on coordinates expressed as the number of grid units @ 1/40th in. = 0.635 mm.
Bottom: Example of a board list, derived from the mechanical specification shown above.

BOARD LIST	POSITION		DIAMETER	
	X	Y	in.	mm
OUTER CONTOUR	100	100		
	100	300		
	400	300		
	400	180		
	500	180		
	500	100		
	100	100		
INNER CONTOUR (CUT-OUTS)	150	210		
	150	250		
	250	250		
	250	210		
	150	210		
MOUNTING HOLES	275	150	0.120	3.0
	325	150	0.080	2.0

internal cutouts, and the coordinates of mechanical mounting holes. See, for example, Figure 9.54.

It is most expedient to express the coordinates in grid units, also called data structure units (DSUs), because this facilitates drawing the board outline on the display screen or entering the data directly into the system. In a few CAD systems, it is also possible to enter metric dimensions.

COMPONENT DESIGNATION	LIBRARY NUMBER	POSITION		ORIENT-ATION	NUMBER OF PINS	
		X	Y		NOMINAL	NOT USED
C8	0002				2	
J1	0101	38	4	0	24	
Q14	0045				14	11.12.13
S3	0036	18	46	1	3	

Fig. 9.55 Example of a component list.

When standardized board sizes are used, e.g., Eurocards, they can be entered as fixed library shapes and called up as needed. In effect, this is a graphic data entry procedure.

The data entry to the CAD system can be performed in several ways. The simplest way is to directly draw the contour on the display screen with the cursor (crosshair); the experienced operator will presumably find it superfluous to prepare a board list. Another way is to enter the coordinates of the board list via the system keyboard. Finally, the data can be entered at the system's control console, which is an alphanumeric display unit with a keyboard, cf. Paragraph 9.5.10c3. The last method relieves the CAD system of considerable work, but it is perhaps neither as fast nor as easy to survey as the direct drawing on the display screen.

9.5.13b Component List

The component list is derived from the schematic diagram and/or the parts list, and it states all the components in reference designation order; for example, R1, R2, R3, etc. See Figure 9.55. For each component, the component list states the corresponding library number, which again defines the library shape to be used. In this way, the component list provides a reference between all the components used and their physical requirements (area occupied and terminal location), which is the first condition for component placement.

If a component must have a certain position, the appropriate coordinates (based on the reference point of the component) should be stated in the component list. It is furthermore expedient to indicate the orientation of the component by means of a code. For example, it could be an LCD (liquid crystal display) device, which must be visible through a window in the front panel, or it could be a switch, the shaft or actuator of which must project through a hole in the front panel. The placement therefore cannot be left to a component placement routine, and furthermore, the location cannot change during the progress of the layout. The best way of assigning a permanent location is through the component list.

The component list, which actually is just a cross-reference between the component designations and the library shapes, can be developed in several ways:

- By manual writing
- By alphanumeric data entry
- By graphic data entry

9.5.13b1 Manual Writing

The fundamental method is a manual transfer of the component designations from the parts list and/or the schematic diagram, whereupon the corresponding library numbers are drawn from the corporate master components file. The method implies considerable referential work unless the operator can remember the library numbers of the most commonly used library shapes. Then the component list is entered manually to the CAD system by means of the keyboard. Obviously, the entire method implies quite a lot of duplicated effort, and a considerable risk of typing errors. To aid checking of the completed component list, it may be expedient to state the nominal number of pins and all unused pin numbers for each component.

9.5.13b2 Alphanumeric Data Entry

The previously described manual method of preparing the component list can, of course, be replaced by a direct entry of data into an alphanumeric display unit via its keyboard. After the entry is complete, the data is transferred to the CAD system, probably by means of a floppy disk.

An improved method is possible when the alphanumeric display unit is connected to a general materials management system, where the data base contains a master file of all components. The parts list is prepared by searching through the component master file for the right components according to the schematic diagram; for example, R12 defined as a resistor with axial leads, 10 kilohms ± 5 %, 1/2 W. The component number stated in the master file and the library number of the corresponding library shape are also displayed.

When the entire schematic diagram has been gone over in this way, it is possible to obtain an automatic printout of the parts list sorted in a sequence corresponding to the component designations, and supplemented with the desired data from the components master file. The possibility of error is negligible, since each item of the parts list except the component designation is derived from the data base (components master file) of the materials management system.

Using the electrical parts list established as described above, a sorting program can print out the desired component list, which now can be entered into the CAD system via the keyboard of the graphics work station.

It is still more advanced when the data from the component list is transferred automatically to the CAD system, either on-line via an electrical connection, or more commonly, off-line via a storage medium, such as a floppy disk.

9.5.13b3 Graphic Data Entry

In some cases the operator prefers to work at the graphics work station. Instead of setting up a component list, the operator simultaneously enters the component

CONDUCTOR WIDTH	COMPONENT		COMPONENT		COMPONENT		COMPONENT	
CODE	NAME	PIN No.	NAME	PIN No.	NAME	PIN No.	NAME	PIN No.
2	J6	2	R2	2	Q1	4	C1	2
	R7	2	Q2	3	C2	2	R8	2
	Q3	11	CR3	2				
0	Q7	9	R10	1	R15	2		
0	Q1	13	R21	2	Q1	5	Q1	12

Fig. 9.56 Example of a wiring list based on manual writing.

designations and the corresponding library numbers. The component shapes with inserted component designations appear on the display screen, i.e., the operator has performed a graphic entry. See Paragraph 9.5.13c for further details.

9.5.13c Wiring List

For the CAD system to wire the board, all the connections between the individual component pins, including the input/output connectors, must be entered according to the schematic diagram. In this way the connections comprise all voltage supplies, whether they occur as conductors or as voltage and ground planes, and all signal conductors. Certain CAD systems also provide connections to possible test points.

A statement of the above connections is usually prepared in the form of a wiring list, also called a net list. Obviously, the wiring list must be complete and error-free. This applies especially to the ICs, where many circuit designers follow the unfortunate practice of not indicating voltage connections, decoupling capacitors, pullup resistors, etc. Here, the operator must add the missing details.

A wiring list can have the appearance shown in Figure 9.56, which shows part of a wiring list. The list can be prepared in any of several different ways, each with its own advantages and disadvantages:

• By manual writing
• By alphanumeric data entry
• By graphic data entry
• By entry from a schematic diagram

9.5.13c1 Manual Writing

During manual writing of the wiring list, the operator records in a blank all connections within the individual circuit or group of the schematic diagram, i.e., all component and connector pins, etc., connected with each other. It is imperative for the operator to work in a systematic way and to tick off or cancel all pins and connections registered, so that none is overlooked.

382

W
;; -5 V, CODE 2
J6-2, R2-2, Q1-4, C1-2, R7-2, Q2-3, C2-2,
R8-2, Q3-11, CR3-2/
;; AMPLIFIER, CODE 0
Q7-9, R10-1, R15-2/
Q1-13, R21-2, Q1-5, Q1-12/

Fig. 9.57 Example of a wiring list appearing as a printout based on alphanumeric data entry.

The wiring list can also contain information about the conductor width required. Here it can be practical to arrange the circuits according to their function; for example, by taking the voltage and ground connections first. Since these connections must be wider than the other ones, it is most expedient to route them first, while there is still room. Then come the signal conductors, which nearly always are narrower so that they are more easily routed.

When the wiring list is finished, it is recommended that the operator check the accuracy by comparing the schematic diagram and the wiring list. Having verified that the wiring list is correct, he can enter it into the CAD system via the keyboard. It is important for the format of the wiring list data to closely follow the CAD system's rules, since the data will otherwise be rejected.

Since it is difficult for an operator to check his own work, some CAD companies recommend that two operators should each prepare a wiring list, and then enter both lists into the CAD system. A special check routine in the CAD system compares the two wiring lists and accepts them only if they are identical. All differences are printed out so that the errors can be corrected.

9.5.13c2 Alphanumeric Data Entry

In the case of alphanumeric data entry, the operator works at an alphanumeric display unit with a keyboard and enters the same data as when working manually. The wiring list can, for example, have the appearance shown in Figure 9.57.

Many of the alphanumeric display units used with CAD systems contain special check routines. These perform a first check of the entered wiring list for certain types of faults and report these faults on the display screen or on a printout. The faults can be:

(a) *Incorrect syntax:* A routine indicates those places in the wiring list where the input is not in the required format. An example is the operator's using a period instead of a comma.

(b) *Unconnected pins:* Another routine lists, for each component, the number of pins connected and unconnected.

(c) *Pins used twice:* A third routine lists, by component designations and pin numbers, all component pins occurring more than once in a single circuit.

Based on the report, the operator can correct a number of typical faults, but the check routines cannot reveal all wiring faults. It is still necessary to check the printed and corrected wiring list, just as in the manual case. The checking, however, is more accurate because a number of faults have been removed beforehand, so that they do not obscure the remaining ones.

Using an alphanumeric display unit for preparing the wiring list implies the great advantage that the wiring list can be entered into the CAD system automatically; for example, from a floppy disk. In this way, the operator does not load the graphics work station as in the manual method. Furthermore, writing and typing errors cannot cause new faults.

During the layout phase it may be necessary to manually change the wiring, which incurs the risk of introducing a few erroneous connections. When the layout is concluded, the operator calls up the manual wiring list from the CAD system; for example, on a floppy disk. On the alphanumeric display unit, he can compare the new wiring list with the initial one. Possible discrepancies will now be shown so that the operator can correct the layout.

9.5.13c3 Graphic Data Entry

In the case of graphic data entry, which is preferred by some operators, the wiring list is prepared immediately after the components are placed as described in Paragraph 9.5.14a. The operator moves the cursor (crosshair) on the display screen from solder pad to solder pad within the individual circuits, so that the wiring list is entered automatically into the CAD system.

As soon as a solder pad is connected, it is marked occupied on the display screen, e.g., by a small cross in the pad, so that the operator does not erroneously use it in another circuit. When the entry is completed, all active solder pads should be marked occupied. In this way the operator can easily see whether there are any missing connections.

The method is very direct and offers the operator a fine view, but the CAD system is tied up with work that could have been performed on the alphanumeric display unit.

9.5.13c4 Entry from the Schematic Diagram

Entry from the schematic diagram always implies that the schematic diagram is drawn and edited by means of a graphics drafting system. When the schematic diagram is completed, all data regarding the component connections are determined, and insofar as the format of the data is correct, the data can automatically be loaded into the CAD system; for example, from a floppy disk.

The obvious advantage is that when the finished schematic diagram is approved by the circuit designer, possibly after a few corrections, the resultant wiring list is automatically correct. It is, however, presumed that there are no missing connections in the schematic diagram, cf. the initial considerations regarding the unfortunate practices of some circuit designers.

9.5.14 Layout

Following the initial data capture described in Paragraph 9.5.13, the operator can commence the layout. Layout consists of the following principal stages:

- Component placement
- Routing
- Cleanup
- Checking the layout

9.5.14a Component Placement

It is neither intended nor possible to describe the internal workings of the component placement routines of the various CAD systems on the market. Since these routines are all variants of the placement routines described in Paragraph 9.5.5, the following description is quite general.

In all systems, the first job is to call up the board contour on the display screen, as described in Paragraph 9.5.13a. The operator should choose a scale that shows the board in a reasonable size on the display screen. If the screen is too small to accommodate a very large board, the board can be divided into two parts. One or more fictitious connectors are placed on each side of the dividing line, and the two halves of the board are then treated separately. After the layout is finished, the two halves are assembled into one board for plotting, and the fictitious connectors are removed.

9.5.14a1 CAD Systems with Component Placement Routines

Most CAD systems on the market have routines for automatic component placement, which require the component list to be entered by one of the methods described in Paragraph 9.5.13b. The components are usually shown stacked in one of the screen corners.

Frequently, a few components must be placed in a certain location on the board. When the placement is defined by coordinates in the component list, the components are automatically placed in the desired locations. Otherwise, the operator picks such components from the stack one by one and uses the cursor to place them as desired.

It is most expedient to let the placement routine place the ICs first and leave the discrete components in the stack. The ICs are now placed by an interaction between the different placement routines described in Paragraph 9.5.5, i.e., cell division, rubber banding, minimization cuts, and successive component swapping. Added to this are auxiliary routines for swapping gates, either within the individual IC or between several ICs, frequently called gate swapping, and auxiliary routines for swapping of terminals, frequently called pin swapping. All such routines seek to reduce the theoretical total length of the internal connections as much as possible.

It should be noted that most manufacturers of CAD systems use designations of their own for the different routines and subroutines, which nearly always can be traced back to the basic algorithms described in Paragraph 9.5.5.

The operator can always follow the progress of the layout by means of a routine that measures the theoretical total length of the internal connections and reports it on the display screen. If the operator, using his general view and intuition, can improve the layout by moving some of the components manually, he uses another routine. The improvement, if any, should have reduced the total length of the internal connections.

When an optimum placement of the ICs has been obtained from the interaction between the different placement routines described, and fortified with a little manual help from the operator, the placement is frozen. The routines for component placement then treat the discrete components in more or less the same way.

9.5.14a2 CAD Systems without Component Placement Routines

A few manufacturers of CAD systems find that the routines for automatic component placement cannot be made to work acceptably, and therefore consider them to be of questionable advantage to the operator. The reason for the difficulty is the overwhelmingly large number of combinations to be evaluated, even when the number is limited by suitable adaptations of the placement algorithms, cf. Paragraph 9.5.5a.

Incidentally, this perception is shared by many operators who use CAD systems with automatic component placement routines. Rather than using these routines, they initially place the components manually.

In the case of CAD systems without routines for automatic component placement, the component list is entered in the usual manner, as described in Paragraph 9.5.13b. Here it can be practical to have the components located individually on the display screen outside the board contour, rather than in a stack. This is easily done by extending the component list with coordinates for the individual components which, for example, are placed above the board. The operator can now pick a component via the cursor and move it to the intended positon. The advantage is that the operator can always survey the number of components and immediately see the size of the component that he will place next.

As a matter of fact, the initial placement phase demonstrates whether the board can be successfully laid out. The operator therefore must utilize both his knowledge of the mutual dependence of the components and his technological knowledge of PCB production conditions.

After the initial component placement, various auxiliary routines optimize the placement. By using the density curves, cf. Paragraph 9.5.5f, one can instantaneously assess possible routing problems arising from the component placement. If there are problems, the operator can experimentally swap various components in the critical areas, and call up new density curves to see whether the conditions have improved. The operator can also use the successive component swapping method described in Paragraph 9.5.5e. Finally, the operator can, just as in the previous case, use the routines for gate swapping, pin swapping, and 180-degree rotation of two-terminal components. The results of all these three auxiliary routines are evaluated on the basis of the theoretical total length of the internal connections; when it is not possible to further reduce the length by swapping components, the component placement is considered optimum. The operator can now proceed to the routing phase.

9.5.14b Routing

The next step is to route the board, i.e., to establish all the connections. Again, it is neither intended nor possible to describe the automatic routing routines used by the many CAD systems on the market. Since these routines are all based on the routing routines described in Paragraph 9.5.4, the following treatment is of a general nature.

A CAD system usually contains several automatic routing routines, also called auto-routers, which can be brought into action in a sequence corresponding to the strategy chosen. Unless he uses one of the rarely purchased CAD systems with batch input, cf. Paragraph 9.5.2, the operator can normally stop an auto-router, perform manual changes, and then let the auto-router continue the routing.

The operator can freely choose the routing strategy. The first thing to decide on is the main direction of the conductors on the various board layers. In the case of double-sided boards, the conductors preferably run in the X-axis direction on one side, and in the Y-axis direction on the other side. The layout can possibly be limited to certain areas of the board.

It is then possible to choose the routing sequence for the following three main groups: voltage and ground connections, busses, and signal conductors. Usually, it is most advantageous to route the largest conductors first, i.e., the voltage and ground connections requiring relatively wide conductors. Next come the busses, all of which also require considerable space.

By means of the cursor, the operator can indicate the connections he wants routed, and on the keyboard he can enter the conductor width if it does not appear in the wiring list. He can furthermore intervene manually during the auto-routing; for example, in order to improve a bus routing, or possibly to repeat it at subsequent ICs.

> *Note:* In many brochures on CAD systems, the manual intervention is called interactive routing, which of course sounds more impressive.

Finally come all the signal conductors, and again the operator can choose the strategy. The auto-router can route the conductors according to their lengths. This presupposes that the CAD system reorganizes the wiring list so that it corresponds to the length of the conductors determined by the actual component placement. It is normally most expedient to route the shortest conductors first.

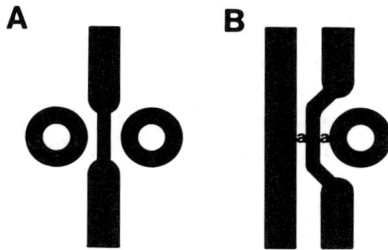

Fig. 9.58 Subroutines for adapting a conductor to crowded conditions.
A: Necking (narrowing) a conductor in a narrow passage.
B: Necking and bending (moving) a conductor.

The conductors can also be routed in a certain direction from a predetermined location. If, for example, the board has an input/output connector located at the left edge, it can be practical to let the routing proceed from left to right to ensure that the connector is connected to the other components in a reasonable manner.

The auto-router can route the signal conductors as chosen by the operator. In complicated cases, it can, for example, be desirable to stop the routing each time a circuit has been routed so that the operator can assess the result, and possibly improve it, before the auto-router starts routing the next circuit. In straightforward cases, this procedure is unnecessary, and the auto-router therefore routes all connections without pausing.

Some CAD systems are provided with subroutines which narrow down the conductor where it passes through a narrow passage; for example, between two IC pads as shown at A, Figure 9.58. This is called "necking." It is also possible to supplement this subroutine with a bending routine, which slightly moves the conductor narrowed by the necking routine, as shown at B, Figure 9.58. This can be important where it is required to meet the minimum spacing a to both sides of the conductor. Meeting the minimum spacing is frequently a precondition of completing the autorouting.

The ability of the auto-router to solve the routing task depends on its effectiveness (cf. Paragraph 9.5.4), the complexity of the board, and the success of the component placement. In general, a solution up to 95 to 98 percent complete can be expected. The missing connections are displayed, for example, as rubber bands or as incomplete conductors which flash or have a higher light intensity. The operator can now intervene manually (interactively) and remove possible obstructions by moving the interfering connections and/or components.

Here the search wave mentioned in Paragraphs 9.5.7 and 9.5.9 will aid the operator, since it very clearly shows where the obstruction is located. Having removed the obstruction, the operator can now route the missing conductor manually or with the auto-router. Then he treats the next missing conductor in the same way until all missing conductors have been routed.

9.5.14c Cleanup

The main task of the CAD system is to establish the required connections according to the schematic diagram. This is, however, not tantamount to achieving the best solution, and there are a number of classic fault types which, in addition to degrading the board's appearance, can also degrade the board's producibility:

- Too many inside corners along the conductors (90° notches)
- Too many sharp bends along the conductors
- Too many insufficiently spaced parallel conductors
- Too many via holes
- Unbalanced board pattern

Most CAD systems come with a number of subroutines for cleaning up (titivating) the board as regards the above faults, and when the primary layout task has been solved, the operator can start these subroutines. The Lee algorithm's penalty system was described in Paragraph 9.5.6, which describes how the functioning of the system can be affected by suitable changes in the penalty conditions. In fact, the cleanup subroutines use modified penalty conditions. Below we shall briefly discuss how to remedy the types of faults described.

When a conductor has a number of undesired 90-degree inner corners—for example, close to a bend—such inner corners can be turned (straightened) by means of a subroutine. See A, Figure 9.59.

It can also be desirable to change 90-degree corners to 45-degree corners. Particularly when a conductor is to be routed to a solder pad within a group of solder pads, e.g., in an input/output connector, it can be an advantage to route the conductor at an angle of 45 degrees. This gives a larger insulation distance than does a 90-degree bend. See B, Figure 9.59. The change is implemented by another subroutine.

A very characteristic trait of many CAD boards is that parallel conductors maintain the minimum spacing, even where there is ample space to spread the conductors. By means of a third subroutine, the system can spread closely spaced conductors where there is room. See C, Figure 9.59.

A minimization routine can reduce the number of via holes. This presupposes that the requirements regarding a pure X- and Y-routing on the two board sides are relaxed. See D, Figure 9.59. The disadvantage is a risk of introducing obstructions to later modifications of the routing; for example, if one wants to add a conductor from a to b.

Many operators prefer to produce the prototype boards with no minimization of the via holes. Experience shows that there will nearly always be a revision stage, and it seems more expedient to perform the minimization as part of the revison.

Another characteristic trait of CAD boards is pattern imbalance. Here also, it seems expedient to wait for the revision stage to balance the pattern. Figure 9.60 shows an example of a CAD board with a balanced pattern.

389

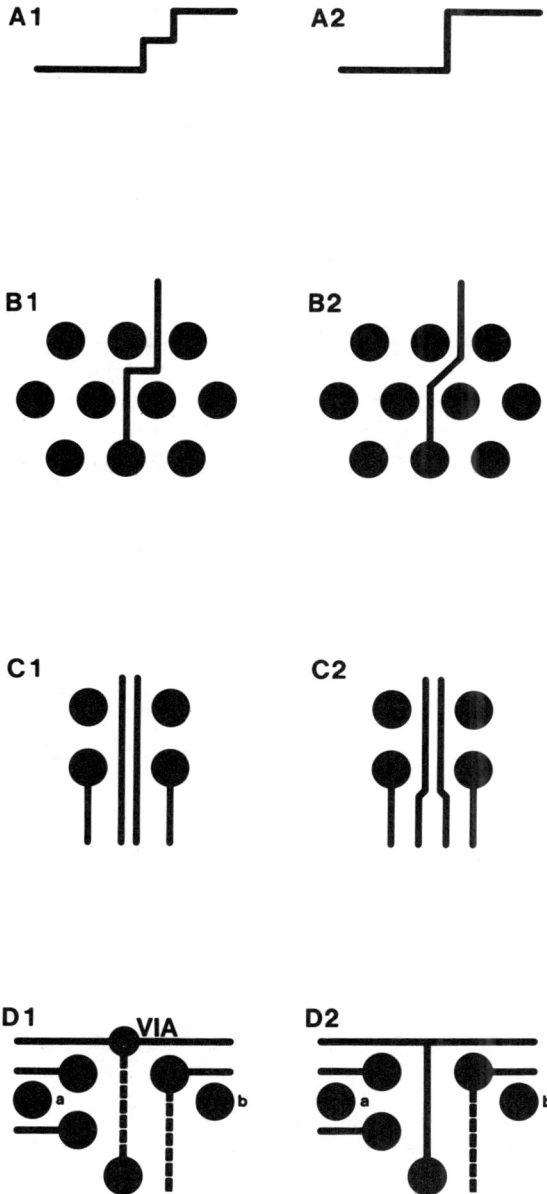

Fig. 9.59 Cleaning up the layout.
A: Straightening unwanted corners.
B: Changing 90° corners to 45° corners.
C: Spreading parallel conductors where space permits.
D: Minimizing the number of via holes

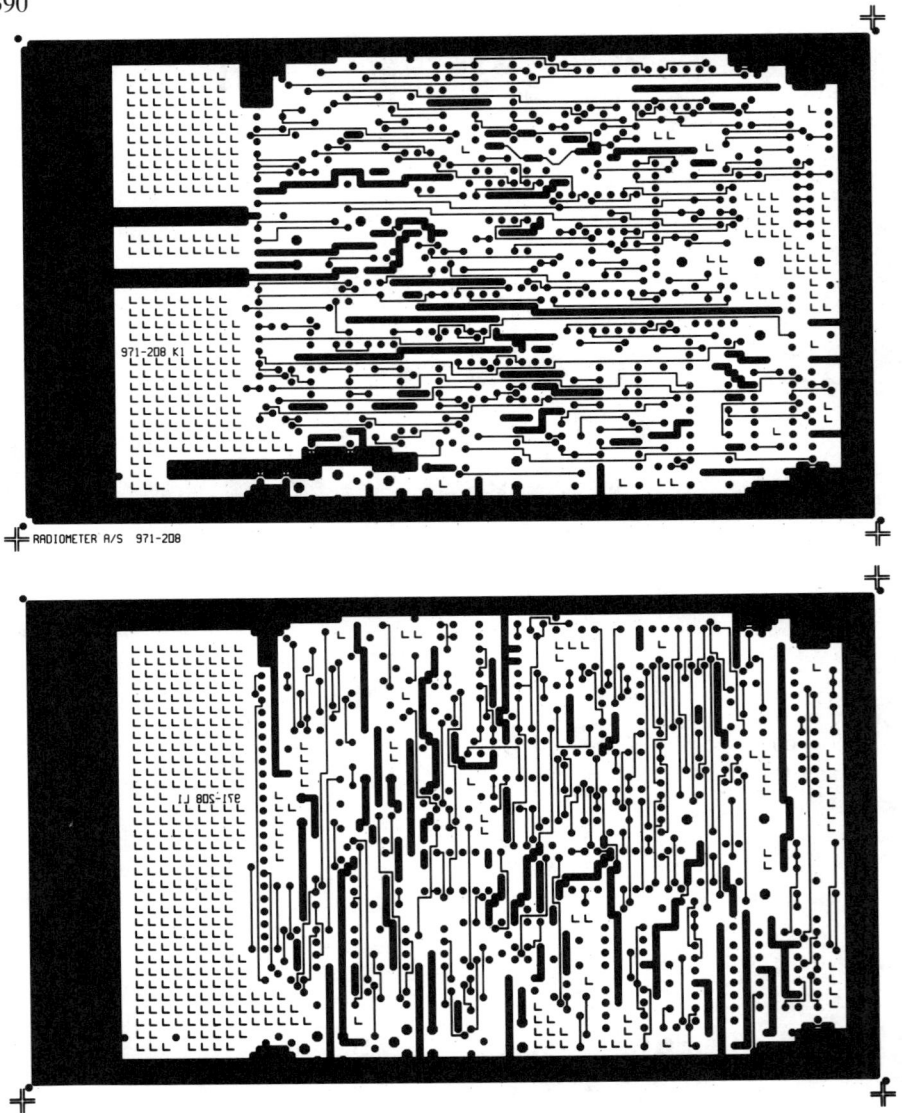

RADIOMETER A/S 971-208

Fig. 9.60 Example of balancing the pattern to achieve a reasonably even density. (Original board size: 100 × 170 mm (3.94 × 6.70 in.)

9.5.14d Checking the Layout

When all conductors have been routed, the operator checks whether the minimum spacings entered into the CAD system have really been met. It should be noted that when the routing is performed by the auto-router, the design rules are always met. Therefore, the checking focuses on conductors that were routed manually (interactively), since the operator inadvertently can violate the design rules.

The check routines examine the layout for violations of the minimum spacing between the various details of the pattern such as conductors, solder pads, via holes, voltage and ground planes, etc. It also verifies that connections stated in the wiring list have been established, and also that there are no undesired connections between independent circuits.

Error indications are displayed both as a code indicating the type of violation—for example, by a letter—and as a marking of the affected area—for example, by a flashing cross on the screen.

Based on the information obtained, the operator can now correct the faults manually and finish the layout with a second check.

9.5.15 Changes

When the layout has been checked via the automatic check routines and been approved, the display can be printed by means of either an electrostatic printer or a pen plotter. This paper printout, which incidentally can be made any time during the layout procedure, is very useful because it allows the operator to discuss layout details with the circuit designer without having to tie up the CAD system in the meanwhile.

The circuit designer can almost always suggest layout improvements while studying a printout. Details which are so obvious to the circuit designer that he has not thought of informing the operator about them, but on the other hand, are so important that they should be corrected, can easily be seen on the printout.

The circuit designer frequently hesitates to suggest numerous or extensive changes to manually taped artwork. But in the case of CAD boards, where circuits can be changed or faults can be corrected quickly, the circuit designer is more inclined to request such changes. Changes which once required hours to complete now require only minutes.

The change routines make it easy to move, add, or delete components as well as conductors. Along with the check routines, this feature is one of the most significant reasons for the high layout quality for which CAD boards are famous.

9.5.16 Post-Processing

Having been through the change phase once or several times, the operator is finally ready to release the layout for post-processing. The CAD system's post-processing software processes the data filed in the memory and converts them to paper tapes, or possibly magnetic tapes, for use in the photoplotter, and to paper tapes for use in the CNC/NC drilling machine.

As previously described, the photoplotter generates master films of the solder and component sides with an accuracy dependent on the photoplotter only. It is also possible to obtain paper tapes which can be used to generate a component notation drawing as well as solder and insulation masks. Finally, it is possible to obtain a paper tape for generating the master drawing which states the board dimensions, hole sizes, etc. The master drawing is frequently drawn by means of a pen plotter.

9.5.17 Later Design Changes

Boards which are completed and in production can easily be changed by means of the CAD system. All data regarding the board are stored on a data medium, e.g., a floppy disk, so that they can easily be loaded into the CAD system. The operator can call up the layout on the display screen, introduce all the desired changes, and generate tapes for new master films and a new drill tape.

The very short turnaround time can be measured in hours instead of days, which frequently would be the case for manually laid-out and taped artwork. A further advantage is that the corrected films are new and undamaged, whereas introducing changes and corrections in preexisting hand-taped artwork inevitably reduces the quality and results in a lower-quality master film.

9.5.18 Example of a CAD Layout

Figures 9.61A through 9.61L show the progress of the layout from contour determination to finished result. Since photographs of displays are often a little blurred and difficult to study in detail, the author has illustrated the layouts from pen plots, which clearly reproduce the images shown on the display screen. The captions contain all the descriptive text.

9.5.19 Capacity of the CAD System

In the previous description, we elaborated the theoretical grounds and the practical modes of operation of the CAD system. By doing so, we obtained an idea of the possibilities of a rational and error-free result from the various routines. The practical use of a CAD system also depends, however, on the amount of circuitry it can handle, and different brands naturally vary in this respect.

It is not expedient in this book to include a survey of available systems because the available systems change so rapidly that the data would become obsolete very quickly. Instead, the most important parameters of interest when assessing a CAD system are stated below:

Primary Design Data:
- Number of components*
- Number of connections
- Number of terminals per component
- Number of layers in multilayer boards
- Resolution and/or size of grid unit(s)
- Maximum board size for a given grid unit size

Secondary design data:
- Number of conductor widths
- Number of solder pad/solder hole sizes
- Number of text sizes

---

*Measured as the total number of discrete components and/or the number of ICs.

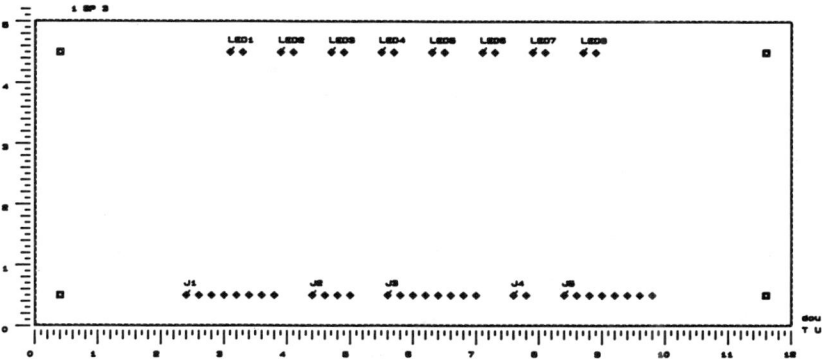

Fig. 9.61A Determining the board contour, fixed component positions (LEDs and I/O connectors), and mounting holes.

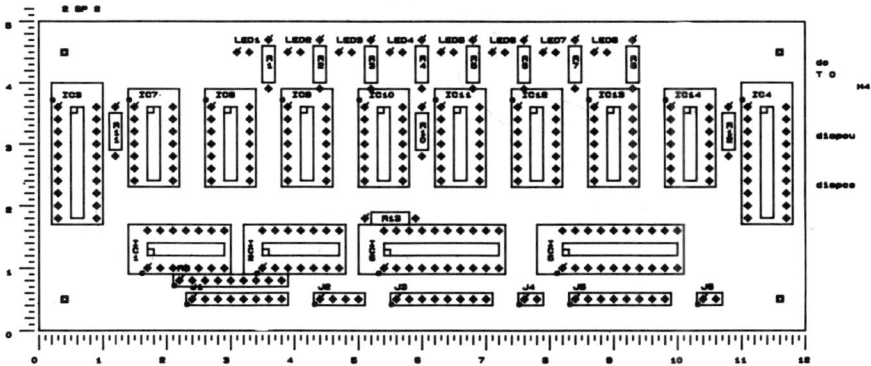

Fig. 9.61B Manual placement of components.

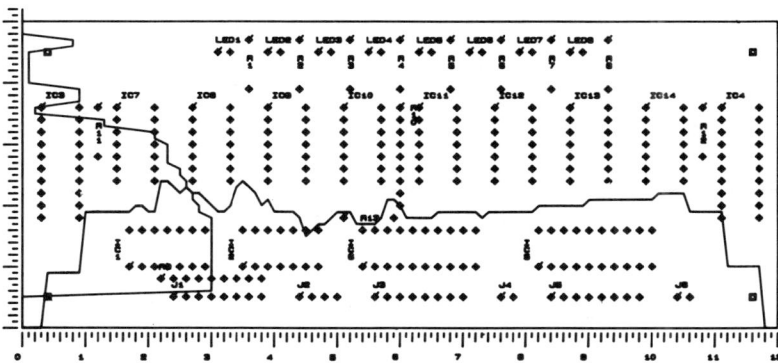

Fig. 9.61C Example of density curves. At the Y-level = 47 grid units, the curve indicates that there are 8 conductors running from the right solder pad of each LED (1 to 8) to the top of resistors R1 to R8. This appears clearly in Figure 9.61H.

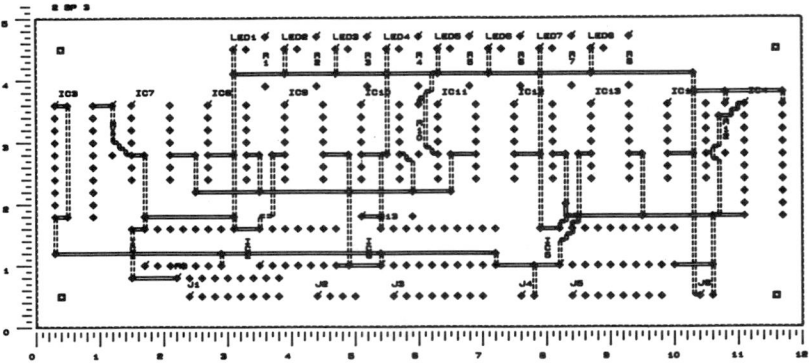

Fig. 9.61D Wiring the wide power supply conductors.

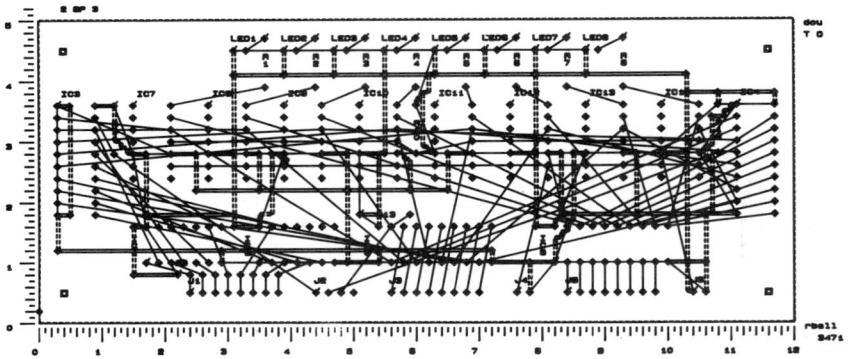

Fig. 9.61E Showing the missing conductors as rubber bands.

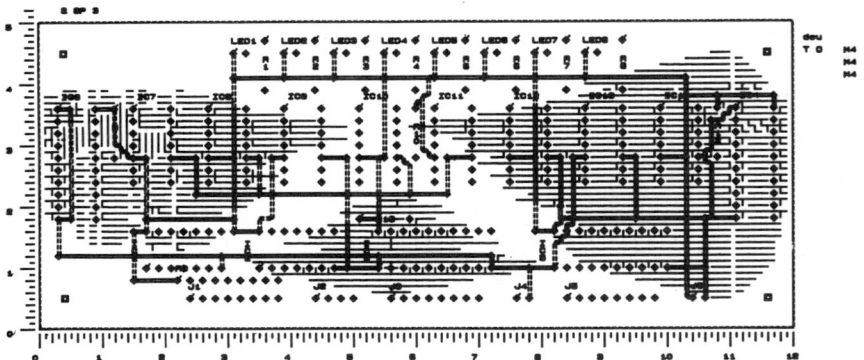

Fig. 9.61F Propagation of the search wave from three different points. The propagation is stopped artificially to simulate obstructions in the middle of the board.

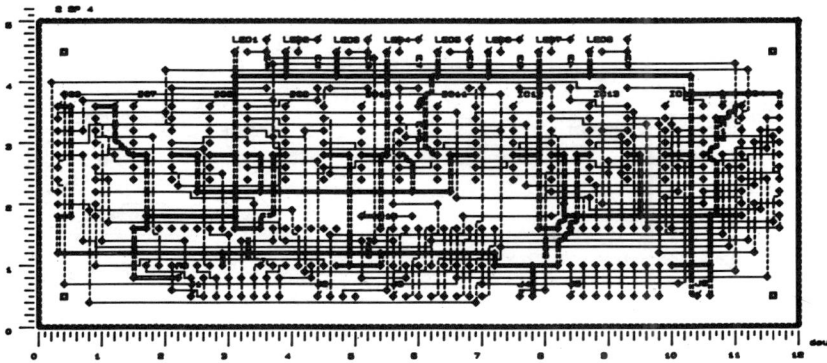

Fig. 9.61G The wiring is finished.

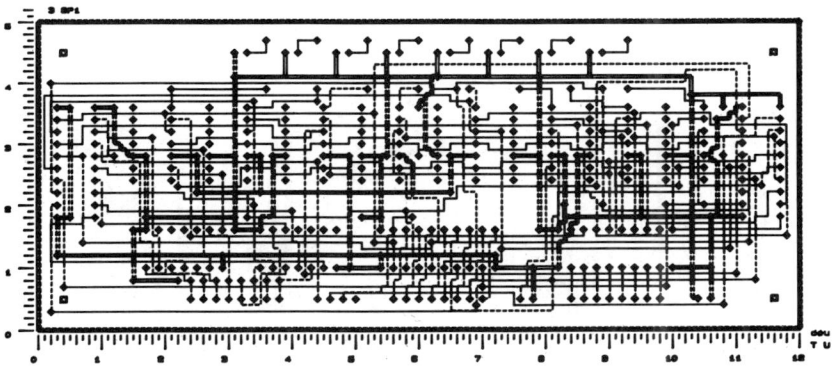

Fig. 9.61H Minimizing the number of via holes on the basis of the wiring of Figure 9.61G.

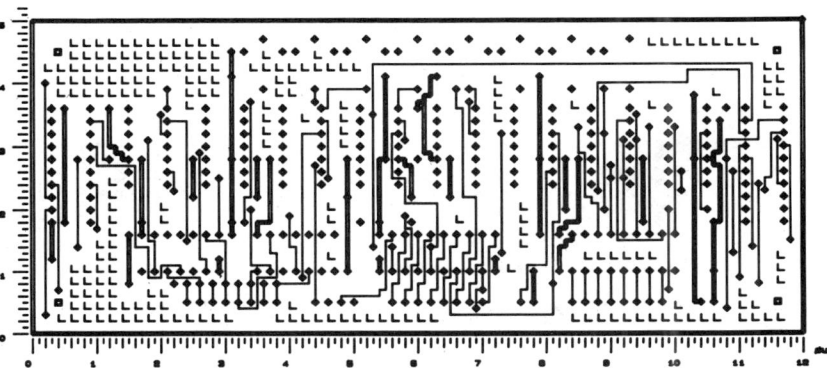

Fig. 9.61I Balancing the pattern by introducing a number of L-shaped dummy conductors.

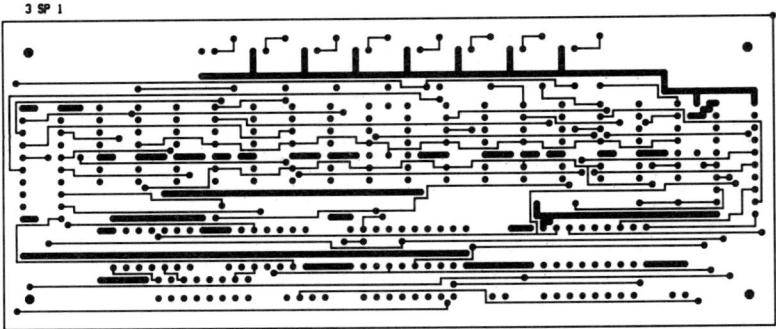

Fig. 9.61J A Gerber plot of the top side of the board. The Gerber plot is a 1:1 pen plot of the pattern as it will look when photo-plotted. The accuracy is a little lower than that of the photo-plotted filmwork, but very convenient for checking the layout before spending money on photo plotting.

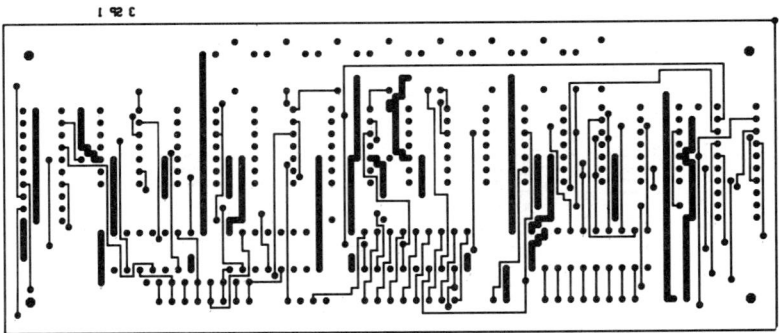

Fig. 9.61K A Gerber plot of the underside of the board. The balancing is omitted here and in Figure 9.61J.

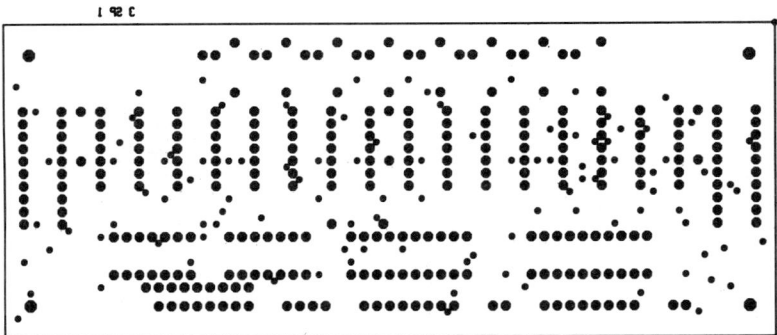

Fig. 9.61L A Gerber plot of the solder mask. The plot is in a negative execution to avoid filling the whole area.

Design Rules:
- Automatic component placement
- Automatic routing of double-sided boards
- Automatic routing of multilayer boards
- Automatic gate assignment
- Automatic gate and pin swapping
- Automatic conductor necking and bending

Design Subroutines:
- Minimization of the number of via holes
- Straightening of conductor inner corners
- Truncation of corners to 45 degrees
- Spreading of parallel conductors
- Balancing of pattern

When assessing a CAD system, one should give the most consideration to its design capacity. The primary design data decide the size and density of boards that the system can handle.

The CAD system's capacity can also be based on the following data:

- The maximum number of cells it can accommodate
- The maximum board area it can handle
- The choice of grid units

A statement can be, for example, 100,000 cells, a board area up to 250 sq. in. (1613 cm^2), and grid units of 1/20 and 1/40 in. (1.27 mm and 0.635 mm). We shall now see what this means:

With a grid unit of 1/20 in., each square inch contains $20 \times 20 = 400$ cells. The board area becomes

$$\frac{100,000}{400} = 250 \text{ sq. in. } (1613 \text{ cm}^2)$$

which agrees with the statement.

When the grid unit is changed to 1/40 in., each square inch contains $40 \times 40 = 1600$ cells, and the board area becomes

$$\frac{100,000}{1600} = 62.5 \text{ sq. in. } (403 \text{ cm}^2)$$

This is a considerably smaller board area than stated.

It should be understood that the area statement must be interpreted correctly in order to know the size of boards which can be laid out with a grid unit of 1/40 in.

Furthermore, there can be a limit to the maximum board dimension, e.g., 20 in. (508 mm). The board sizes which the CAD system can handle appear in Figure 9.62. The curve is a hyperbola characterized by $XY = A$, where X is the board length, Y is the board width, and A is the maximum area. In the preceding example, A is 250 and 62.5 sq. in., respectively. Depending on the system's grid unit chosen, the curve can be entered from the board length axis, which gives the corresponding maximum width.

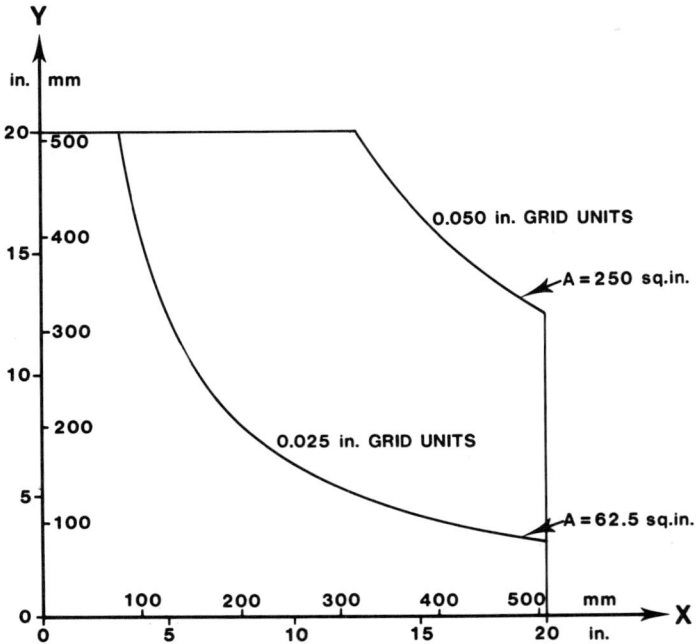

Fig. 9.62 Assessment of the largest board size a CAD system can handle as a function of the grid unit chosen, either 1/20th in. or 1/40th in.

It is also important to assess the response time of the system, i.e., the time elapsed from when a command is given until the response is displayed. If the operator, for example, wants to zoom, a waiting time of one second (perhaps two seconds at the most) is acceptable. But if the time is 15 or 20 seconds the operator loses his patience, or even worse, the ability of concentrating on the job. Furthermore, long waiting times reduce the productivity of the system.

It certainly is very difficult to assess the working speed of the system from the information given in brochures and manuals, so the author strongly recommends the system be subject to a benchmark test, preferably based on the customer's own PCB design. See also Paragraph 9.6.

9.6 Using a CAD Service Bureau

The number of CAD service bureaus established during recent years greatly exceeds the number of drafting service bureaus established within the same period. The latter bureaus offer design and plotting based on automated drafting systems. Although the automated drafting systems are being extended with more and more processing power so that they in many respects approach the CAD systems, the difference in numbers clearly indicates the superiority of the CAD system.

Quite a large number of CAD service bureaus exist. Their customers include both smaller companies that, because of the limited number of board designs per year, cannot justify buying their own CAD system, and larger companies that wish to reduce peak loads on their own CAD system(s).

When a company considers buying a CAD system, it is expedient to gain practical knowledge of this technology by letting a CAD service bureau perform some board layouts. It is usually possible to participate in this work, either as a passive onlooker or as a more or less active operator.

Many CAD service bureaus willingly sell equipment time with or without operator assistance, so that the future CAD user can operate the CAD system and learn for himself. It should be noted, however, that because even a moderate command of a CAD system requires a fairly long learning time, a comparison between the time spent for manual artwork generation versus computer-aided design should not be made too soon.

9.6.1 Documentation Package for the CAD Service Bureau

Since the design problem fundamentally is solved on the basis of the customer's schematic diagram and parts list, the efforts of the CAD service bureau can be compared with a full bureau design from a bureau based on automated drafting. Nevertheless, the bureau requires more information than just the schematic diagram and parts list, and the documentation package must therefore include all the information listed in Paragraph 9.4.1, Subparagraphs 1 through 5.

9.6.2 Output of the CAD Service Bureau

On the whole, the result is the same as in the case of a full bureau design from a drafting service bureau. See Paragraphs 9.4.1 to 9.4.3. The customer should not expect the same degree of cleanup as achieved when he performs his own layout on his own CAD system. The CAD service bureau is frequently fully booked up, and the customer may not appreciate the time expended on cleaning up the board, especially when he has to pay for it.

9.6.3 Price Considerations

The reader is again referred to the discussion, in Paragraph 9.4.4, of the price conditions valid for a service bureau for automated drafting. The tendency is to not require a price quotation beforehand, so that the billing is based on the actual operator time. Obviously, the better and clearer the documentation package is, the less operator time is required. On the other hand, the documentation should not be overly comprehensive and state data already known to professional operators; for example, that a type 7400 IC has 14 pins and that the distance between rows of pins is 0.3 in.

9.6.4 Design Changes Performed by a CAD Service Bureau

It was previously mentioned that the execution of design changes is quite simple. The turnaround time and the cost, however, complicate matters when the changes are performed by a CAD service bureau. What, in the case of hand-taped artwork, seemed to be a simple matter of a few hours' work now seems a little more complicated.

The intended changes must be described in detail, and a new documentation package prepared and sent to the CAD service bureau. The turnaround time can easily amount to some days or even more, depending on the service bureau's current workload. The change itself requires expensive computer time and plotter time. Therefore, the change becomes relatively expensive, even in the case of a small and harmless change.

In practice, it is not uncommon for the costs of changes for five or six revisions to add up to the price paid for the first issue of the board. One should therefore include the costs of changes when assessing the economic aspects of using a CAD service bureau.

10

Design of Printed Circuit Boards
for Automatic Component Insertion

10.1 Introduction

In Section 5.1 it was pointed out that printed circuit boards display a high degree of uniformity, which in turn implies that they lend themselves to automatic component insertion. This means, however, that certain design rules must be observed by the PCB designer when making the layout. Even small deviations from such rules can cause difficulties and possibly complete failure in the automatic insertion process, and the aim of this chapter is to give a brief survey of the design parameters that must be taken into consideration.

10.2 Inserting Machine

For the reader who is unfamiliar with automatic component insertion, a short description is given below. Automatic insertion is used mainly for the assembly of axial lead components, transistors, and dual-in-line ICs, these components being equally com-

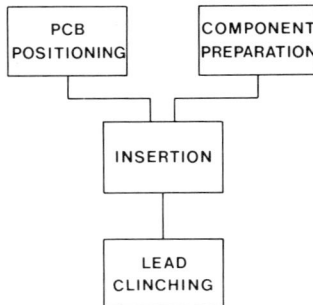

```
  ┌──────────┐      ┌───────────┐
  │   PCB    │      │ COMPONENT │
  │POSITIONING│      │PREPARATION│
  └──────────┘      └───────────┘
        │                 │
        └────────┬────────┘
            ┌──────────┐
            │INSERTION │
            └──────────┘
                 │
            ┌──────────┐
            │  LEAD    │
            │ CLINCHING│
            └──────────┘
```

Fig. 10.1 Flow chart for an automatic component-insertion system.

402

patible with the mechanical operation of the inserting machine. A flow chart for the process is shown in Fig. 10.1. The components must be prepared for insertion: Axial lead components and three-terminal transistors are supplied from lead tape reels and dual-in-line ICs from slide magazines. Usually, the axial lead components, i.e. resistors, capacitors, diodes or transistors with a special lead form, which are required, are of different types and of different electrical values and physical sizes. They must therefore be sequenced in accordance with the assembly program. This is performed by a sequencing machine operated under program control (see Fig. 10.2).

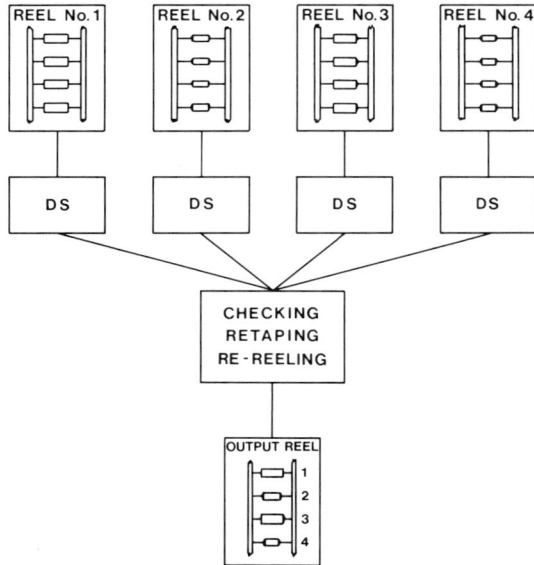

Fig. 10.2 Block diagram for an automatic sequencing machine. DS, dispensing station.

The lead taped components are fed to dispensing stations where the components are cut loose from the tapes in a sequence that minimizes the cycle time of the inserting machine. They are then transferred to a checking station for verification of their presence or absence against the program, and finally re-taped and re-reeled. Dual-in-line ICs are collected in individual slide magazines for each type, and a sequencing

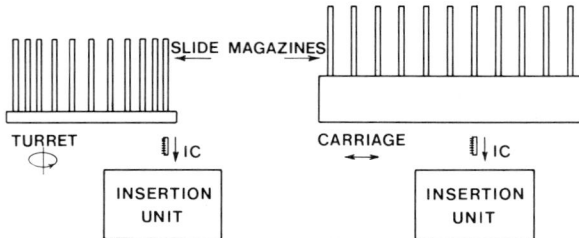

Fig. 10.3 IC sequencing and insertion machine. Left, turret type; right, carriage type.

and inserting machine can have either a turret or a carriage loaded with magazines (see Fig. 10.3). The turret is loaded in proper order for insertion, and it releases one or more ICs to the insertion unit after advancing one slide magazine position. The carriage is also loaded with slide magazines and positioned under program control to release the desired type of IC for insertion.

The circuit board is now placed beneath the insertion unit and positioned, either manually, by pantographic control or by fully automatic control (see Fig. 10.1). In the latter case, all components must lie on a grid system to permit the automatic control to be programmed. Conveniently this may be a 0.05-in (1.27-mm) grid. When the board is positioned, the automatic insertion starts. An axial lead component is cut free of the supporting tapes, gripped around the leads, formed to the desired lead span and moved downwards, the leads being guided down to the solder holes. The leads are pushed through the solder holes until the component body is seated, and then cut to a precise length and clinched so that the component is held in place during the subsequent handling (see Fig. 10.4). An IC is treated in nearly the same way: after

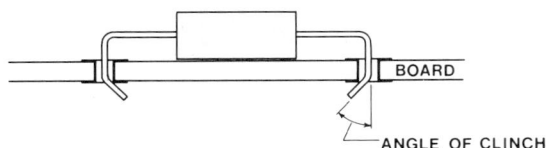

Fig. 10.4 Clinching of a component. The angle of clinch is adjustable from 0° to 90°.

automatic selection of the proper IC from the turret or the carriage, the leads are formed with a slight outward bend. The IC is then gripped and clamped between two fingers and moved downwards so that the leads enter the solder holes. When the IC is seated, the leads are cut to the desired length and clinched so that it is held in place during the subsequent handling.

The brief description above serves to give the PCB designer an idea of the working principles of automatic component inserting machines so that he is in a better position to understand the design considerations attending automatic component insertion.

Very large and complex insertion systems can be built, but except for an illustration of such a system (see Fig. 10.5) these installations will not be dealt with in this book.

10.3 Design Considerations

10.3.1 Introduction

It stands to reason that printed circuit boards intended for automatic component insertion must be designed with this in view in order to achieve satisfactory operation. This imposes some restrictions on the PCB designer, and the following sub-sections serve to draw his attention to important design parameters. The layout must, of course, be adapted to the insertion system installed in the assembly department, and although all makes have features in common in their method of operation, the recommendation given should be regarded as a guide only. The PCB designer should

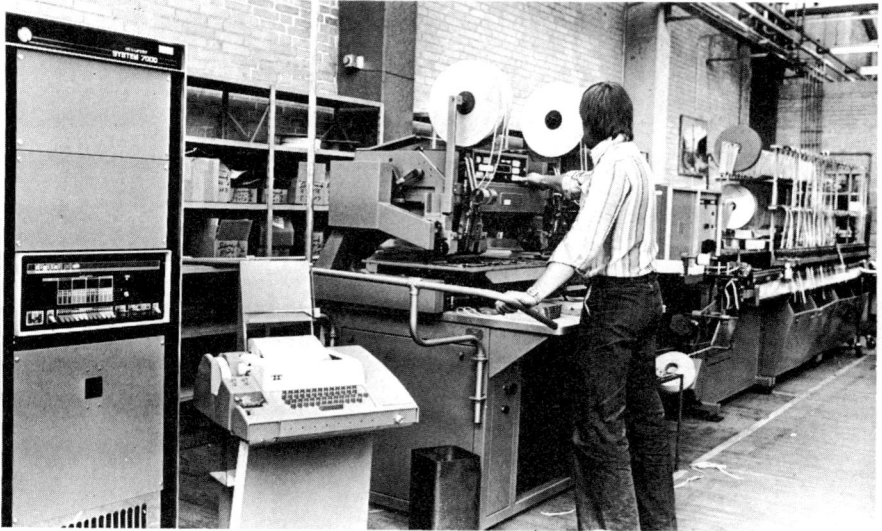

Fig. 10.5 Automatic component-inserting system used by Bang & Olufsen A/S, Denmark.

always consult the production assembly engineer regarding the specifications of the inserting machines actually used before setting up design standards and rules for his own work.

10.3.2 Board Size

Two important limitations are the maximum board size and the insertion area. These differ from make to make, but typical values are 20×19 in (508×482 mm) for the maximum board size, and 18×18 in (457×457 mm) for the insertable area. The PCB designer must, however, consider also uninsertable areas caused by the holding fixture which locates the board on the worktable. See Fig. 10.6, which shows a board placed on a holding fixture, uninsertable areas around tooling holes and supporting edges being indicated by borderlines. Such areas are caused by the operation of the cut and clinch unit which, being located beneath the board, requires a certain clearance with respect to the holding fixture. Uninsertable areas are stated below, based on typical values of R and W indicated in Fig. 10.6. The PCB designer should, however, consult the production assembly engineer to obtain information on the exact clearance. In some cases it is possible to reduce the uninsertable areas by chamfering the cut-out edge of the holding fixture.

> *Axial lead components:*
> Around a tooling hole: $R = 0.375$ in (9.5 mm)
> Along a supporting edge: $W = 0.250$ in (6.4 mm)
>
> *Dual-in-line ICs:*
> Around a tooling hole $R \times 0.500$ in (12.7 mm)
> Along a supporting edge: $W = 0.250$ in (6.4 mm)

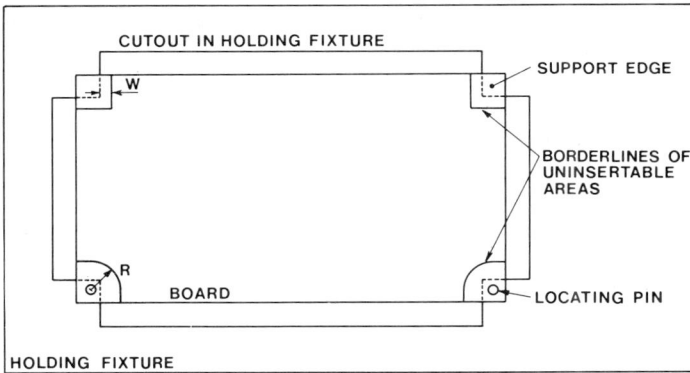

Fig. 10.6 Holding fixture with board.

10.3.3 Tooling Holes

Tooling holes are used for accurate location of the board on the holding fixture. In order to achieve a well-defined relation between the pattern and the tooling holes, the latter should be drilled or punched in the same operation as the solder holes. In view of the long tolerance chain between pattern and contour, the board contour must never be utilized for locating the board on the holding fixture. Misinsertion or damage to adjacent components could be the result of excessive displacements. It is recommended that tooling holes be placed as far as possible from each other, preferably with the centre-to-centre line parallel to the longest edge and at a distance of 5–10 mm (0.2–0.4 in) from the edge (see Fig. 10.7). This makes it a little easier to mount the board on and remove it from the holding fixture, than would be the case if the tooling holes were placed diagonally. One of the tooling holes should preferably coincide with the datum of the reference system, which in turn should fall on a grid-line

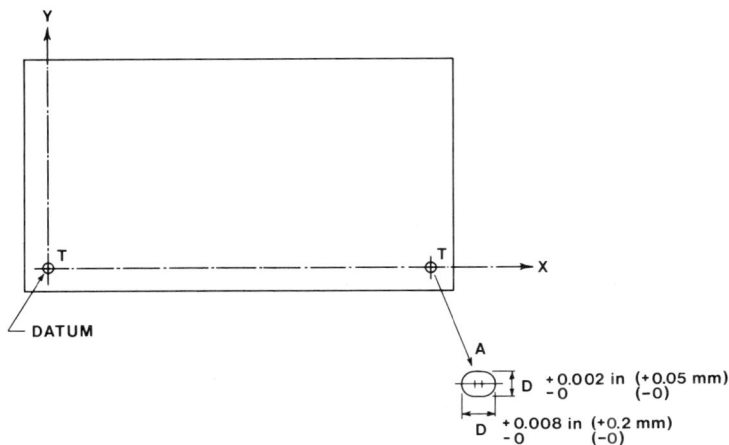

Fig. 10.7 Tooling hole locations. T, tooling hole; A, alternative execution of second tooling hole.

intersection in order to facilitate the dimensioning of the solder holes. In the case of punched boards, the second tooling hole can be made slightly oblong, which further facilitates mounting and removal of the board. (See the enlarged drawing inserted in Fig. 10.7.) If the tooling holes are drilled, the same effect can be achieved by giving the second locating pin an oblong cross-section.

The tooling holes should not be too small, and a diameter of 0.125 in is generally recommended. The tolerances on diameters of the tooling holes and the locating pins are indicated in Fig. 10.8, while the tolerances on the dimensions of oblong tooling

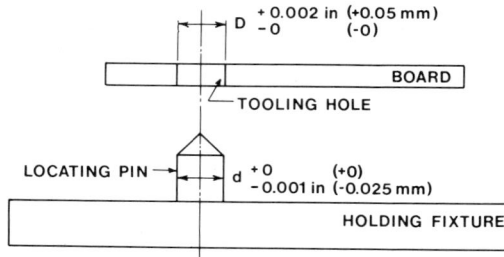

Fig. 10.8 Tolerances on tooling hole and locating pin.

holes are indicated in Fig. 10.7. Except for the length of oblong tooling holes, all tolerances must necessarily be very close as excessive play could cause a displacement of the board which might result in misinsertions. The stated tolerances permit a worst case play of 0.075 mm (0.003 in), which is usually considered acceptable.

10.3.4 Component Types and Sizes

As already mentioned, automatic inserting machines can handle axial lead components. Resistors, capacitors and diodes must be delivered by the component manufacturer as lead taped components on standard reels in compliance with either the IEA Specification No. RS296-C, 'Reel Packaging of Components with Axial Leads' and spaced on a 0.2-in or 0.375/0.400-in pitch, or the IEC Recommendation No. 286, 'Packaging of Components on Continuous Tapes' and spaced on a 5-mm or 10-mm pitch.

The maximum axial lead component size varies from make to make, but typical values are:

Maximum body length:	0.700 in (17.8 mm)
Optionally:	1.000 in (25.4 mm)
Maximum body diameter:	0.250 in (6.4 mm)
Optionally:	0.400 in (10.2 mm)
Lead diameter, copper:	0.016–0.036 in (0.4–0.9 mm)
Component span:	0.300–1.300 in (7.6–33.0 mm)
(centre-to-centre distance)	

Only seldom are transistors inserted automatically as this calls for a lead forming as shown in Fig. 10.9 with subsequent taping on conventional lead tapes. As no stan-

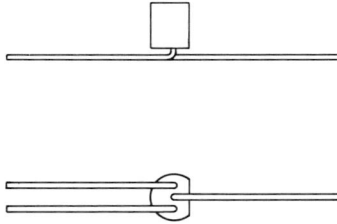

Fig. 10.9 Forming of transistor leads for automatic insertion (TO-92).

dards exist, the transistor manufacturer must be consulted before a decision is made regarding automatic insertion.

Dual-in-line ICs conforming to TO-116 configurations with 6–18 leads intermixed can be handled by most inserting machines:

Body length:	0.340–1.000 in (8.6–25.4 mm)
Body width:	0.220–0.270 in (5.6–6.9 mm)
Body thickness:	0.080–0.220 in (2.0–5.1 mm)
Lead length: (from top of body)	0.420 in (10.7 mm)

10.3.5 Layout

Careful layout of the board is of paramount importance to the success of automatic component insertion, and below some recommendations for proper layout will be given.

10.3.5a Grid-based Layout

Whenever possible, component should be located on a 0.05-in precision grid with all insertion holes lying over grid-line intersections. Not only does this facilitate the practical layout work as described in the previous chapters, it also offers some specific advantages as regards the preparatory work for automatic component insertion. In the first place, great savings can be obtained in the case of inserting machines requiring manual pantograph control. Due to the grid-based solder holes, a master template can be used instead of individual templates, the master grid template being adapted to the board by means of a cheap mask overlay. In the second place, preparation of data for numerically-controlled inserting machines is facilitated. Owing to the fact that the solder holes are grid based, the programmer can easily and accurately determine their location and enter the data in the NC program.

In the preceding chapters, it was pointed out that the layout sketch should be made on the basis of a precision grid, and that great care should be taken in the taping stage to ensure that the solder pads do actually lie concentrically on the grid-line intersections. Because they are better able to satisfy these requirements, automatic artwork/filmwork-generating methods are becoming more attractive, accuracies of ±0.04 mm (±0.0016 in) being obtainable.

10.3.5b Component Orientation

For the sake of efficient automatic insertion, all components should preferably have the same orientation as shown in Fig. 10.10. This places very rigorous constraints on

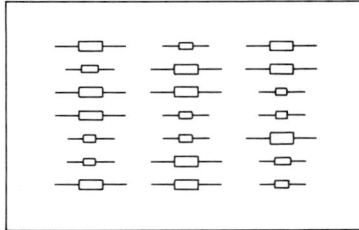

Fig. 10.10 Single-axis orientation of components.

the PCB designer and, in practice, he is often compelled to use a two-axis X–Y orientation as illustrated in Fig. 10.11. The effect is, however, a more complicated insertion process, as the board must be handled twice, i.e. components along one axis are

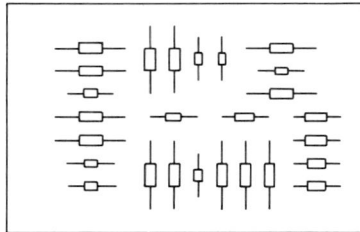

Fig. 10.11 Two-axis orientation of components.

inserted first, the board is then turned 90°, and the remaining components inserted along the other axis. Although large in-line insertion systems allow for multiple-axis orientation by providing rotatable tooling (insertion units and cut and clinch units) or rotatable holding fixtures, the PCB designer should avoid this solution as the efficiency of the automatic insertion is impaired.

10.3.5c Lead Span

A further complication lies in the variety of lead spans. In manual assembly, it is well known to standardize on as few lead spans as possible. This applies with even more force to automatic component insertion. Some inserting machines are set up with fixed tooling to handle components having identical lead spans, a range of tooling heads being available to accommodate for different spans. Obviously, the board must be handled as many times as different span sizes render it necessary. More advanced inserting machines have variable tooling which, under program control, can be adjusted to handle components with different lead spans.

The conclusion is that the PCB designer should always limit the number of different lead spans as far as possible.

The minimum span depends primarily on the length of the component body and on how close to the body the leads can be bent, as discussed in Section 5.10. Since dimension 'a' indicated in Fig. 5.10 depends, in the case of automatic inserting machines, on the width of the inside lead former of the insertion unit, the PCB designer should consult the production assembly engineer when working out standards for lead span. Regardless of the width of the inside lead former, dimension 'a' should never be less than 1.5 mm (0.060 in). The total span should always be rounded upwards to the nearest grid module, 0.050 in (1.27 mm) or 0.100 in (2.54 mm).

10.3.5d Component Spacing

Printed circuit boards of today are often characterized by a high-packaging density, which complicates automatic insertion of components as this—as opposed to manual assembly—requires a certain tooling clearance around the insertion holes. The PCB designer must therefore consider some design requirements as regards minimum tooling clearances in order to avoid damage to components already mounted. Figures 10.12–10.14 show some typical configurations for axial lead com-

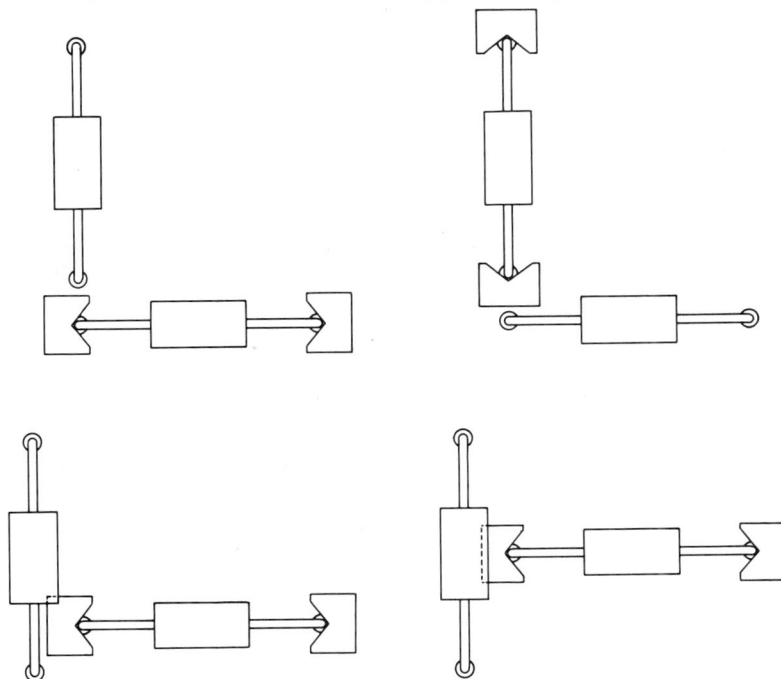

Fig. 10.12 End-to-side insertion.

410

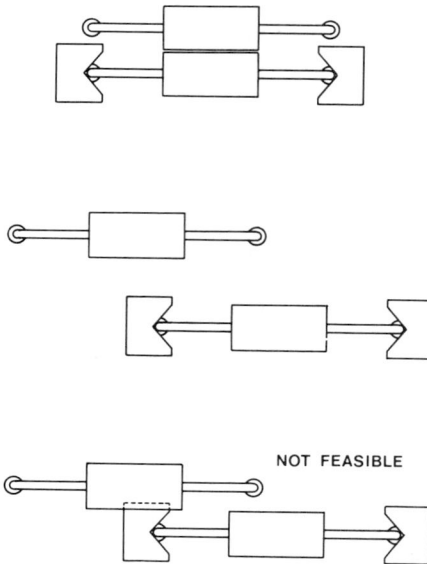

Fig. 10.13 Side-by-side insertion.

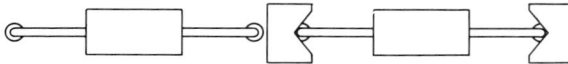

Fig. 10.14 End-to-end insertion.

ponents, the calculation of the exact allowable spacings being left to the PCB designer as this depends on the type of machine and its tooling. Directions including some values for guidance are, however, stated below:

A. Body Clearance Axial lead components can be mounted side by side with practically no body clearance, i.e. the centre-line spacing depends on the body diameters including the diameter tolerance, rounded upwards to the nearest 0.100-in (2.54-mm) module. In the case of end-to-end mounting, the spacing between adjacent insertion holes must in general not be less than 0.100 in (2.54 mm).

Dual-in-line ICs can be mounted side by side with a centre-line spacing of 0.400 in (10.16 mm), i.e. 0.100 in (2.54 mm) between the centre-lines of adjacent rows of insertion holes. A nearly zero clearance is allowed when ICs are mounted end to end, tolerances on body lengths being taken into consideration.

B. Minimum Tooling Clearance The tooling requires a certain clearance around the solder holes (insertion holes). This appears from Fig. 10.15, which shows a simplified drawing of the insertion tool. The component is nearly seated in the board, and the leads are guided downwards and through the insertion holes by the outside formers and the driver. The necessary clearance depends on the dimensions of the

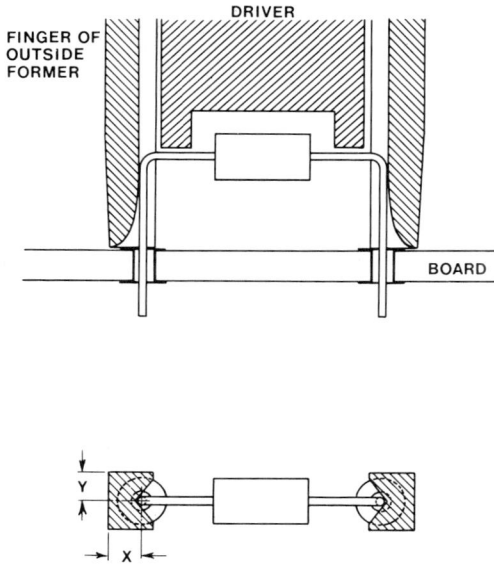

Fig. 10.15 Simplified drawing of insertion unit.
Top, front view; bottom, top view.

fingers of the outside former, which should be checked by the PCB designer before setting up design rules. Values for guidance of dimensions X and Y (Fig. 10.15) are 0.085 in (2.2 mm) and 0.060 in (1.5 mm), respectively, X depending slightly on the lead diameter.

Having determined the exact values of X and Y, it is fairly easy to determine how closely a component can be inserted without damaging adjacent components.

Note: The actual spacing must, of course, also take the body clearances into account. The spacing is then rounded upwards to the nearest 0.100-in (2.54-mm) module.

C. Insertion Sequence   Obviously, the insertion sequence is important in the case of a high-packaging density, if interference between adjacent components is to be avoided. This is illustrated in Fig. 10.12 where the configurations shown at the bottom require that the components drawn horizontally are inserted first. It should be noted, however, that the configuration shown at the bottom of Fig. 10.13 cannot be realized. Regardless of which component is inserted first, insertion of the second component will interfere with the first, the only feasible solution being an increase of the spacing or a horizontal displacement.

10.3.5e Diameter of Solder Hole

Owing to the long tolerance chain from artwork or filmwork to inserting machine, the oversize of the solder hole (insertion hole) must be larger than when the printed circuit board is manually assembled. A rule of thumb indicates an oversize of 0.25 mm (0.010 in) with respect to the lead diameter. In order to ensure absolutely reliable insertion, a further clearance of 0.1 mm (0.004 in) is sometimes added, especially in the case of ICs with 16 leads to introduce simultaneously, where worst-case conditions for insertion are approached. In the author's opinion, the said oversize is valid only for photo-plotted artworks, manually-taped artworks requiring a further 0.1 mm (0.004 in) increase of the oversize if absolutely reliable insertion is to be expected. This is discussed in detail in Section 10.3.6. As a rule, axial lead components do not present any design problems as regards the finished hole diameter, and thereby the outer pad diameter, owing to the high design freedom. The difficulties are usually found in connection with ICs, as the pad configuration is predetermined. In Section 6.5.4a we found the diagonal dimension of the lead cross-section, which in practice has a maximum value of 0.59 mm (0.023 in). In accordance with the above rule of thumb, the diameter of a plated through hole becomes 0.94 mm (0.037 in) which often is rounded up to 1.0 mm (0.040 in). In that case, this value applies for photo plotting and for manual taping.

The hole diameter found is in conflict with the normally required 0.3-mm (0.012-in) wide annular rings, when the outer diameter of the solder pad is limited to 1.4 mm (0.055 in), or even to 1.27 mm (0.050 in) as discussed in Chapter 9. It was shown in Section 6.5.5 that it is not practicable to increase the diameter of circular pads or the width of oblong pads beyond 1.4 mm (0.055 in) if it still is to be possible to route a 0.30-mm (0.012-in) wide conductor transversely through an IC package. Some PCB designers have accepted the full consequences of this and use solder pads with virtually no annular rings (see Fig. 10.16).

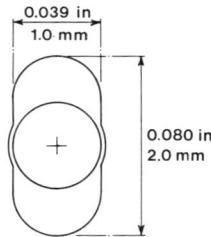

Fig. 10.16 Solder
pad with no annular
ring.

10.3.5f Grouping of Components

In Section 10.3.5c it was concluded that the PCB designer should always limit the number of different component spans as far as possible. He must, however, also take care that components are grouped in accordance with their span as this increases the efficiency of the inserting machine by reducing the travel of the worktable. This is illustrated in Fig. 10.17 which shows three different insertion paths.

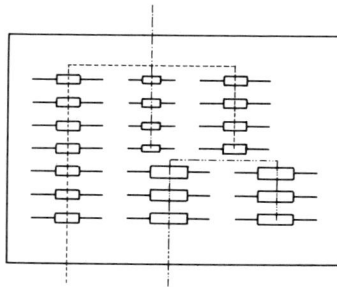

Fig. 10.17 Insertion paths. Grouping of components in order to optimize insertion paths.

10.3.6 Tolerance Considerations

In Section 10.3.5e, a certain oversize of the insertion hole was indicated. In order to gain a fuller understanding of the tolerance chain, we shall now assess the various contributions to the total tolerance. The calculation of the oversize necessary to ensure proper insertion is based on a determination of the radial deviation between a component lead and the matching insertion hole, allowances for diameter deviations being included:

Radial Deviations:

This is illustrated in Fig. 10.18, the same diameter of lead and hole being assumed. As the deviation a can occur in any direction, it is evident that proper insertion requires an increase of the hole diameter by $2a$.

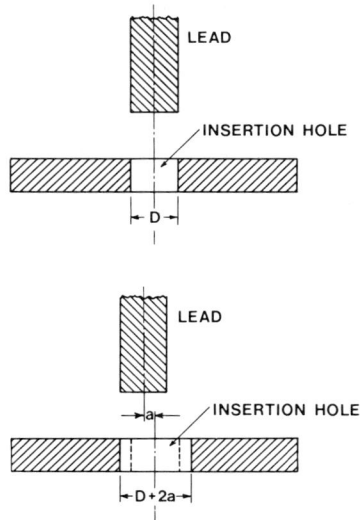

Fig. 10.18 Radial deviations. Cross-section of board and component lead.

Diameter Deviations:

If the diameter of the lead is increased by *b*, the diameter of the insertion hole must be increased by the same amount.

Calculation of Oversize of Insertion Hole:

When the tolerance chain comprises a large number of deviations arising from the various processes to which the board has been subjected ever since artwork preparation and manufacture up to the moment of component insertion, it is not reasonable just to add the contributions arithmetically. Instead we use the formula

$$\Delta D = \sqrt{(2x_1)^2 + (2x_2)^2 + \ldots + y_1^2 + y_2^2 + \ldots}$$

where ΔD indicates the oversize, $2x_1, 2x_2, \ldots$ indicate radial deviations, and $y_1, y_2, \ldots$ indicate diameter deviations as defined above.

Another interesting point is the comparative suitability of manual artwork preparation and photo plotting, when automatic component insertion is to be utilized.

When the master film is derived from a taped artwork, and the taping is done on a 4 : 1 scale, the pads will deviate more or less from the intersections of the grid lines, and the grid itself will be inaccurate, especially if aged.

Positional accuracy of pads (see Section 6.4.1a)	±0.15 mm (±0.006 in)
Accuracy of grid (estimated) (see Section 8.2.3)	±0.075 mm (±0.003 in)
Positional accuracy of pads, referred to a 1:1 scale (approximate)	±0.05 mm (±0.002 in)
Tolerance on photo reduction	±0.05 mm (±0.002 in)
Total positional accuracy of pads (approximate)	±0.10 mm (±0.004 in)

The drill tape is derived by digitizing the pad positions, whereas the control tape for the inserting machine is based on true grid positions, i.e. the full deviation will appear when the components are being inserted automatically.

In the case of photo plotting, the specified plotting accuracy of ±0.04 mm (±0.0016 in) will not affect the insertion. The drill tape used by the PCB manufacturer and the control tape for the component inserting machine are derived from the same data base, so if no other sources of error existed, all component leads should match their respective insertion holes. There are, however, other sources of error, and a complete survey of the individual contributions is provided below.

Contributed error

a1 Positional accuracy of pads of master film
(Manually taped artwork)
Total positional accuracy: ±0.1 mm (±0.004 in)   0.20 mm   (0.008 in)

a2	Deviation between drill tape and control tape (Automated draughting/CAD)	0	0
b	Positional tolerance on drilled holes: Board length: $\leqslant$200 mm: $\pm$0.08 mm ($\pm$0.003 in)	0.16 mm	(0.006 in)
c	Positional tolerance on tooling holes: $\pm$0.08 mm ($\pm$0.003 in)	0.16 mm	(0.006 in)
d	Maximum play between tooling hole and locating pin, see Fig. 10.8	0.075 mm	(0.003 in)
e	Positional tolerance on locating pin of holding fixture: $\pm$0.025 mm ($\pm$0.001 in)	0.05 mm	(0.002 in)
f1	Holding fixture positioning: $\pm$0.025 mm ($\pm$0.001 in)	0.05 mm	(0.002 in)
f2	Lead forming and positioning: $\pm$0.025 mm ($\pm$0.001 in)	0.05 mm	(0.002 in)
g	Tolerance on lead diameter: $\pm$0.025 mm ($\pm$0.001 in)	0.025 mm	(0.001 in)
h1	Tolerance on diameter of drilled or punched insertion hole: $\pm$0.05 mm ($\pm$0.002 in)	0.05 mm	(0.002 in)
h2	Tolerance on diameter of plated through insertion holes: $\pm$0.10 mm ($\pm$0.004 in)	0.10 mm	(0.004 in)

We shall now determine the oversize ΔD of the insertion hole for the following four cases:

Manually taped artwork, drilled or punched holes:

$$\Delta D = \sqrt{a_1^2 + b^2 + c^2 + d^2 + e^2 + f_1^2 + f_2^2 + g^2 + h_1^2}$$
$$\Delta D = 0.33 \text{ mm } (0.013 \text{ in})$$

Manually taped artwork, plated through holes:

$$\Delta D = \sqrt{a_1^2 + b^2 + c^2 + d^2 + e^2 + f_1^2 + f_2^2 + g^2 + h_2^2}$$
$$\Delta D = 0.34 \text{ mm } (0.013 \text{ in})$$

Automated draughting/CAD, drilled or punched holes:

$$\Delta D = \sqrt{a_2^2 + b^2 + c^2 + d^2 + e^2 + f_1^2 + f_2^2 + g^2 + h_1^2}$$
$$\Delta D = 0.26 \text{ mm } (0.010 \text{ in})$$

Automated draughting/CAD, plated through holes:

$$\Delta D = \sqrt{a_2^2 + b^2 + c^2 + d^2 + e^2 + f_1^2 + f_2^2 + g^2 + h_2^2}$$

$$\Delta D = 0.27 \text{ mm } (0.011 \text{ in})$$

The above figures do not include the additional diameter increase which is often used in the case of multiple-lead components such as ICs.

A surprising result of the investigation is the large insertion holes necessary for setting off the unavoidable inaccuracies caused by manual taping if completely reliable automatic component insertion is to be achieved. It is easily understood that this particular assembly method calls for the inherent advantages of automatic draughting or computer-aided design.

11

Design of Boards
for Automatic Testing

11.1 Introduction

The use of automatic test equipment (ATE) has become quite common. The two main reasons are the ever-increasing board complexity and the growing problem of finding qualified personnel to perform the testing and troubleshooting. An automatic test equipment as shown in Fig. 11.1 is primarily intended to test assembled boards. The objective is to ensure that components which might have been damaged during assembly, including mass soldering, are identified so that they can be replaced before the board reaches the functional test stage.

The test equipment usually contains a guard system which serves to electrically isolate any components from all the other components on the board. Therefore, it is possible to ascertain whether the component value is within the required tolerance. The components are tested in sequence according to a test program written for the particular type of board. The designations of possible defective components are automatically printed on paper, so that such components can be identified and replaced.

In exactly the same way, the ATE can test the pattern itself for short circuits and interruptions, permitting a lower inspection level during the incoming inspection. It should be noted, however, that near-shorts and near-opens cannot be detected by the ATE.

The earlier in the production process a defective component or PCB can be found, the cheaper and easier it is to replace. This is particularly true if a component failure in normal service would damage other components, requiring that the additional defective components also be found and replaced.

The fact that automatic test equipment will undoubtedly increase in use and importance is good reason to design the boards so that they can be tested automatically, without delay or significant complications.

Experience shows furthermore that the actual test time can be halved when the board is designed with proper consideration for the special conditions of automatic testing. Since the PCB designer greatly influences the success of the tests, we shall therefore consider some of the problems occurring in practice and state various design rules.

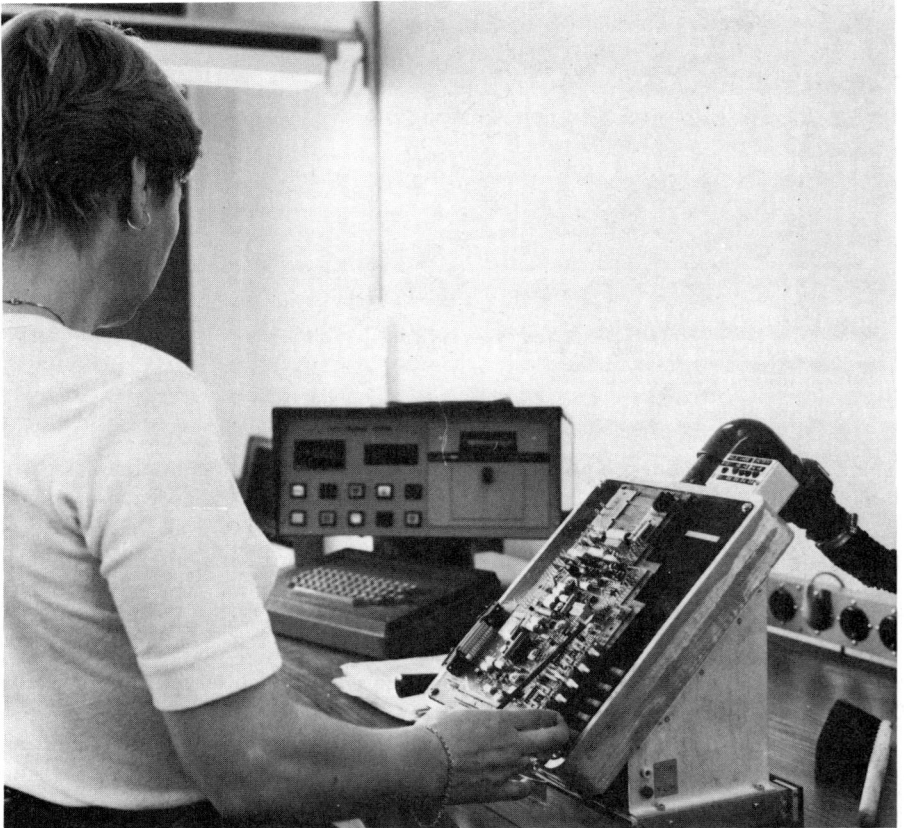

Fig. 11.1 Example of automatic test equipment. The "bed of nails" fitted with a board is at the right.

11.2 Test Equipment

An automatic test equipment will usually be based on one of the two designs described below.

11.2.1 ATE with Movable Base Plate

The test equipment and the assembled board are connected via a bed of nails, on which the board is placed. See Figs. 11.2 and 11.3A. A base plate carries a number of fixed and movable test contact probes. The movable probes are controlled by a fixed guide plate. By raising the base plate, the ATE moves the contact probes upwards to make contact with selected test points.

The contact pressure itself is determined by a small spring inside the test contact probe. The base plate can be raised in several different ways: mechanically,

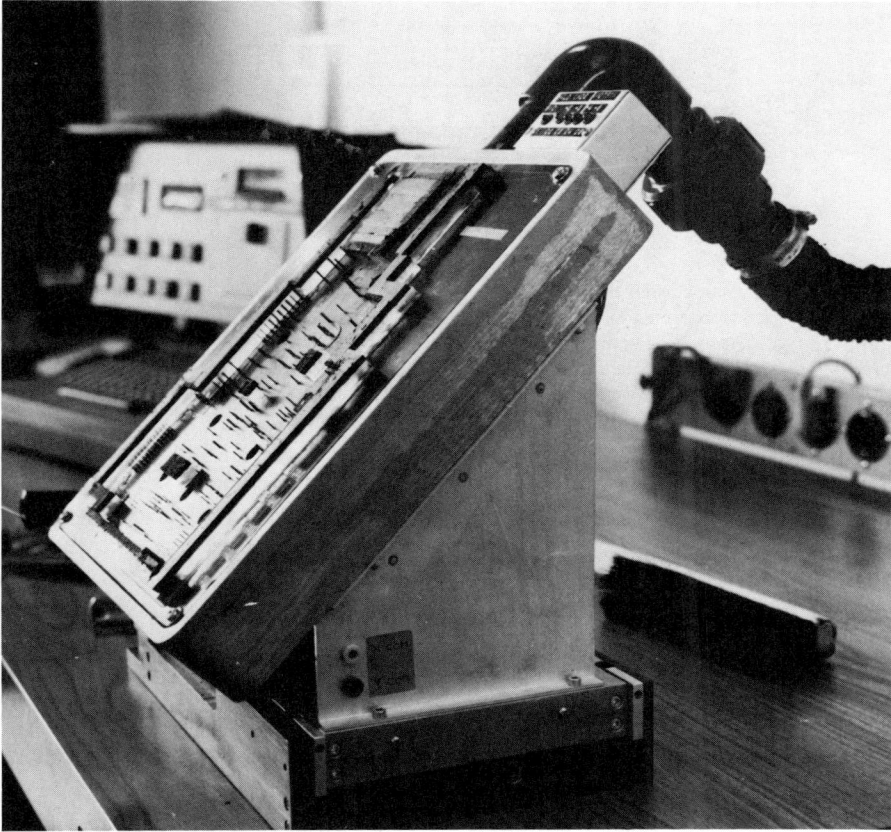

Fig. 11.2 Close-up of "bed of nails". The movable base plate, cf. Fig. 11.3A, has been blocked in its topmost position so that the test contact probes are visible.

pneumatically, or with a vacuum, the latter being most common today. Fig. 11.3A shows how the plate is raised by a vacuum pump's evacuating the chamber between the guide plate and the base plate.

In order to build up and maintain the vacuum, all leakages must be reasonably small. The board is therefore sealed by a gasket along the board edges. Cutouts and larger holes in the board must be sealed by suitably shaped gaskets glued to the guide plate. For bare-board testing, a plastic sheet placed on the top side of the board can seal the smaller holes, but cutouts and larger holes must still be sealed by gaskets.

The board must, of course, be oriented correctly with respect to the test contact probes. The simplest way is to use the board's tooling holes and to mount guide pins at the right places in the guide plate.

420

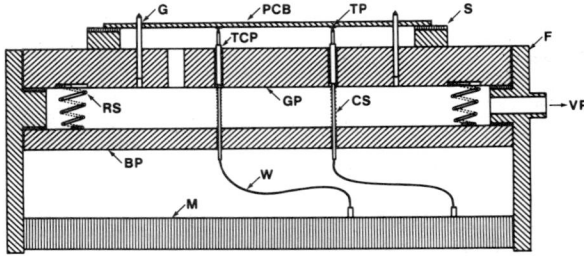

Fig. 11.3A Cross-sectional drawing of older type of "bed of nails." The movable base plate is shown in its uppermost position. PCB = printed circuit board under test. TP = test pad. G = guide pin. S = sealing along board edge. F = frame. TCP = test contact probe. GP = guide plate. RS = retainer spring. CS = contact spring. VP = vacuum pump. EP = movable base plate. W = wiring. M = matrix for connection to test equipment.

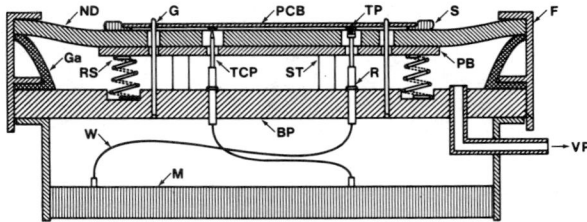

Fig. 11.3B Cross-sectional drawing of a "bed of nails" of comparatively recent design. The "bed of nails" is shown in an actuated condition. PCB = printed circuit board under test. TP = test pad. G = guide pin. S = sealing along board edge. F = frame. ND = neoprene diaphragm. PB = phenolic paper board. Ga = gasket. RS = retainer spring. TCP = test contact probe. R = receptacle. ST = stop. BP = fixed base plate. VP = vacuum pump. W = wiring. M = matrix for connection to test equipment.

11.2.2 ATE with Fixed Base Plate

In a newer model of the fixture, the test contact probes are stationary as shown in Fig. 11.3B because of their mounting in the fixed base plate. The fixed guide plate shown in Fig. 11.3A has been replaced by a neoprene diaphragm, and the center part made rigid by a phenolic paper board glued to the under side of the diaphragm. Since the phenolic paper board is somewhat smaller than the diaphragm, the latter can move downward when the vacuum actuates the fixture (bed of nails). Because of the vacuum, the board under test follows the diaphragm so that the probe tips hit the test pads, suitable clearance holes being provided. The board is sealed and positioned as described above.

The newer design has several advantages: The vacuum is higher because the fixed base plate does not cause sealing problems as in the older equipment. The test contact probes are inserted into wired receptacles mounted in the base plate, so that they are readily removed for cleaning or replacement. Finally, the wiring from the receptacles is not subject to breakage because it does not move when the fixture is actuated.

11.2.3 Contact Problems

The system appears to be very simple, but in practice contact problems can be more or less detrimental to the test, particularly in the case of assembled boards. Solder pads are commonly used as test points because this practice requires no changes to the board.

When the component lead soldered to the pad is a little too long, and perhaps also is bent outwards slightly, the probe tip may be deflected so much that it misses the pad as shown at A, Fig. 11.4.

The PCB designer should note that in dense boards the nominal width of the annular ring can be as little as 0.012 in. (0.30 mm), or even less. Furthermore, if the drilling is eccentric, the actual width of the annular ring can be as little as 0.002 to 0.004 in. (0.05 to 0.10 mm) on one of the sides. It is therefore possible that the contact is made between the end of the component lead and the side of the probe tip, and not between the pad and the end of the probe tip. For several reasons, the former type of contact is very undesirable: The contact pressure can be much too low, the lead end can be oxidized where it has been trimmed, or the flux residue on the lead end can be so thick that electrical contact is impossible.

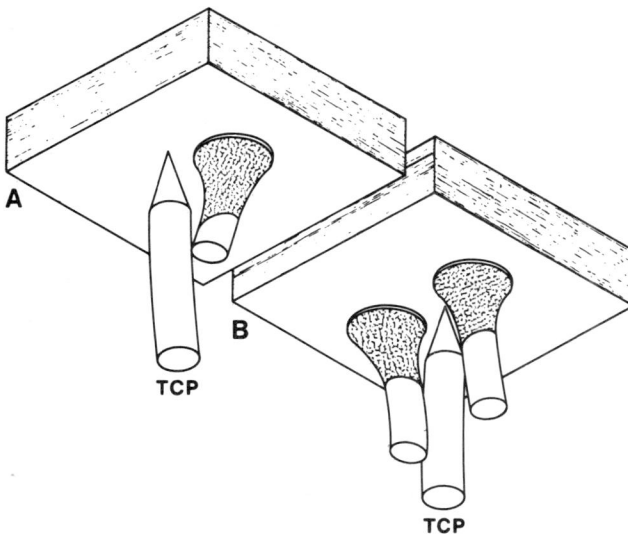

Fig. 11.4 Deflection of test contact probe (TCP).
    A: Probe tip deflected because of long and bent component lead.
    B: Probe tip short-circuits two long and bent component leads.

The result is frequently either an intermittent contact or no contact at all, and the printout will therefore indicate an error.

Another type of fault is the probe tip's short-circuiting two component leads. See B, Fig. 11.4. There can be two reasons for this type of fault: The component leads may have been cut a little too long and bent towards each other so that the probe tip touches both leads, or the probe tip itself can be so wide that a short circuit is inevitable. The latter fault occurs most frequently when using probes with crown tips or similar wide tips.

11.3 Obtaining a Good Contact

The erroneous measurements described in Paragraph 11.2.3 are, of course, very annoying to the operator. He can possibly improve the situation by vacuum-actuating the bed of nails a few times in succession to "massage" the board, but he risks contact failure at other places. The situtation can easily become so unbearable that it demoralizes the operator.

11.3.1 Probe Tip Designs

Several different versions of probe tips are now available, each type having its own advantages and disadvantages. See Fig. 11.5. The tendency to deflect the probe tip can be reduced by using a crown-shaped probe tip, which is more easily captured by the lead end. See B, Fig. 11.5. Practice shows, however, that in time the crown tip fills with dust, flux residues, etc., which can cause contact problems. The same problems can also result from the lead end's oxidization or contamination with flux residues. Newer types of test contact probes are installed in receptacles so that they can easily be removed for cleaning or replacement without rewiring.

Finally, it should be noted that there is a difference in price between the various types of test contact probes shown in Fig. 11.5.

11.3.2 Contact Design for Automatic Testing

Because of the aforementioned conditions, it seems expedient to mention another aspect of the philosophy about printed circuit boards as carriers and interconnecting devices for electronic components: A board should be designed to permit automatic testing without any contact problems. The reader should realize that effective measures in that respect mean quite a lot of extra work for the PCB designer and draftsman. On the other hand, this is only a one-time cost that is very quickly recovered because of the problem-free test procedure.

The above point of view is particularly valid in the case of boards with surface-mounted components, i.e., chip carriers and chip components, which frequently are located on the board's solder side. Because of the small dimensions of the solder pads, it usually is impossible for the probe tip to hit the pad. It is inadvisable to let the probe tip hit the terminal area of the component, partly because the

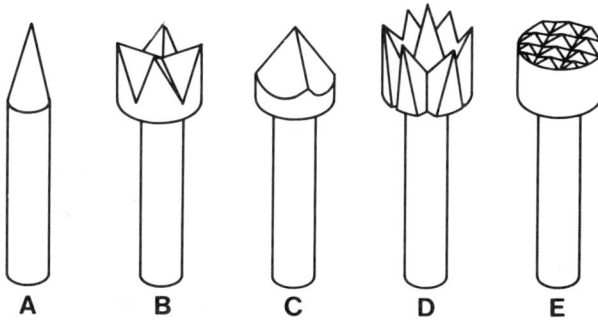

Fig. 11.5 Various types of probe tips.
  A: Pointed tip.
  B: Crown tip.
  C: Chisel tip.
  D: Tulip tip.
  E: Serrated tip.

component can be damaged, and partly because the contact can be hampered by a layer of insulating lacquer. Finally, it should be realized that an interruption caused by a cracked solder joint cannot always be detected because the terminal is pressed against the pad by the probe tip during the test.

A more reliable contact can be achieved by the introduction of special test pads connected to selected test points. Such test pads are distinguished from the usual solder pads by having just one function, namely to serve as contact points for the probe tips. See Fig. 11.6. Since the test pads are not drilled, they do not increase the manufacturing costs. A section of a board provided with such special pads is shown in Fig. 11.7

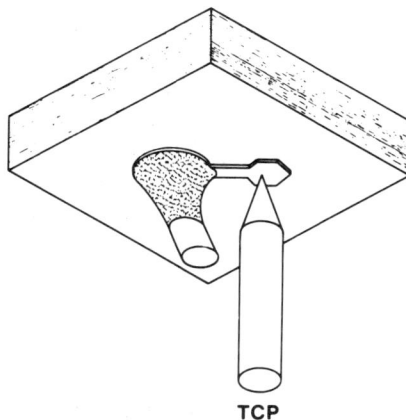

TCP

Fig. 11.6 Using a special test pad prevents the problems illustrated in Fig. 11.4.

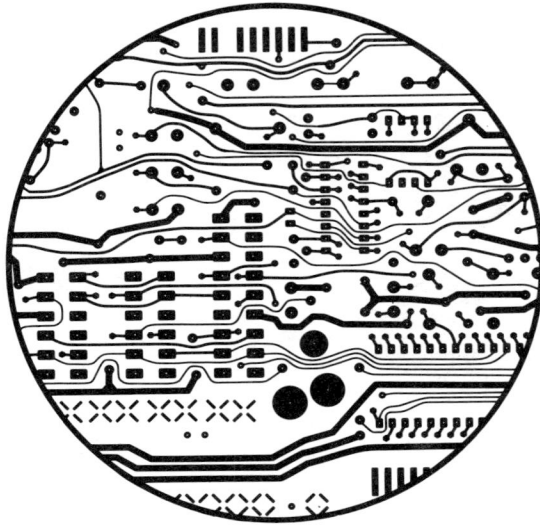

Fig. 11.7 Section of board provided with test pads.

The use of test pads implies an immediate advantage by defining the various test points very precisely. For the sake of the later drilling of the base plate and the guide plate/diaphragm of the bed of nails, the test pads are usually digitized so that an NC drilling machine can perform the drilling.

An important advantage of using special test pads is that changes in the circuitry do not necessarily result in changes to the drilling pattern of the base and guide plate/diaphragm. Even when some of the components are moved a little, the test pads can in many cases remain untouched.

11.3.3 Use of Conductors as Contact Points

Another idea could be to use a short section of a reasonably wide conductor as a contact point. If so, determination of the exact position implies that the master film of the solder mask is laid over the master film on the solder side, since the contact point is defined only by the aperture in the solder mask. This is not a particularly practical procedure, and is impossible when the board has been designed with no solder mask. In such cases, the contact point must be determined in another way, e.g., by marking on an additional film of the solder side. To avoid mistakes during future changes, this film should be part of the documentation package in exactly the same way as the set of artwork, i.e., the film must be defined by its issue level.

11.4 Establishing Test Points

Most ATE (automatic test equipment) systems use a special guarding technique which makes it possible to measure a component soldered in place on the board, without having to interrupt conductors or to desolder the component. It is also possible to test the PCB pattern itself for interruptions and short-circuits, even if all components have been soldered in place.

The guarding technique implies the use of various methods of connecting the equipment, depending on the type of component and the adjacent circuitry. The technique of programming the test equipment, however, requires such extensive knowledge of the equipment's modes of operation that the subject is outside the PCB designer's field of activity. Below we will therefore only elaborate on establishing the necessary test points in the form of special test pads, and state various design rules.

In the following discussion, certain theoretical design rules for establishing test points are given. In the extreme, a very large number of test points can be necessary, which implies both much work during the layout phase and a risk of side effects, e.g., short circuits during mass soldering. Finally, the component density can be so high that there is not room for all the test points. Limitations in the number of test points is therefore given in the design rules of Paragraphs 11.4.2b and 11.4.3d.

It is highly recommended that the PCB designer collaborate with the test engineer regarding which test points to choose, preferably based on the schematic diagram. The test engineer marks on the schematic diagram the necessary test points, which in the layout phase are treated in the same way as all other components.

In the case of computer-aided design, the designer should indicate, by suitable numbering, the test points and component pins which are to be connected, whereupon the test points are introduced automatically. It is important to select a routine which ensures that a test pad does not occur on both sides of the board, i.e. be treated as ordinary solder pads by the CAD system. To facilitate the later drilling of the base plate and guide plate/diaphragm used in the bed of nails, it is beneficial to derive a special drill tape.

11.4.1 General Considerations

The general considerations regarding the size and physical location of the test pads on the board are dealt with below, whereas the actual establishment of test points in the form of test pads are addressed in Paragraphs 11.4.2 and 11.4.3.

11.4.1a Shape and Size of Test Pads

In the nature of the case, all test pads are placed on the solder side of the board. Since this side frequently is used when the PCB manufacturer digitizes the board, there is a certain risk of mistaking the test pads for the usual solder pads. This could cause an unnecessary and unwanted drilling of the test pads. The risk can be reduced considerably by giving the test pads a different shape than the normal solder pads.

The test pads must, of course, be large enough so that the probe tips can hit them reliably. On the other hand, they must also be so small that their presence does not increase the risk of solder bridging. This could happen if a test pad has been placed so close to a conductor that the solder mask aperture around the test pad reaches the conductor edge.

The minimum size of the test pads depends on the hit accuracy of the bed of nails, particularly when test contact probes with pointed tips are used. The hit accuracy depends upon a number of factors, which can be related both to the board itself and to the mechanical design and execution of the bed of nails.

The first category comprises a possible permanent dimensional change of the board and the positional accuracy of the board when placed in the bed of nails, including the location and diametral tolerance of the board's tooling holes.

The other category comprises the positional accuracy of the test contact probes, including the drilling accuracy of the fixture's base plate, the positional accuracy of the guide pins, and the play in the test contact probes themselves. Since it is difficult for the PCB designer to survey these conditions, he should procure this information from the test engineer.

All things considered, a hit accuracy corresponding to $+0.008$ to $+0.016$ in. ($+0.2$ to $+0.4$ mm) can normally be expected.

Incidentally, the hit accuracy should be considered in conjunction with the minimum spacing of the board pattern. When the minimum spacing is 0.010 in. (0.25 mm), and a conductor passes by a pad at this distance, a crown tip having a diameter of 0.054 in. (1.4 mm) can easily cause a short circuit, particularly if the solder mask inadequately covers the conductor. On the other hand, a pointed tip will hit within the test pad, provided the size of the test pad is a little more than twice the accuracy of the hit.

These considerations lead to the following compromise regarding the shape and size of the test pads:

Manual Taping: Hexagonal test pads have the dimensions shown in Fig. 11.8. Most suppliers of drafting aids stock only hexagonal pads with a center hole, but this is unimportant when the draftsman closes the hole with tape. The hole diameter is usually 0.01 in. (0.25 mm) in a 1:1 scale.

Computer-Aided Design: Here the problem of choosing the size and shape of the test pads is greater, since many photoplotters use a symbol disc with a relatively modest selection of pads. Since the cost of modifying a symbol disc is usually very high, the designer has to accept one of the existing pads. The pads will usually be circular, and the diameter should be reasonably small, e.g., 0.04 in. (1.0 mm) or 0.05 in. (1.27 mm).

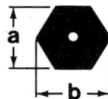

Fig. 11.8 Hexagonal test pad. a = 0.050 in. (1.27 mm). b = 0.058 in. (1.47 mm).

Quite another problem is that the CAD equipment must be able to place test pads on the solder side of the board only, and not on both sides as usual. Finally, the usual routine for checking the spacing (insulation distance) shall also check the test pads.

11.4.1b Number of Test Points

The total number of test points, and with that the total number of test contact probes, must be adapted to the force available for actuating the test fixture (bed of nails). When the test fixture is vacuum-actuated, the theoretical force is 14 lb/sq. in. (1 kg/cm^2), corresponding to a complete vacuum. A certain leakage is always present in the vacuum pump and the test fixture, and also in the board because of open holes, e.g., via holes not soldered due to partial blockage by solder mask ink. Add to this a number of retainer springs, cf. Figs. 11.3A and 11.3B.

When determining the maximum number of test contact probes, the PCB designer should assume no more than 50 percent of the theoretical force; i.e., approximately 7 lb/sq. in. (0.5 kg/cm^2). In this way, he also allows for the fact that the test fixture must be actuated rapidly to help the probe tips to penetrate a possible layer of oxide and flux.

Test contact probes are available with many different spring resistances. It should be noted that the spring resistance varies with the travel, e.g., from 1 oz. (approximately 30 g) in the rest position to 4 oz. (approximately 120 g) at full travel, which, for example, implies a travel of 0.15 in. (approximately 4 mm).

A rule of thumb states that a reasonably reliable contact can be achieved at a contact pressure of approximately 3½ oz. (100 g), so no more than 32 test contact probes per square inch (5 test contact probes per cm^2) should be used. This number is an average value valid for the total board area, and the PCB designer can therefore easily calculate the maximum number of test contact probes allowable.

Large concentrations of test contact probes, e.g., in the center area of the board, can cause an undesirable warp of the board. Particularly in the case of surface-mounted leadless chip components, warpage can cause the solder joints to crack. The test engineer, however, can compensate for an uneven distribution of the test contact probes by means of the retainer springs previously mentioned.

11.4.1c Location of Test Pads

It is most practical to place the test pads according to the grid, and because of the physical size of the test contacts, the test pads should not be spaced closer than 0.1 in. (2.54 mm). Since common types of test contact probes have a tip diameter of 0.054 in. (1.4 mm) or 0.062 in. (1.6 mm), spacings closer than 0.1 in. (2.54 mm) are inadvisable. A still smaller type of test contact probe has a diameter of 0.031 in. (0.8 mm), so that a center-to-center distance of 0.05 in. (1.27 mm) seems possible. Nevertheless, the risk of short circuits between the test contact probes should not be underestimated.

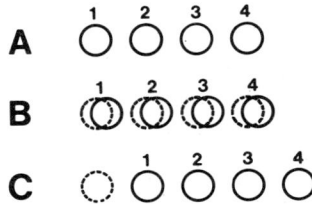

Fig. 11.9 Drilling of the base plate should always be taken into consideration when moving test pads.

  A: Original positions of test pads.
  B: Nonmodular shifting of pad positions makes drilling of base plate difficult because of converging holes. Preferred method of moving test pads requires no redrilling of original holes in base plate.

The test pads should be placed at a reasonable distance from the board edges, particularly in the case of a vacuum-actuated bed of nails. For the seal to ensure the vacuum, the distance between the board edge and a test pad should not be less than 0.12 to 0.20 in. (3 to 5 mm). For the same reasons, all test pads shoud be kept at a distance of 0.12 to 0.20 in. (3 to 5 mm) from larger internal holes and cutouts.

To protect the seal, no solder holes with soldered component leads should occur in a zone at least 0.12 in. (3 mm), preferably 0.20 in. (5 mm) along the board edges or along larger internal holes and cutouts.

11.4.1d Moving of Test Pads

In the event of board changes, the test pads should, as far as possible, remain at the same place to avoid the inconvenience of modifying the base plate and guide plate/diaphragm of the test fixture. It can, however, become necessary to move a test pad. In such cases, the test pad should preferably be moved at least 1 module = 0.1 in. (2.54 mm) in order to avoid the new hole's overlapping the rim of the old one. Should it, in spite of everything, be necessary to move the hole less than 0.1 in. (2.54 mm), the old hole must be filled with epoxy before the new one is drilled. When the test pads are placed in a row, a general shifting of the entire row along the center line and assigning new functions to the test pads is recommended in order to retain as many test pads as possible in the original location. See Fig. 11.9.

11.4.1e Guiding the Board in the Bed of Nails

The board must be guided with respect to the bed of nails so that the test contacts can hit the test pads. The best way of achieving the guiding action is to use two guide holes located diagonally. Such holes must, of course, observe the above directions concerning a distance of 0.12 to 0.20 in. (3 to 5 mm) from the board edge. If the board already has tooling holes, it will be advantageous to use these as guide holes.

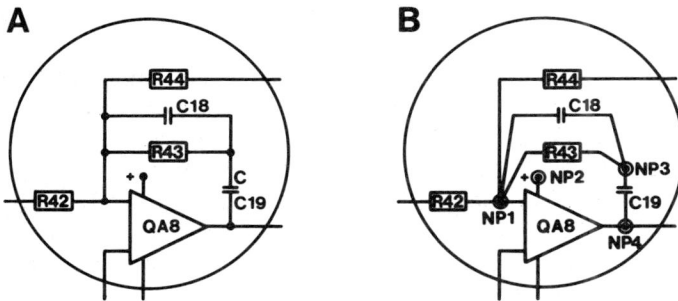

Fig. 11.10 Test points for testing components.
    A: Section of schematic diagram.
    B: Redrawn schematic diagram with indication of nodal points (NP).

11.4.1f Board Size

The bed of nails must, of course, be of a size to accommodate the board. Beds of nails can be delivered in many sizes, depending upon the vendor, so it is difficult to state more than a few examples of available sizes, which are understood as the maximum board size:

Small bed of nails: $8\frac{3}{4} \times 14\frac{1}{2}$ in. (222×368 mm)
Large bed of nails: 15×20 in. (381×508 mm)

11.4.2 Test Points for Testing the Components

To auto-test a printed circuit board, the ATE must establish electrical contact to the terminals of each component. Since the components will necessarily be connected with each other in various configurations, they have a number of points in common.

In the schematic diagram, a common point is frequently represented by a conductor line interconnecting one side of a number of components. This point is usually called a nodal point. See A, Fig. 11.10, which shows a section of a schematic diagram. B shows a redrawn diagram which elucidates the nodal points. Any component can now be made accessible for testing by a suitable choice of nodal points. In the board itself, the nodal point occurs as a conductor which connects a number of pads with each other.

11.4.2a Connection of Test Points

The most practical procedure is for the PCB designer or the test engineer to indicate the above nodal points on the schematic diagram, and to treat the points as a type of component to be placed on the solder side of the board during the layout phase.

In accordance with the definition of a nodal point, basically only one test point, in the form of a test pad, should be connected to the conductor that represents the nodal point, and consequently not to the individual conductor sections.

It must be realized that it can be physically impossible to make room for all test pads when the board pattern is very dense. The test equipment can also have a limited capacity that restricts the number of test points, e.g., to 1000. In such cases it is necessary to reduce the number of test pads. Since the PCB designer usually is not responsible for the test program, he has to collaborate with the test engineer, who can divide the test points into two or more catagories, e.g., absoutely necessary test points and desirable but dispensable test points.

11.4.2b Design Rules

Deciding where to place the test points physically can be difficult. The design rules stated below serve to aid the PCB designer:

1. A test pad is connected to each absolutely necessary test point (nodal point). Fundamentally, the connection location is of minor importance, but when it is desirable to perform a test for interrupted conductors, cf. Paragraph 11.4.3b, the test pad should be connected to the extreme end of the conductor.
2. When the space available permits, a test pad is connected to every desirable but dispensable test point (nodal point). See Item 1 above regarding the location.
3. In the case of a power bus, two test pads should be connected. The reason is that a common terminal for both power and test signals can give misleading results during the test.
4. A test pad should be connected to functional but unused IC terminals, e.g., a gate output. A short-circuited and unused gate can adversely affect other parts of the IC. It is therefore recommended to include such unused components in the test.

11.4.3 Test Points for Testing the Pattern

As mentioned earlier, it is also possible to test the pattern itself for short circuits and interruptions. Below we will address reasonable criteria for assigning test pads.

The production department must understand that the introduction of test points in itself does not increase the board reliability, but serves exclusively to facilitate the testing of the board for pattern faults. Therefore, it does not seem reasonable to put pressure on the PCB designer to introduce test access to all individual parts of the pattern. Faults in the form of short circuits and interruptions will affect the circuitry, so such faults will always be recognized and corrected even if no test pads have been introduced. This contrasts with the proper component test where tolerances and polarities play a decisive part. Here a conventional test access to at least the most important nodal points is of far greater consequence.

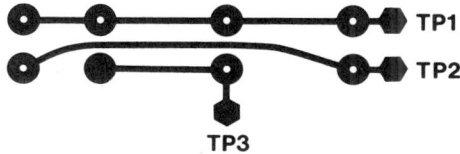

Fig. 11.11 Test for short-circuits between conductors. TP = test point.

11.4.3a Short Circuits

Short circuits between the individual parts of the pattern can always occur, either as etching faults or as solder bridges. Since the incoming inspection of the boards usually is based on a sampling system, the possibility always exists that a few defective boards, e.g., with etching faults, slip through the inspection so that they are not found before reaching automatic testing.

Mass soldering can cause solder bridging on boards with no solder mask, and even boards with solder masks can exhibit a few solder bridges. This happens particularly when the spacing between the conductors and the pads is so small that it prevents a perfect screen printing of the solder mask, i.e., with full coverage of the conductor edges.

Special programming of the test equipment makes it possible, based on resistance measurements, to determine between which conductors the short circuits occur. In principle, the test is carried out by successive measurements between one conductor and all other conductors which, during the measurement, are connected to ground. Conductors exhibiting short circuits are indicated by a printout of their number so that they can be identified and the fault corrected.

As shown in Fig. 11.11, it is not necessary to connect more than one test point to each conductor in order to perform the test. The test points are therefore identical with the test points for components.

When there is a large distance between a conductor and the remaining parts of the pattern, the probability of short circuits in the form of etching faults and solder bridging is practically zero. In such cases, there is no need for testing for short circuits, but for the sake of component testing, a test pad should nevertheless be connected to the conductor (at the nodal point). When programming the test equipment, it is better to include such conductors, partly because an omission requires the operator to assess the total conductor configuration, and partly because no appreciable time is saved during the test.

It should be noted that the test cannot reveal near-shorts as shown in Fig. 11.12. The resistance of even a very small gap, e.g., 0.002 in. (0.05 mm) is so high that nothing unusual will be found.

11.4.3b Interruptions

Whether a conductor is interrupted or not can be checked simply by measuring the resistance between its two ends. A test point at each end as shown in Fig.

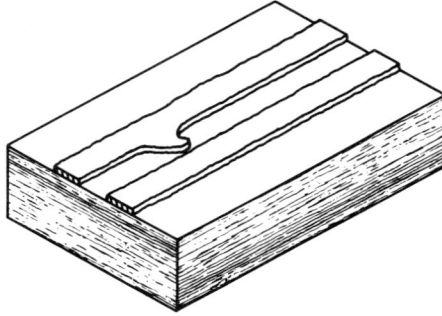

Fig. 11.12 "Near-short" between two conductors because of unwanted projection.

11.13 is therefore necessary. In this way, all partial conductor lengths including possible via holes are checked. The resistance will normally be in the milliohm range, i.e., practically zero. An interruption will not always result in an infinitely high resistance value, e.g., if the components have been connected as shown in Fig. 11.14.

By performing successive resistance measurements, the test equipment can localize interrupted conductors and print out the conductor number. It should be noted that it is also possible to check the conductors for interruptions by using the results of the component test, cf. Paragraph 11.4.2. A printout indicating an open component can in several cases be traced back to an interrupted conductor.

In practice it is, however, difficult for the operator to decide whether it is a component fault or a pattern fault. The result is often that the operator chooses to exchange the component, particularly in such cases where the interruption of the conductor cannot be seen immediately. It can be a nearly invisible crack, or the fault can be hidden under a component. For this reason, it is considered safer to establish direct test access to the conductors, which requires a test pad at each conductor end.

It can, however, be quite an extensive job to place test pads that allow testing of every single conductor for interruptions. In addition, the test pads inevitably increase the pattern density and therefore the risk of solder bridging.

Interrupted conductors are usually caused by mishaps during the pattern imaging: In the case of bare-copper boards, voids in the etch resist can allow the copper to be etched, and in the case of plated-through boards, the plating resist perhaps was not removed completely, so that the tin/lead layer protecting the conductors against etching contained voids. Other reasons can be the overlooking of minor faults in the master film or in the dry film resist, e.g., an unfortunately

Fig. 11.13 Test for interruption of conductor. TP = test point.

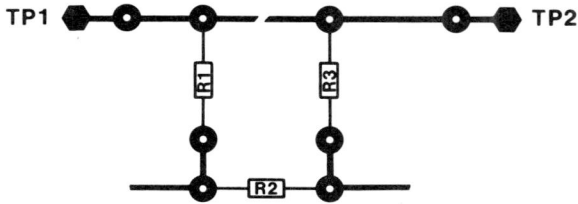

Fig. 11.14 Effect of interrupted conductor. Measured resistance value is $R_1 + R_2 + R_3$.

located pinhole in the emulsion layer or the photosensitive layer. Also scratches in the plated tin/lead layer, which acts as an etching resist, can cause an etching of a short length of a conductor.

The largest risk of interrupted conductors occurs therefore at very narrow conductors, whereas wide conductors exhibit a nick along the edge or perhaps a pinhole at the middle, which is of no importance. In the same way, long conductors are more exposed to interruptions than are very short conductors. The conclusion is therefore that it is superfluous to test wide conductors, regardless of their length. Similarly, it is unnecessary to test narrow conductors which are also very short.

The previous two statements are not quite true when the conductors shift from one side of the board to the other by means of plated-through via holes. To save space, such via holes are frequently covered with the solder mask so that they are not filled with solder after mass soldering. When the copper plating is not sufficiently ductile, i.e., elastic, cf. Paragraph 3.5.2, the thermal shock caused by reflowing and mass soldering can, under unfavourable conditions, cause barrel and/or corner cracking. The more via holes a conductor comprises, the higher the risk of interruptions. In principle, all such conductors should therefore be tested for interruptions. However, when no complaints about interrupted via holes have been received from the production department or from the field, the plating quality achieved by the PCB manufacturer in question can be considered satisfactory. A general test of all via holes is therefore superfluous.

The reader's attention is called to the growing tendency to use smaller and smaller via holes (which could be called "mini via holes") with a drilled hole diameter of 0.012 to 0.016 in. (0.30 to 0.40 mm) or even smaller. In such holes the thickness of the copper plating will frequently be somewhat less than the specified thickness of 0.001 in. (25 μm) valid for ordinary 0.031 in. (0.8 mm) holes, possibly only 0.0006 to 0.0008 in. (15 to 20 μm). The plating thickness can vary a good deal depending on the type of plating bath, the plating current, and the vat geometry. Furthermore, there is a certain tendency to land the conductor directly at the rim of the via hole without using any annular ring, for which reason corner cracking is a significant risk.

In the author's opinion, the reliability of such mini via holes is slightly less than that of the more conventional via holes. Once more it is emphasized that the introduction of test points does not improve the reliability of the board in this respect, but only serves to facilitate the testing.

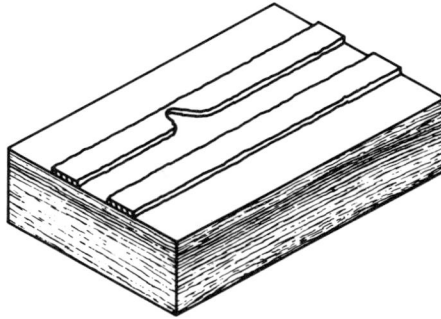

Fig. 11.15 "Nearly-open" conductor because of unwanted indentation.

The conditions are a little different in the case of ordinary solder holes since the component leads, because of the capillary action, improve the filling of the holes and make them stronger.

It should be noticed that the test cannot disclose near-opens as shown in Fig. 11.15, because the conductor resistance does not increase perceptibly when there is a pinhole or a nick at the edge.

11.4.3c Connection of Test Pads

The PCB designer must realize that the introduction of test pads for a complete test of the pattern for interrupted conductors can require considerable board space. A test of the pattern for short-circuited conductors/solder pads is far less demanding since, on the whole, the necessary test pads coincide with the component nodal points.

In the case of bare-board testing, either at the PCB manufacturer or at the incoming goods inspection, introducing additional test pads to test for interrupted conductors and via holes is unnecessary. The disadvantages mentioned in Paragraph 11.3 do not occur, so the ordinary solder pads can be used with no contact problems.

In the absence of bare-board testing, the PCB designer usually must compromise regarding the extent to which he prepares the pattern for testing for interrupted conductors and via holes. Design rules are therefore stated in Paragraph 11.4.3d.

11.4.3d Design Rules

Below we shall deal with some design rules which can aid the PCB designer when deciding to what extent test pads for testing the pattern for short circuits and interruptions should be introduced.

Test for Short Circuits:

1. A test pad should be connected to one end of each conductor, regardless of its length and width. The test points coincide with the component test points.
2. When the component test points have been divided into absolutely necessary test points and desirable but dispensable test points, cf. Paragraph 11.4.2b, it is enough to introduce test pads of the former catagory only. Special conditions, e.g., two or more conductors running between a pair of IC pads, may justify additional test pads.

Test for Interruptions:

3. An additional test pad is connected to the other end of narrow and long conductors (width < 0.012 in. (0.30 mm) and length > 8 in. (200 mm)).
4. Narrow conductors (width < 0.012 in. (0.30 mm)), which run under IC packages or other difficult-to-remove components, are treated as described in Item 3.
5. Conductors changing side by means of mini via holes (drilled hole diameter < 0.016 in. (0.40 mm)), are treated as described above in Item 3, but when experience shows such holes to be reliable, the requirement can be withdrawn.

When a conductor branches as shown in Fig. 11.16, it is necessary to connect a test pad at the end of each branch so that the test for interruptions can be carried out. Since branching can occur on both sides of the board, the PCB designer or draftsman must provide access to both sides, preferably during the layout stage.

11.4.3e Special Considerations when Placing Test Pads

All test pads must necessarily be placed on the solder side of the board. Because conductors, on the whole, act as connection lines between the solder pads, con-

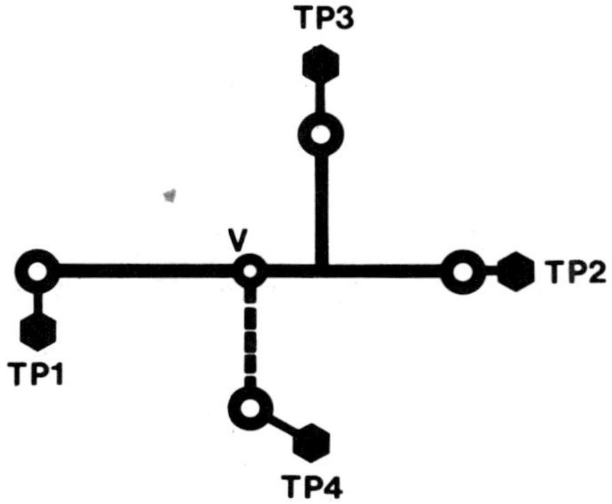

Fig. 11.16 Conductor with branches. Dotted line = conductor on component side. V = via hole. TP = test point.

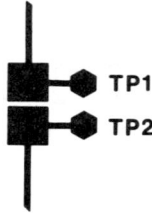

Fig. 11.17 Circuitry dividing points. TP = test points.

ductors on the component side are accessible from the solder side through the plated-through holes. In the case of a double-sided, nonplated-through board, there will be either a component lead, a piece of wire, or an eyelet soldered to both sides of the board. It will therefore always be possible to achieve contact with conductors on the component side by connecting the test pads to the solder pads on the solder side. Certain cases, however, require special precautions:

1. Conductors connected to edge connector contacts on the component side: If the conductor is to be tested for interruption, a via hole should be connected to the edge connector contact as close to the contact as possible. The via hole is then used as a test pad.

    In some types of test fixtures it is, however, possible to establish test access to certain points on the top side of the board, e.g., an edge connector.

2. Circuitry dividing points: It is frequently desirable to test the condition of a circuitry dividing point, particularly when the connection is established by soldering (as a desired solder bridge) between two adjacent solder areas. Such areas are often designed as square pads with a spacing of 0.01 in. (0.25 mm). See Fig. 11.17.

    A fully efficient test must be performed directly across the dividing points, to which test pads are connected. This can easily be established on the solder side, whereas the component side can mean a small complication. Insofar as the pattern of the solder side allows, the dividing points are provided with plated-through holes, i.e., matching solder pads are placed on the solder side and used as test pads. The plated-through holes should, however, not be so large that the solder suction can prevent the intended solder bridges. A hole diameter of 0.025 in. (0.6 mm), possibly also 0.031 in. (0.8 mm) is acceptable.

    Other types of circuitry separation, e.g., a switch or a soldered strap, imply solder pads on the solder side. The condition of the dividing point can be checked in the same way as described above, test pads being connected to the solder pads on the solder side.

3. Guard or shielding conductors: These conductors start from solder pads or from a ground area and dead-end. When such a conductor is to be tested for interruptions, it is necessary to connect a test pad to the dead-end. If

the conductor is located on the component side, the test pad is replaced by a via hole, which is used as a test pad on the solder side.

Note: Via holes should not be covered with the solder mask when used as test points. For correct drilling of the base plate and the guide plate/diaphragm of the bed of nails, such via holes, as well as contact fingers used as test points on the solder side, should be marked on a special issue of the master drawing.

It seems most expedient to let the test pads be assigned during the layout procedure. The reason is that at this stage it is much easier to pay attention to possible congested areas.

11.4.4 Bare-Board Testing

Bare-board testing for short circuits and interruptions can be performed as part of the incoming goods inspection. Since the problems discussed in Paragraph 11.2 are not present, for obvious reasons, the use of special test pads is unnecessary. The reader is referred to Subparagraphs 1 to 4 of Paragraph 11.4.3d. When the components are to be tested according to Paragraph 11.4.2 at a later production stage, it will of course be necessary to connect component test points as described in the design rules given above.

Index

of electrical components, 109
of mechanical parts, 110
Deflection, of PCBs, 96
Deformation, under load, 12
Delamination, 64
of gold plating, 39, 82
Delivery time, confirmation of, 49
Density curves, 335
Design changes, by a CAD service
    bureau, 400
Design modifications
by CAD, 391
by CAD service bureau, 400
Design Rule
Number 1, minimum spacing
    (conductor/pad), 161
Number 2, minimum spacing (parallel
    conductors), 162
Number 3A, minimum spacing
    (secondary voltage), 164
Number 3B, minimum spacing (mains
    voltage), 165
Number 4, minimum spacing
    (soldering), 166
Number 5, size of solder hole, 168
Number 6, size of pad (plated
    through holes), 170
Number 7, size of pad (non-plated
    through holes), 171
Number 8, solderability of pads, 176
Number 9, conductor width, 179
Number 10, width of current-carrying
    conductors, 183
Number 11, solder mask clearances,
    203
Number 12, overprinting of solder-
    mask pads, 207
Design rules
for automatic component testing, 430
for automatic pattern testing, 434
for drafting systems
    pattern, 322
    solder masks, 323
Designation, standardized location
    of, 200
Despatch, of secondary artwork, 223
Development capacity, 290
Development
of diazo films, 147, 254
of negative diazo films, 280
Development of PCBs, 288
correction stage, before pilot
    production, 289
correction stage, after pilot

production, 289
functional model stage, 288
prototype stage, 288
Development projects
low-priority, 290
parallel, 290
Dewetting, 71
Diazo contact printing
avoidance of enlargement, 217
comparisons between methods A, B,
    and C, 217
determination of exposure time, 252
development of prints, 254
distance of light source, 210
enlargement of prints, 215
exposure monitor, 211
exposure of films, 209
light source, 210
method A, 212
method B, 214
method C, 215
of secondary artwork, 219
of tertiary artwork, 220
optimum pad and film thicknesses,
    217
pad master for component side, 253
pad master for master drawing, 254
pad master for solder mask, 254
pad master for solder side, 253
parallax error, 210
sequence of, 212
use of secondary artwork, 219
warning, 253
Diazo film
development of, 147
for component side, 221
for contact printing, 209
for pad master, 221
for secondary artwork, 150
for secondary artwork to be filed, 220
for secondary artwork to be sent to
    manufacturer, 220
for solder side, 221
influence of daylight, 147
misregistration, 150
negative, 280
overdevelopment, 148
permanent artwork, 150
printing turnround time, 149
purchase of, 220
resolution of, 148
shelf life of unexposed, 220
stability of, 147
wavelength of light, 148